FUNDAMENTALS IN MODELING AND CONTROL OF MOBILE MANIPULATORS

AUTOMATION AND CONTROL ENGINEERING
A Series of Reference Books and Textbooks

Series Editors

FRANK L. LEWIS, Ph.D.,
Fellow IEEE, Fellow IFAC
Professor
Automation and Robotics Research Institute
The University of Texas at Arlington

SHUZHI SAM GE, Ph.D.,
Fellow IEEE
Professor
Interactive Digital Media Institute
The National University of Singapore

FUNDAMENTALS IN MODELING AND CONTROL OF MOBILE MANIPULATORS

Zhijun Li
Shuzhi Sam Ge

CRC Press
Taylor & Francis Group
Boca Raton   London   New York

CRC Press is an imprint of the
Taylor & Francis Group, an **informa** business

CRC Press
Taylor & Francis Group
6000 Broken Sound Parkway NW, Suite 300
Boca Raton, FL 33487-2742

First issued in paperback 2017

© 2013 by Taylor & Francis Group, LLC
CRC Press is an imprint of Taylor & Francis Group, an Informa business

No claim to original U.S. Government works

Version Date: 20130502

ISBN 13: 978-1-4665-8041-1 (hbk)
ISBN 13: 978-1-138-07436-1 (pbk)

Visit the Taylor & Francis Web site at
http://www.taylorandfrancis.com

and the CRC Press Web site at
http://www.crcpress.com

To our parents, for bringing us to the world and raising us to explore;

To our teachers and professors, for educating us about rights and instilling us with principles; and

To our families, for their love showered upon us and their grace and companionship in the journey!

Contents

xii

Preface

In the robotics and automation literature, mobile manipulators (manipulators mounted on mobile bases, platforms or vehicles) have received much attention since the beginning of the 1990s for their extended space of operation and extra degrees of freedom offered by the vehicle for accomplishing specific (typically, "long range") manipulative tasks that otherwise could not be executed and accomplished.

With a mobile platform, mobile manipulators can move around and reach places unfriendly to human beings, and perform operations over a large space or long range. Indeed, they can find a wide range of applications including (i) service or security applications at factories, offices, or homes, (ii) search and rescue operations in a harsh, demanding and/or dangerous environment, (iii) remote operations or explorations in remote places or spaces, among others.

Such systems combine the complementary advantages of mobile platforms and robotic arms to extend operational ranges and functionality, and at the same time to bring along the complexity and difficulty in dynamic modelling and control system design for such a class of systems. One the one hand, mobile robotic manipulators possess strongly coupled dynamics of mobile platforms and robotic manipulators, not to mention the nonholonomic constraints introduced by the wheeled/tracked mobile platforms, which present many difficulties in the motion planning, control, coordination, and cooperation of such systems. On the other hand, many advances in nonlinear system analysis and control system design offer powerful tools and fundamental concepts for the control of mobile manipulator systems. The study of mobile manipulators provides yet another class of practical systems where intelligent control, adaptive control, and robust control are called upon.

In the course of research on mobile manipulators, we see an exponential growth in the literature where many fundamental concepts and power tools have been developed in understanding and appreciation of the scientific problems and solutions involved. This motivated us to write the current monograph to give the subject of interest a systematic treatment yet in a timely manner.

The main objective of this book is to give a thorough theoretical treatment of several fundamental problems for mobile robotic manipulators, and some major issues that the authors have been analyzing over the past ten years. By integrating fresh concepts and state-of-the-art results to give a systematic treatment for kinematics and dynamics, motion generation, feedback control, coordination and cooperation, a basic theoretical framework is formed toward a mobile robotic manipulator which not only extends the theory of nonlinear control, but also applies to more realistic problems.

A second objective is to write a book that only encompasses the fundamentals of mobile manipulators, the first specialized book on this topic. The book is primarily intended for researchers and engineers in the system and control community. It can also serve as complementary reading for nonlinear system theory at the post-graduate level.

The goal of this book is to investigate the fundamental issues including modeling, motion planning, control, coordination, and cooperation for single and multiple mobile manipulators.

The book contains eight chapters, which exploit several independent yet related topics in detail.

Chapter 1 introduces the system description, background, and motivation of the study, and presents several general concepts and fundamental observations which provide a sound base for the book.

Chapter 2 describes the kinematics and dynamics equations of manipulators and mobile platforms separately, and derives the kinematics and dynamics equations for the mobile manipulator in order to investigate the dynamic interaction between the manipulator and the platform. Deriving the entire equations of one 2-DOF mobile manipulator and 3-DOF mobile manipulator in explicit forms, which can be found in the Appendix, we can conduct the simulations to verify the proposed controller in the following chapters.

Chapter 3 investigates the motion generation approach for both single nonholonomic mobile manipulators and multiple nonholonomic mobile manipulators in the coordination under conditions of obstacles.

Chapter 4 presents the model-based control for the mobile manipulator with the precision known dynamic model and external disturbance, which is the basis for the following chapters.

Chapter 5 systematically investigates adaptive robust control for mobile manipulators with unknown system parameters and external disturbances by three aspects, including the motion/force control and output feedback control. Finally, the position stabilization and constraint force control for mobile manipulators with uncertain holonomic constraints is systematically investigated with unknown system parameters and external disturbances.

Chapter 6 presents developing the dynamics of a mobile manipulator with an underactuated joint, and deriving the motion/force control for mobile manipulators interacting with the environments.

Considering multiple mobile manipulators carrying a common object in a cooperative manner, Chapter 7 studies two coordination control approaches: centralized coordination and decentralized coordination with unknown inertia parameters and disturbances. Both proposed coordination controls are robust not only to system uncertainties such as mass variation but also to external disturbances. Simulation studies on the control of coordinated two-wheel driven mobile manipulators show the effectiveness of the proposed scheme.

Chapter 8 investigates centralized robust adaptive controls for two cooperating mobile robotic manipulators manipulating an object with relative motion in the presence of uncertainties and external disturbances. The proposed adaptive controls are robust against relative motion disturbances and parametric uncertainties and validated by simulation studies.

Acknowledgements

For the creation of the book, we are very fortunate to have received many helpful suggestions from our colleagues, friends, and coworkers, through many stimulating and fruitful discussions. First of all, we would like to express our sincere appreciation to our co-workers who have contributed to the collaborative studies of mobile manipulators.

For the final completion of the book, we gratefully acknowledge the unreserved support and constructive comments from and fruitful discussions with Martin D. Adams, University of Chile; Hugh Durrant-Whyte, The University of Sydney; Javier Ibanez Guzman, Renault; Frank F. Lewis, University of Texas at Arlington; Mou Chen, Nanjing University of Aeronautics and Astronautics; Rongxin Cui, Northwestern Polytechnical University; Zhuping Wang, Tongji University; Yanjun Liu, Liaoning University of Technology; Wijerupage Sardha Wijesoma, Nanyang Technological University; Hongbo Li, Tsinghua University; Beibei Ren, University of California, San Diego; C.-Y. Su, Concordia University; Pey Yuen Tao, SimTech; Loulin Huang, Auckland University of Technology; Danwei Wang, Nanyang Technological University; Tianmiao Wang, Beijing University of Aeronautics and Astronautics; Yuanqing Xia, Beijing Institute of Technology; and Jie Zhao, Harbin Institute of Technology.

Much appreciation goes to Qun Zhang, Sibang Liu, Zhongliang Tang, Qian Zhao, Xiaoming Sun, and Zhen Zhong for their time and effort in proofreading, and providing numerous useful comments and suggestions to improve the readability of the book.

This work is partially supported by (i) Intelligent Control of Unmanned Vehicles, Temasek Young Investigator Award, Defence Science and Technology Agency (DSTA), Singapore; (ii) Collaborative Autonomous Systems for Built Environments (CARSyB), A-Star/SERC research project, Singapore; (iii) the National Natural Science Foundation of China (Nos. 60804003, 61174045, and 60935001); (iv) International Science & Technology Cooperation Program of China (No. 2011DFA10950); (v) the Fundamental Research Funds for the Central Universities under Grant 2011ZZ0104; and (vi) National High Technology Research and Development Program of China (863, 2011AA040701).

Zhijun Li

Wushan, Guangzhou

Shuzhi Sam Ge

Kent Ridge Crescent, Singapore / Clear River, Chengdu

1

Introduction

CONTENTS

1.1 Mobile Manipulator Systems

Traditionally, robotic manipulators are usually bolted on the fixed base. The workspace of such a fixed base manipulator is a limited volume of the operation space that can be reached by the end-effector of the manipulator. The workspace is limited and tasks must be carefully structured so that the manipulator can reach parts to be assembled. This is typically achieved by means of conveyor belts or other transporting devices. In the recent several decades, there has been a great deal of interest towards mobile robots [9], [8], [21], [154]. A mobile robot typically refers to a mobile platform or vehicle, equipped with computing units and various sensors. The study of mobile robots is mostly concentrated on a central question: how to move from here to there in a structured/unstructured environment. It involves many issues such as kinematics, dynamics, motion planning, navigation, control and coordination.

Different from mobile robots, a mobile manipulator consists of one or several manipulators and a mobile platform (or a mobile robot). The manipulators are mounted on the top of the mobile platform. A mobile manipulator combines the dextrous manipulation capability offered by fixed-base manipulators and the mobility offered by mobile platforms. A mobile manipulator has a considerably larger operation workspace than a fixed-based one and more manipulation than mobile robots. Mobile manipulators have many potential applications in manufacturing, nuclear reactor maintenance, construction, and

FIGURE 1.1
Weichun earthquake, Sichuan, China.

planetary exploration [58], [59], [60]. The possible applications of mobile manipulators can be found in hazardous areas after earthquakes (See. Fig. 1.1), nuclear/chemical hazardous areas (See Fig. 1.2), space exploration, and military applications (including the mine-clearing during peace-keeping/holding missions).

Some illustration examples of mobile manipulators are shown in Fig. 1.3–Fig. 1.9.

It is obvious that mobile manipulators offer a tremendous potential for performing very wide tasks. However, they bring about a number of challenging problems rather than simply increasing the structural complexity. The following fundamental issues are listed and shall be addressed in this book:

(i)   What are the exact kinamatic and dynamic models of mobile manipulators since the mobile manipulator is subject to both holonomic and nonholonomic constraints?

(ii) How can we plan the effective motion trajectory of a mobile manipulator under both holonomic and nonholonomic constraints?

FIGURE 1.2
Nuclear accidents, Fukushima, Japan.

FIGURE 1.3
The Stanford assistant mobile manipulator [1].

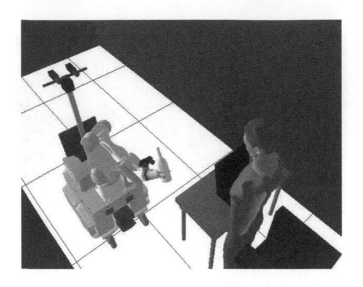

FIGURE 1.4
The mobile manipulator in Washington University [2].

FIGURE 1.5
DLR's humanoid robot called Justin [3].

FIGURE 1.6
NASA's Mars Rover [4].

FIGURE 1.7
Autonomous robot based on iRobot platform in National University of Singapore.

FIGURE 1.8
RAS-1 mobile manipulator by South China University of Technology.

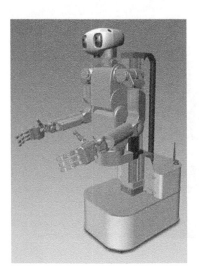

FIGURE 1.9
M1 mobile manipulator robot made by Meka Robotics [6].

(iii) How do we design the hybrid motion/force control and hybrid position/force for mobile manipulators since the mobile manipulator needs to interact with the environments?

(iv) How can we coordinate and cooperate multiple mobile manipulators in order to efficiently fulfill the desired tasks, since the coordination and cooperation could improve the manipulation performance of the multiple robots?

The objective of this book is to investigate the fundamental problems including modeling, motion generation, control, coordination and cooperation for mobile manipulators. The emphasis will be systematic presentation of the fundamental approaches for the mobile manipulator.

1.2   Background and Motivations

In this subsection, we briefly discuss the background and motivation of investigating mobile manipulator systems and review the previous works related to the above issues.

Investigation of the modeling, control, coordination and cooperation of mobile manipulators spans several different research domains. Some of them have been extensively studied while others are new and little research has been done. Fundamental issues related to the topic include the kinematic and dynamic modeling, the control of nonholonomic systems, the path-planning considering motion and manipulation, the hybrid motion/force control and hybrid position/force control if the mobile manipulator is required to interact with environments, the coordination and cooperation strategy of multiple mobile manipulators. However, there is only a limited literature available on the fundamental issues, although the advantage of a mobile manipulator over a conventional fixed-base manipulator has been widely acknowledged.

In physics and mathematics, a wheeled mobile manipulator is fundamentally a nonholonomic system, which is a system described by a set of parameters subject to differential constraints, such that when the system evolves along a path in its parameter space, where the parameters vary continuously (in the mathematical sense) and return to the identical values they held at the start of the path, the system itself may not have returned to its original state. A classical example of nonholonomic systems is a rigid disk rolling on a horizontal plane without slippage in [135], which is equivalent from the control perspective to a wheeled cart driven by two wheels. As a matter of fact, a car-like system in general is a nonholonomic system except a few examples of omnidirectional vehicles [40], [33], [34], [35]. Other examples of nonholonomic systems can be seen in underwater vehicles [36], [37], underactuated robotic manipulators [38], [139], underactuated mobile manipulators [183], [184]. A good survey of the recent development in terms of nonholonomic motion planning is given by Li and Canny [28]. There are many studies on motion planning of mobile robots using various approaches, e.g., potential field [45], graph search algorithms, the A* algorithm [41], Bellman-Ford algorithms [42], the wavefront algorithm [43], and visibility graph approaches [44]. For the motion generation plan planning for mobile manipulators, since mobility is the main concern, the approaches are similar to motion planning for mobile robots. Therefore, we briefly review motion planning for mobile robots.

Dynamic window approaches [46] search the velocity space for a heading close to the goal direction, without hitting obstacles within several command intervals. The curvature-velocity method [47] searches the velocity space for a point that satisfies the velocity and acceleration constraints and maximizes an objective function. In [48], a model for planning the shortest path in configuring the space of a mobile robot was proposed, based on a Lagrange method for optimizing a function. In [49], an iterative algorithm was proposed for motion generation for parking a mobile robot. In [50], a motion planning model for mobile robots used a bubble method to find the locally reachable space and a parameterization method to satisfy the kinematic constraint. In [51], a time-optimal motion planning method was proposed for a robot with kinematic constraints, which consist of three stages: (i) planning for a point robot; (ii) planning for a robot with size and shape; and (iii) optimizing cost functions for a time-optimal solution. In [52], a path planner was proposed for a nonholonomic mobile robot using a search based algorithm, which requires a local collision-checking procedure and the minimization of cost functions. In [53], a multi-level approach to motion planning of a nonholonomic mobile robot was proposed, where at the first level, a path is found that disrespects the nonholonomic constraints; at each of the next levels, a new path is generated by transformation of the path generated at the previous level; at the final level, all nonholonomic constraints are respected. In [54], a search based model was proposed for path planning with penetration growth distance, which searches the collision paths instead of searching the free workspace as most other models do. In [55], a learning method was proposed for path planning of a robot in a cluttered workspace where the dynamic local minima can be detected. In [56], a probabilistic learning approach was proposed for the motion planning of a mobile robot, which involves a learning phase and a query phase and uses a local method to compute feasible paths for the robot. In [57], a multilayer reinforcement learning model was proposed for path planning of multiple mobile robots. However, the planned robot motion using learning-based approaches is not efficient and is computationally expensive, especially in its initial learning phase.

Due to Brocketts theorem [135], it is well-known that a wheeled mobile robot with restricted mobility cannot be stabilized to a desired configuration (or posture) via differentiable, or even continuous, pure-state feedback, although it is controllable. A number of approaches have been proposed for the problem, which can be classified as (i) discontinuous time-invariant stabilization [83], (ii) time-varying stabilization [84], [9] and (iii) hybrid stabilization [87], [86]. See the survey paper [88] for more details and references therein.

One commonly used approach for controller design of the mobile robot is to convert, with appropriate state and input transformations, the original systems into some canonical forms for which controller design can be easier [10], [89], [90]. Using the special algebra structures of the canonical forms, various feedback strategies have been proposed to stabilize mobile robots [83], [108], [10], [92]. Recently, adaptive control strategies were proposed to stabilize the dynamic mobile robot systems with modeling or parametric uncertainties [108]. Hybrid control based on supervisory adaptive control was presented to globally asymptotically stabilize a wheeled mobile robot [93]. Adaptive state feedback control was considered in [94] using input-to-state scaling. Output feedback tracking and regulation were presented in [95] for practical wheeled mobile robots. In [96], robust exponential regulation for nonholonomic systems with input and state-driven disturbances was presented under the assumption that the bounds of the disturbances are known. However, these studies consider neither vehicles with manipulators nor a system's dynamics.

For the mobile manipulators, the base degrees-of-mobility are treated equally with the arm degrees-of-manipulation, and solve the redundancy by introducing a user-defined additional task variable [97]. In [98], a weighted multi-criteria cost function is defined, which is then optimized using Newton's algorithm and the coordination of mobility and manipulation is formulated as a nonlinear optimization problem. A general cost function for point-to-point motion in Cartesian space is defined and is minimized using a simulated annealing method. In [99], a controller design was proposed for a mobile manipulator. The controller consists of a feedforward part which executes off-line optimization along the desired trajectory and a feedback part which realizes decoupling and compensation of the tracking errors. In [100], a decentralized robust controller was described for a mobile robot by considering the platform and the manipulator as two separate systems with which two interconnected subsystems are stable if the unknown interconnections are bounded. Their model used for simulation consists of a two-link manipulator attached to a planar base, in which the angular motion of the base is excluded.

Input-output feedback linearization was investigated to control the mobile platform such that the manipulator is always positioned at the preferred configurations measured by its manipulability [101]. Similarly, through nonlinear feedback linearization and decoupling dynamics in [61], force/position control of the end-effector along the same direction for mobile manipulators was proposed and applied to nonholonomic cart pushing. In [102], the effect of the dynamic interaction between the arm and the vehicle of a mobile manipulator was studied, and nonlinear feedback control for the mobile manipulator was developed to compensate for the dynamic interaction. In [62], coordination and control of mobile manipulators were presented with two basic task-oriented controls: end-effector task control and platform self posture control. In [63], the concept of manipulability was generalized to the case of mobile manipulators and the optimization criteria in terms of manipulability were given to generate the controls of the system.

However, control of mobile manipulators with uncertainties is essential in many practical applications, especially for the case when the force of the end-effector should be considered. To handle unknown dynamics of mechanical systems, robust and adaptive controls have been extensively investigated for robot manipulators and dynamic nonholonomic systems. Robust controls assume the known boundaries of unknown dynamics of the systems, nevertheless adaptive controls could learn the unknown parameters of interest through adaptive tuning laws.

Under the assumption of a good understanding of dynamics of the systems understudy, model-based adaptive controls have been much investigated for dynamic nonholonomic systems. In [103], adaptive control was proposed for trajectory/force control of mobile manipulators subjected to holonomic and nonholonomic constraints with unknown inertia parameters, which ensures the motion of the system to asymptotically converge to the desired trajectory and force. In [105], adaptive state feedback and output feedback control strategies using state scaling and backstepping techniques were proposed for a class of nonholonomic systems in chained form with drift nonlinearity and parametric uncertainties. In [64], the nonholonomic kinematic subsystem was first transformed into a skew-symmetric form, then a virtual adaptive control designed at the actuator level was proposed to compensate for the parametric uncertainties of the kinematic and dynamic subsystems.

Because of the difficulty in dynamic modeling, adaptive neural network control, a non-model based approach, has been extensively studied for different classes of systems, such as robotic manipulators [111, 112, 79] and mobile robots [113]. In [108], robust adaptive control was proposed for dynamic nonholonomic systems with unknown inertia parameters and disturbances in which adaptive control techniques were used to compensate for the parametric uncertainties and sliding mode control was used to suppress the bounded disturbances. In [66], adaptive robust force/motion control was presented systematically for holonomic mechanical systems and a large class of nonholonomic mechanical systems in the presence of uncertainties and disturbances. In [80], adaptive neural network control for robot manipulator in the task space was proposed, which neither requires the inverse dynamical model nor the time-consuming off-line training process. In [114], the unidirectionality of the contact force of robot manipulators was explicitly included in modeling and the fuzzy control was developed. In [81], adaptive neural fuzzy control for function approximation was investigated for uncertain nonholonomic mobile robots in the presence of unknown disturbances. In [115], adaptive neural network controls were developed for the motion control of mobile manipulators subject to kinematic constraints.

Coordinated controls of multiple mobile manipulators have attracted the attention of many researchers [68], [172], [173], [174]. Interest in such systems stems from the greater capability of the mobile manipulators to carry out more complicated and dextrous tasks which may not be accomplished by a single mobile manipulator. It is an important technology for applying cooperative mobile manipulators to modern factories for transporting materials, and dangerous fields for dismantling bombs or moving nuclear infected objects. Coordinated controls of multiple mobile manipulators have attracted the attention of many researchers [68], [172], [173], [174]. Interest in such systems stems from the greater capability of the mobile manipulators in carrying out more complicated and dextrous tasks which may not be accomplished by a single mobile manipulator. It is an important technology for applying cooperative mobile manipulators to modern factories for transporting materials, and dangerous fields for dismantling bombs or moving nuclear infected objects.

Controls of multiple mobile manipulators present a significant increase in complexity over the single mobile manipulator case; moreover, it is more difficult and challenging than the controls of multiple robotic manipulators [170] and [171]. The difficulties of the control problem lie in the fact that when multiple mobile manipulators coordinate each other, they form a closed kinematic chain mechanism. This imposes a set of kinematic and dynamic constraints on the position and velocity of coordinated mobile manipulators. As a result, the degrees of freedom of the whole system decrease, and internal forces are generated which need to be controlled.

Until now, a few control methods have been proposed for solving this problem. These controls include: (i) hybrid position/force control, where the position of the object is controlled in a certain direction of the workspace and the internal force of the object is controlled in a small range of the origin [68], [174], [177], [170], [169]; and (ii) leader-follower method, where one or a group of mobile manipulators plays the role of the leader, which tracks a preplanned trajectory, and the rest of the mobile manipulators form the follow group which are moved in conjunction with the leader mobile manipulators [172], [175], [176].

Most previous studies on the coordination of multiple mobile manipulators systems only deal with motion-tracking control [68], [172], [174], [175], and [176], on the assumption of known complex dynamics of the system. If there exit uncertain dynamics and disturbances from the environments, the controls so designed may give degraded performance and may incur instability. Moreover, the large scale tasks, such as manufacturing and assembly in automatic factories and space explorations, often include situations where multiple robots are grasping an object in contact with the environment, for example, scribing, painting, grinding, polishing, contour-following on a larger scale. The purpose of controlling a coordinated system is to control the contact forces between the environment and object in the constrained direction and the motion of the object in unconstrained directions. The internal forces are produced within the grasped object, which do not contribute to system motion; the larger internal forces would damage the object. We have to maintain internal forces in some desired values. Moreover, the parameter uncertainties and external disturbances existing between the robots and the environment would disrupt the interaction.

1.3   Outline of the Book

The book contains seven chapters which exploit several independent yet related topics in detail.

Chapter 1 introduces the system description, background and motivation of the study, and presents several general concepts and fundamental observations which provide a sound base for the book.

Chapter 2 describes the kinematics and dynamics equations for manipulators and mobile platforms separately. There are two reasons for treating the two subsystems separately: (i) the mobile platform has very unique control properties due to its nonholomomic nature, the modeling of the mobile platform should be addressed independently of the manipulator which is holonomic for the sake of clarity; and (ii) the mobile manipulators have different types, several examples are listed in the chapter. Chapter 2 further derives the kinematics and dynamics equations for the mobile manipulator in order to investigate the dynamic interaction between the manipulator and the platform. Deriving the entire equations of a 2-DOF mobile manipulator and a 3-DOF mobile manipulator in explicit forms, which can be found in the appendix, we can conduct the simulations to verify the proposed controller in the following chapters.

Chapter 3 investigates the motion generation for both single nonholonomics mobile manipulators and multiple nonholonomic mobile manipulators in the coordination under the consideration of the obstacles. A collision-free motion generation approach is proposed for nonholonomic mobile manipulators and coordinated nonholonomic mobile manipulators in the highly cluttered environments, which employ smooth and continuous polynomial functions. The approach generates a collision-free initial path for a mobile manipulator. While following this path, the detected obstacles can be avoided. The current path is iteratively deformed in order to get away from obstacles and satisfy the nonholonomic constraints and yields admissible input trajectories that drive both the manipulator and the platform to a desired configuration. The core idea of the approach is to deform the normal vector along the current path in order to modify this path, making the collision constraints decrease. Illustrative examples demonstrate the planning methodology in obstacle-free and obstructed environments.

Chapter 4 presents the model-based control for the mobile manipulator with the precision of a known dynamic model and external disturbance, which is the basis for the following chapters.

Chapter 5 systematically investigates the control for mobile manipulators. It describes the effective adaptive robust control strategies to control a class of holonomic constrained noholonomic mobile manipulators in the presence of uncertainties and disturbances. The system stability and the boundedness of tracking errors are proved using Lyapunov synthesis. All control strategies have been designed to drive the system motion to converge to the desired manifold and at the same time guarantee the boundedness of the constrained force. Moreover, adaptive robust output-feedback force/motion control strategies are then presented for mobile manipulators under both holonomic and nonholonomic constraints in the presence of uncertainties and disturbances. The controls are developed on structural knowledge of the dynamics of the robot and actuators and in conjunction with a linear observer. The proposed controls are robust not only to parametric uncertainty such as mass variations but also to external ones such as disturbances. The system stability and the boundedness of tracking and observation errors are proved using Lyapunov stability synthesis. Simulation results validate that not only the states of the system converge to the desired trajectory, but also the constraint force converges to the desired force. In the third section, hybrid position stabilization/force tracking control is invetigated for nonholonomic mobile manipulators with unknown parameters of interest and disturbances under uncertain holonomic constraints. The nonholonomic mobile manipulator is transformed into a reduced chained form, and then, robust adaptive force/motion control with hybrid variable signals is proposed to compensate for parametric uncertainties and suppress bounded disturbances. The control scheme guarantees that the outputs of the dynamic system track some bounded auxiliary signals, which subsequently drive the kinematic system to the desired trajectory/force.

Chapter 6 investigates adaptive motion/force control by dynamic coupling and output feedback for nonholonomic mobile manipulators with an under-actuated joint, in the presence of parametric and functional uncertainties. It is obvious that the constraints of the system consist of kinematic constraints for the mobile platform and dynamic constraints for the under-actuated joint. Through using dynamic coupling property of nonholonomic mobile under-actuated manipulators, adaptive output feedback control is investigated for the system by using a high-gain observer to reconstruct the system states, whose states and time derivatives of the output are unavailable. Moreover, the nonholonomic constraint force between the wheels and the ground is also considered in the control design such that the slipping or slippage is avoided during the motion. It is shown that output tracking errors of motion and force converge to adjustable neighborhoods of the origin for the output feedback control.

Chapter 7 investigates two coordination control approaches. First, the centralized robust adaptive coordination controls of multiple mobile manipulators carrying a common object in a cooperative manner with unknown inertia parameters and disturbances are presented. A concise dynamics consisting of the dynamics of mobile manipulators and the geometrical constraints between the end-effectors and the object is developed for multiple mobile manipulator coordination. Subsequently, we design centralized robust adaptive controls where parametric uncertainties are compensated by adaptive update techniques and the disturbances are suppressed. The controls ensure that the output tracking errors of the system converge to zero whereas the internal force tracking error remains bounded and can be made arbitrarily small. Feedback control design and stability analysis are performed via explicit Lyapunov techniques. Simulation studies on the control of coordinated two wheels driven mobile manipulators show the effectiveness of the proposed scheme. Then, a decentralized robust adaptive coordination control version of multiple mobile manipulators cooperatively carrying a common object interacting with nonrigid environments is proposed. First, the decentralized dynamics of system coupled with the physical interactions are developed, which includes the dynamics of mobile manipulators, the internal force between end-effectors and the object, and interaction force between the object and environments. Then, a decentralized adaptive robust control based on impedance approach for coordinated multiple mobile manipulators is designed and analyzed using Lyapunov synthesis. The proposed controls are robust not only to system uncertainties such as mass variation but also to external disturbances. Simulation results are presented to validate that the position tracking errors converge to zero whereas the impedance-based internal force tracking error can be made arbitrarily small.

Chapter 8 presents the coupled dynamics for two cooperating mobile robotic manipulators manipulating an object with relative motion in the presence of uncertainties and external disturbances. Centralized robust adaptive controls are introduced to guarantee the motion and force trajectories of the constrained object converge to the desired manifolds with prescribed performance. The stability of the closed-loop system and the boundedness of tracking errors is proved using Lyapunov stability synthesis. The tracking of the constraint trajectory/force up to an ultimately bounded error is achieved. The proposed adaptive controls are robust against relative motion disturbances and parametric uncertainties and validated by simulation studies.

2

Kinematics and Dynamics

CONTENTS

2.1   Introduction

Kinematics are the velocity relationships relating the linear and angular velocities of the task space end-effector to the joint space, while dynamics are concerned with the relationship between the forces acting on them and the positions, the velocities and the accelerations they produce.

Mathematically, the forward kinematic equations define a function between the task space and the joint space. The velocity relationships are then determined by the Jacobian which is a matrix that can be thought of as the vector version of the ordinary derivative of a scalar function. The Jacobian is one of the most important quantities in the analysis and control of robot motion. It arises in virtually every aspect of robotic manipulation: in the planning and execution of smooth trajectories, in the determination of singular configurations, in the execution of coordinated anthropomorphic motion, in the derivation of the dynamic equations of motion, and in the transformation of forces and torques from the end-effector to the manipulator joints.

The dynamic equations describe the chains time evolution of interconnected rigid body chains under a given set of internal and external forces and/or desired motion specifications, and provide a means of designing robot prototype. We can test control approaches without building the actual robots. The theoretical principle behind rigid body dynamics is rather straightforward: one extends Newton's laws for the dynamics of a point mass to rigid bodies, which are indeed nothing else but conglomerations of point masses that keep constant distances with respect to each other. However, the interconnection of rigid bodies by means of (prismatic, revolute, etc.) joints gives rise to new physical properties that do not exist for one single point mass or one single rigid body. More in particular, the topology of the kinematic chain determines to a large extent the minimal complexity of the computational algorithms that implement these physical properties.

Kinematics and dynamics deal with the mathematical formulation of the dynamic equations of robot motion. Most of the introductory material can be found in textbooks on classical physics and mechanics, e.g., [102], [143], [185]. Model building could help us understanding the physical meaning and allow real-time execution in a robot controller, i.e., the calculations require less time than the real physics. In practice, this means that only ideal kinematic chains are considered: rigid bodies interconnected via ideal joints. Introducing flexible joints and joint friction increases the computational costs, although real-time execution remains possible.

All dynamic algorithms discussed in this chapter assume that the physical parameters of the robot are known: dimensions of links, relative positions and orientations of connected parts, mass distributions of links, joints and motors. In practice, it's not straightforward to find realistic values for all these parameters in a given robot. In addition, this chapter assumes ideal systems, i.e., perfectly rigid bodies, joints without backlash and with perfectly modelled flexibilities and friction. For the easy computation of kinematics model for mobile robotic manipulators, we decomposed mobile robotic manipulators into the mobile platform and the robotic manipulators.

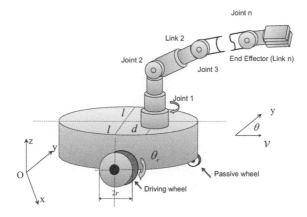

FIGURE 2.1

Differential-driven mobile manipulator.

2.2 Kinematics of Mobile Platform

In this section, we consider two mobile platforms—the differential-driven mobile platform and car-like mobile platform. Since the procedures for the derivation of the kinematic model for the differential driven mobile robot and mobile manipulator are similar, we use a common symbolic representation. If the mobile platform satisfies nonholonomic constraints without slipping, then the following constraint holds for the above two types:

$$A(q)\dot{q} = 0 \tag{2.1}$$

where $A(q) \in \mathbf{R}^{n_v \times n}$ is the matrix associated with the constraints.

2.2.1 Differential-driven Mobile Platform

The mobile robot shown in Fig. 2.1 is a typical differential-driven mobile manipulator. It consists of a vehicle with two driving wheels mounted on the same axis, and a passive front wheel. The motion and orientation are achieved by independent actuators, e.g., dc motors providing the necessary torques to the rear wheels. The nonholonomic constraint states that the robot can only move in the direction normal to the axis of the driving wheels, i.e., the mobile

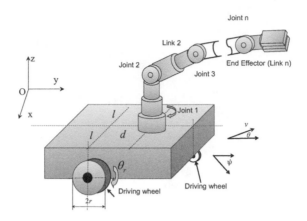

FIGURE 2.2

Car-like driven mobile manipulator.

base satisfies the conditions of pure rolling and nonslipping, therefore, the mobile platform is generally subject to three constraints. The first one is that the mobile robot cannot move in the lateral direction, i.e.,

$$\dot{y}_o \cos\theta - \dot{x}_o \sin\theta = 0 \tag{2.2}$$

since $\dot{x}_o = v\cos\theta$ and $\dot{y} = v\sin\theta$ where (x_o, y_o) is the center point of the driving wheels, and v is the velocity of the platform, and θ is the heading angle of the mobile robot measured from X-axis.

Eq. (2.2) is a nonholonomic constraint and cannot be integrated analytically to result in a constraint between the configuration variables of the platform, namely x_o, y_o, and θ . As is well known, the configuration space of the system is three-dimensional (completely unrestricted), while the velocity space is two-dimensional. This constraint becomes

$$\dot{x}_o \cos\theta + \dot{y}_o \sin\theta + l\dot{\theta} = r\dot{\theta}_r \tag{2.3}$$

$$\dot{x}_o \cos\theta + \dot{y}_o \sin\theta - l\dot{\theta} = r\dot{\theta}_l \tag{2.4}$$

where θ_r and θ_l are the angular positions of the two driving wheels, respectively, and $2l$ is the platform width.

Let the generalized coordinates of the mobile platform be $q =$

$(x_o, y_o, \theta, \theta_r, \theta_l)$. The three constraints lead to matrix $A(q)$ as follows

$$A(q) = \begin{bmatrix} -\sin\theta & \cos\theta & 0 & 0 & 0 \\ \cos\theta & \sin\theta & l & -r & 0 \\ \cos\theta & \sin\theta & -l & 0 & -r \end{bmatrix} \tag{2.5}$$

Let m rank matrix $S(q) \in \mathbf{R}^{n \times m}$ formed by a set of smooth and linearly independent vector fields spanning the null space of $A(q)$, i.e.,

$$S^T(q)A^T(q) = 0 \tag{2.6}$$

According to (2.1) and (2.6), it is possible to find an auxiliary vector time function $v(t) \in \mathbf{R}^{n-m}$ such that, for all t

$$\dot{q} = S(q)v(t) \tag{2.7}$$

where $v(t) = [\dot{\theta}_r, \dot{\theta}_l]$.

For the two wheels mobile robot as shown in Fig. 2.1, we can define the matrix $S(q) \in R^{5 \times 2}$ as follows

$$S(q) = [s_1(q) \ s_2(q)] = \begin{bmatrix} \frac{r}{2}\cos\theta & \frac{r}{2}\cos\theta \\ \frac{r}{2}\sin\theta & \frac{r}{2}\sin\theta \\ \frac{r}{2l} & -\frac{r}{2l} \\ 1 & 0 \\ 0 & 1 \end{bmatrix} \tag{2.8}$$

It is obvious that the matrix $S(q)$ is in the null space of matrix $A(q)$, that is, $S^T(q)A^T(q) = 0$. A distribution spanned by the columns of $S(q)$ can be described as

$$\Delta = \text{span}\{s_1(q), s_2(q)\} \tag{2.9}$$

Remark 2.1 *The number of holonomic or nonholonomic constraints can be determined by the involutivity of the distribution Δ. If the smallest involutive distribution containing Δ (denoted by Δ^*) spans the entire five-dimensional space, all the constraints are nonholonomic. If $dim(\Delta^*) = 5 - k$, then k constraints are holonomic and the others are nonholonomic.*

Example 2.1 *To verify the involutivity of Δ, we compute the Lie bracket of $s_1(q)$ and $s_2(q)$.*

$$s_3(q) = [s_1(q) \ s_2(q)] = \frac{\partial s_2}{\partial q}s_1 - \frac{\partial s_1}{\partial q}s_2 = \begin{bmatrix} -\frac{r^2}{2l}\sin\theta \\ \frac{r^2}{2l}\cos\theta \\ 0 \\ 0 \\ 0 \end{bmatrix} \tag{2.10}$$

which is not in the distribution Δ spanned by $s_1(q)$ and $s_2(q)$. Therefore, at least one of the constraints is nonholonomic. We continue to compute the Lie bracket of $s_1(q)$ and $s_3(q)$

$$s_4(q) = [s_1(q) \; s_3(q)] = \frac{\partial s_3}{\partial q} s_1 - \frac{\partial s_1}{\partial q} s_3 = \begin{bmatrix} -\frac{r^3}{4l^2} \cos \theta \\ \frac{r^3}{4l^2} \sin \theta \\ 0 \\ 0 \\ 0 \end{bmatrix} \tag{2.11}$$

which is linearly independent of $s_1(q)$, $s_2(q)$, and $s_3(q)$. However, the distribution spanned by $s_1(q), s_2(q), s_3(q)$ and s_4 is involutive. Therefore, we have

$$\Delta^* = \text{span}\{s_1(q), s_2(q), s_3(q), s_4(q)\} \tag{2.12}$$

It follows that two of the constraints are nonholonomic and the other one is holonomic.

To obtain the holonomic constraint, we subtract Equation (2.3) from Equation (2.4)

$$2l\dot{\theta} = r(\dot{\theta}_r - \dot{\theta}_l) \tag{2.13}$$

Integrating the above equation and properly choosing the initial condition of $\theta(0) = \theta_r(0) = \theta_l(0)$, we have

$$\theta = \frac{r}{2l}(\theta_r - \theta_l) \tag{2.14}$$

which is obviously a holonomic constraint equation. Thus θ may be eliminated from the generalized coordinates.

The two nonholonomic constraints are

$$\dot{x}_o \sin \theta - \dot{y}_o \cos \theta = 0 \tag{2.15}$$
$$\dot{x}_o \cos \theta + \dot{y}_o \sin \theta = \frac{r}{2}(\dot{\theta}_r + \dot{\theta}_l) \tag{2.16}$$

The second nonholonomic constraint equation in the above is obtained by adding Eq. (2.3) from Eq. (2.4). It is understood that θ is now a shorthand notation for $c(\theta_r - \theta_l)$ rather than an independent variable. We may write these two constraint equations in the matrix form

$$A(q)\dot{q} = 0 \tag{2.17}$$

where the generalized coordinate vector q is now defined as

$$q = \begin{bmatrix} q_1 \\ q_2 \\ q_3 \\ q_4 \end{bmatrix} = \begin{bmatrix} x_o \\ y_o \\ \theta_r \\ \theta_l \end{bmatrix} \tag{2.18}$$

and $A(q)$ is given by

$$A = \begin{bmatrix} -\sin\theta & \cos\theta & 0 & 0 \\ -\cos\theta & -\sin\theta & \frac{r}{2} & \frac{r}{2} \end{bmatrix} \tag{2.19}$$

The kinematics of this mechanism can be written as

$$\begin{bmatrix} \dot{x}_o \\ \dot{y}_o \\ \dot{\theta} \\ \dot{\theta}_r \\ \dot{\theta}_l \end{bmatrix} = \begin{bmatrix} \frac{r}{2l}(l\cos\theta - d\sin\theta) & \frac{r}{2l}(l\cos\theta + d\sin\theta) \\ \frac{r}{2l}(l\sin\theta + d\cos\theta) & \frac{r}{2l}(l\sin\theta - d\cos\theta) \\ \frac{r}{2l} & -\frac{r}{2l} \\ 1 & 0 \\ 0 & 1 \end{bmatrix} \begin{bmatrix} \dot{\theta}_r \\ \dot{\theta}_l \end{bmatrix} \tag{2.20}$$

2.2.2 Car-like Mobile Platform

Consider a mobile manipulator whose platform includes front and rear wheels, as shown in Fig. 2.2. The rear wheels are parallel to the main axis of the car and used for driving the platform, while the front wheel is used for steering the platform. We also assume no-slip on the wheels. For simplicity, the manipulator is mounted at point m, where the steering wheel is located on the point m. For this point the nonholonomic constraint is written as

$$\dot{x}_m \sin\theta + \dot{y}_m \cos\theta + d\dot{\theta} = 0 \tag{2.21}$$

where $\dot{x}_m$ and $\dot{y}_m$ are the x and y components of the velocity v of point m respectively, and d is the distance between the point m and the back wheel axis.

The differential kinematics of the car-like mobile platform are described by the following equations

$$\dot{x}_m = v\cos(\theta + \psi) \tag{2.22}$$

$$\dot{y}_m = v\sin(\theta + \psi) \tag{2.23}$$

$$\dot{\theta} = \frac{v}{l}\sin\psi = \frac{\omega r}{l}\sin\psi \tag{2.24}$$

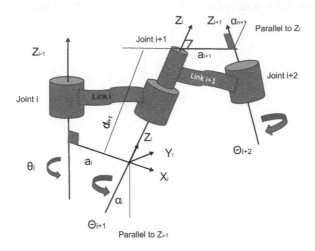

FIGURE 2.3
Denavit-Hartenberg frame assignment.

where ψ is the steering angle, $v = \omega r$ is the velocity at the point m, ω is the front wheel angular rate, and r is its radius. Eqs. (2.22)-(2.24) can be written

$$
\begin{bmatrix} \dot{x}_m \\ \dot{y}_m \\ \dot{\theta} \end{bmatrix} = \begin{bmatrix} \cos(\theta + \psi) & 0 \\ \sin(\theta + \psi) & 0 \\ l^{-1} \sin \gamma & 0 \end{bmatrix} \begin{bmatrix} v \\ \dot{\gamma} \end{bmatrix} \tag{2.25}
$$

Eq. (2.25) maps the two input velocities, v and $\dot{\gamma}$, to the three output velocities, $\dot{x}_m$, $\dot{y}_m$ and $\dot{\theta}$. If one eliminates the input velocities, the nonholonomic constraint given by Eqs. (2.22)-(2.24) results. From Eq. (2.25), we know that one of its columns is zero, therefore, if the mobile platform is not moving $v = 0$ then neither the position nor the orientation of the platform can be changed using the steering wheel.

2.3   Kinematics of Robotic Manipulators

Mobile manipulators consist of a sequence of rigid links connected by either revolute or prismatic joints mounted on a mobile base. The mobile base could

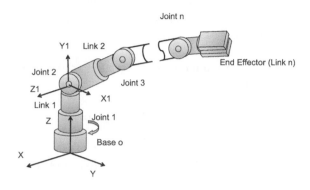

FIGURE 2.4
Serial manipulators with n-DOF.

be decomposed as several wheels and a platform. Each wheel and the platform could be treated as one joint-link pair, and one link. The mobile platform holds three degrees of freedom. For the manipulators, each joint-link pair constitutes one degree of freedom. For an n degrees of freedom robot, there are n joint-link pairs with link 0 attached to a supporting mobile base. The joints and links are numbered outwardly from the base to the arm; and the i-th joint of the arm is the point of connection between the i-th link and the $(i-1)$-th link.

In order to represent a relative, kinematic relationship precisely between two adjacent links, we follow the Denavit-Hartenberg (D-H) convention, which is commonly used for selecting frames of reference in robotic systems where each homogeneous transformation matrix representing each link's co-ordinate system at the joint with respect to the previous link's co-ordinate system. Figure 2.3 will be useful for understanding the definition of Denavit-Hartenberg frame. If the ith joint is revolute, the homogeneous transformation matrix from the ith frame to the $(i-1)$th frame is given as

$$T_i^{i-1}(\theta_i) = \begin{bmatrix} \cos\theta_n & -\cos\alpha_i \sin\theta_i & \sin\alpha_i \sin\theta_i & a_i \cos\theta_i \\ \sin\theta_i & \cos\alpha_n \cos\theta_n & -\sin\alpha_i \cos\theta_n & a_n \sin\theta_n \\ 0 & \sin\alpha_i & \cos\alpha_n & d_i \\ 0 & 0 & 0 & 1 \end{bmatrix} \quad (2.26)$$

- a_i is the length of the common normal, equal to the shortest distance between the z_{i-1} axis and the z_i axis.

- d_i is the offset, the distance from the origin of the $i-1$ coordinate frame to the intersection point of the z_{i-1} axis.

- α_i is the twist, the angle between the z_{i-1} axis and the z_i axis about the x_i axis in the right-hand sense.

- θ_n is the angle between the x_{i-1} axis and the x_i about the z_{i-1} axis in the right-hand sense.

If the i-th joint is prismatic, the homogeneous transformation matrix is

$$T_i^{i-1}(\theta_i) = \begin{bmatrix} \cos\theta_i & -\cos\alpha_i \sin\theta_i & \sin\alpha_i \sin\theta_i & 0 \\ \sin\theta_i & \cos\alpha_i \cos\theta_n i & -\sin\alpha_i \cos\theta_i & 0 \\ 0 & \sin\alpha_i & \cos\alpha_i & d_i \\ 0 & 0 & 0 & 1 \end{bmatrix} \quad (2.27)$$

where θ_i, α_i and a_i are constants and d_i is the variable distance. In what follows, the variable quantity, i.e., θ_i for revolute joints and d_i for prismatic joints will be denoted by the generalized co-ordinate q_i.

Therefore, we define r_i^i as the co-ordinate of point i on the link with respect to the i-th frame attached to i-th link, so the inertia frame co-ordinate of the point is

$$r_i = r_i^0 = T_i^0 r_i^i = T_1^0 T_2^1 \ldots T_i^{i-1} r_i^i \quad (2.28)$$

Consider the mobile manipulators shown in Figs. 2.1 and 2.2, where the four principal coordinate frames are shown: world frame O_w, platform frame O_p, manipulator base frame O_m and end effector frame O_e. Then, the manipulator's end effector position/orientation with respect to O_w is given by:

$$T_e^w = T_p^w T_b^p T_e^b \quad (2.29)$$

where T_p^w is determined by the position of the platform, T_b^p is a fixed matrix decided by the mounted point of the manipulators, and T_e^b is determined by the joint variables' vector $q = [q_1, q_2, \ldots, q_m]^T$ with m denoting the DOF of the robotic manipulator.

The end effector's position/orientation vector $x_e^w(q)$ is a nonlinear function $f(\cdot)$ of the mobile manipulator's overall configuration (posture) vector $q = [p^T, \theta^T]^T \in \mathbf{R}^n$ with $n = 3 + n_m$, i.e.

$$x_e^w = f(q) \qquad (2.30)$$

If x_e^d is the end effector's desired space vector, then one must have

$$\dot{x}_e^d = J(q)\dot{q} \qquad (2.31)$$

where $J(q)$ is the $m \times n$ Jacobian matrix of the mobile manipulator.

2.4 Dynamics of Mobile Manipulators

In general, the dynamics of mobile manipulator can be derived by two different formulations: the closed-form Lagrange-Euler formulation and forward-backward recursive Newton-Euler formulation. The Lagrange-Euler approach treats the mobile manipulator as a whole and performs the analysis using the Lagrangian function (the difference between the kinetic energy and the potential energy of the mobile robotic system), which compose each link of the mobile manipulator. The Newton-Euler approach describes the combined translational and rotational dynamics of a rigid body with respect to the each link's center of mass. The dynamics of the whole mobile manipulator can be described by the forward-backward recursive dynamic equations. Therefore, two different kinds of formulations provide different insights to the physical meaning of dynamics. Dynamic analysis is to find the relationship between the generalized co-ordinates q and the generalized forces τ. A closed-form equation like the Lagrange-Euler formulation is preferred such that we can conduct the controllers to obtain the time evolution of the generalized co-ordinates.

Thus, in the following section, only the Lagrange-Euler formulation will be discussed in detail from Section 2.4.1 to Section 2.4.4, which follows the description of the previous work [111]. Section 2.4.5 comes from lots of the previous works, such as [23], [111], etc. In this chapter, we further extend the derivation of dynamics for the manipulators presented in mobile manipulators.

2.4.1   Lagrange-Euler Equations

We briefly introduce the principle of virtual work since the Lagrange-Euler equations of motion are a set of differential equations that describe the time evolution of mechanical systems under holonomic constraints.

Consider a system consisting of l particles, with corresponding co-ordinates r_1, r_2, $\ldots$, r_l is subject to holonomic constraints as follows

$$f_i(r_1 \ldots r_l) = 0, i = 1, 2 \ldots m \tag{2.32}$$

Presence of constraint implies presence of a force (called constraint force), that forces this constraint to hold. The system under constraint (2.32) has m fewer degrees of freedom than the unconstrained system, then the co-ordinates of the l constraints are described in term of n generalized co-ordinates q_1, q_2, $\ldots$, q_n as

$$r_i = r_i(q), i = 1, 2, \ldots, l \tag{2.33}$$

where $q = [q_1, q_2, \ldots, q_n]^T$ and $q_1, q_2, \ldots, q_n$ are independent. To keep the discussion simple, l is assumed to be finite.

Differentiating the constraint function $f_i(\cdot)$ with respect to time, we obtain new constraint

$$\frac{d}{dt} f_i(r_1, r_2, \ldots, r_l) = \frac{\partial f_i}{\partial r_1} \frac{dr_1}{dt} + \cdots + \frac{\partial f_i}{\partial r_l} \frac{dr_l}{dt} = 0 \tag{2.34}$$

The constraint of the form

$$\omega_1(r_1, \ldots, r_l) dr_1 + \cdots + \omega_k(r_1, \ldots, r_k) dr_k = 0 \tag{2.35}$$

is called nonholonomic if it cannot be integrated back to $f_i(\cdot)$.

Given (2.34), by definition a set of infinitesimal displacements Δr_1, $\ldots$, Δr_l, that are consistent with the constraint

$$\frac{\partial f_i}{\partial r_1} \Delta r_1 + \cdots + \frac{\partial f_i}{\partial r_l} \Delta r_l = 0 \tag{2.36}$$

are called virtual displacements, which can be precisely defined as follows with Eq.(2.33) holding

$$\delta r_i = \sum_{j=1}^{n} \frac{\partial r_i}{\partial q_j} \delta q_j, \ i = 1, 2, \ldots, l \tag{2.37}$$

where $\delta q_1, \delta q_2, \ldots, \delta q_n$ of the generalized co-ordinates are unconstrained.

Consider a system of l-particles with the total force F_i, and suppose that

- the system has a holonomic constraint, that is, some of the particles are exposed to constraint force f_{ic};

- there are the external force f_{ie} to the particles; and

- the constrained system is in equilibrium;

then the work done by all forces applied to ith particle along each set of virtual displacement is zero,

$$\sum_{i=1}^{l} F_i^T \delta r_i = 0 \qquad (2.38)$$

If the total work done by the constraint forces corresponding to any set of virtual displacement is zero, that is

$$\sum_{i=1}^{l} f_{ic}^T \delta r_i = 0 \qquad (2.39)$$

Substituting Eq. (2.38) into (2.39), we have

$$\sum_{i=1}^{l} f_{ie}^T \delta r_i = 0 \qquad (2.40)$$

which expresses the principle of virtual work: if satisfying (2.39), the work done by external forces corresponding to any set of virtual displacements is zero. Suppose that each constraint will be in equilibrium and consider the fictitious additional force $\dot{p}_i$ for each constraint with the momentum of the i-th constraint p_i. By substituting p_i with F_i in Eq. (2.38), and the constraint forces are eliminated as before by using the principle of virtual work, we can obtain

$$\sum_{i=1}^{l} f_{ie}^T \delta r_i - \sum_{i=1}^{l} \dot{p}_i \delta r_i = 0 \qquad (2.41)$$

The virtual work by the force f_{ie} is expressed as

$$\sum_{i=1}^{l} f_{ie}^T \delta r_i = \sum_{j=1}^{n} \left(\sum_{i=1}^{l} f_{ie}^T \frac{\partial r_i}{\partial q_j} \right) \delta q_j = \sum_{j=1}^{n} \psi_j \delta q_j = \psi^T \delta q \qquad (2.42)$$

where $\psi = [\psi_1, \psi_2 \ldots \psi_n]$, $\psi_j = \sum_{i=1}^{k} f_{ie}^T \frac{\partial r_i}{\partial q_j}$ is called the j-th generalized force.

Considering $p_i = m_i \dot{r}_i$, the second summation in Eq. (2.41) becomes

$$\sum_{i=1}^{l} \dot{p}_i^T \delta r_i = \sum_{i=1}^{l} \sum_{j=1}^{n} m_i \ddot{r}_i^T \frac{\partial r_i}{\partial q_j} \delta q_j = \sum_{j=1}^{n} \eta_j \delta q_j = \eta^T \delta q \tag{2.43}$$

where $\eta = [\eta_1, \eta_2 \dots \eta_n]^T$, and $\eta_j = \sum_{i=1}^{k} m_i \ddot{r}_i^T \frac{\partial r_i}{\partial q_j}$.

Using the chain-rule, we can obtain

$$\frac{\partial v_i}{\partial \dot{q}_j} = \frac{\partial r_i}{\partial q_j} \tag{2.44}$$

since

$$v_i = \dot{r}_i = \sum_{j=1}^{n} \frac{\partial r_i}{\partial q_j} \dot{q}_j \tag{2.45}$$

We can further obtain

$$\frac{d}{dt} \left[\frac{\partial r_i}{\partial q_j} \right] = \sum_{l=1}^{n} \frac{\partial^2 r_i}{\partial q_j \partial q_l} \dot{q}_l = \frac{\partial v_i}{\partial q_j} \tag{2.46}$$

Based on the product rule of differentiation, we have $m_i \ddot{r}_i^T \frac{\partial r_i}{\partial q_j} = \frac{d}{dt}[m_i \dot{r}_i^T] - m_i \dot{r}_i^T \frac{d}{dt}[\frac{\partial r_i}{\partial q_j}]$, and considering the above three relations, we can rewrite η_j as

$$\eta_j = \sum_{i=1}^{l} m_i \ddot{r}_i^T \frac{\partial r_i}{\partial q_j} = \sum_{i=1}^{l} \left(\frac{d}{dt} \left[m_i v_i^T \frac{\partial v_i}{\partial \dot{q}_j} \right] - m_i v_i^T \frac{\partial v_i}{\partial q_j} \right) \tag{2.47}$$

Let $K = \sum_{i=1}^{l} \frac{1}{2} m_i v_i^T v_i$ be the kinetic energy, considering (2.47). We can obtain $\eta_j = \frac{d}{dt} \frac{\partial K}{\partial \dot{q}_j} - \frac{\partial K}{\partial q_j}$, $j = 1, 2, \ldots, n$. Rewriting the above equation in a vector form, we have

$$\eta = \frac{d}{dt} \frac{\partial K}{\partial \dot{q}} - \frac{\partial K}{\partial q} \tag{2.48}$$

Considering the equations from (2.41), (2.42), (2.43), and (2.48), we have

$$\left[\frac{d}{dt} \frac{\partial K}{\partial \dot{q}} - \frac{\partial K}{\partial q} - \psi \right]^T \delta q = 0 \tag{2.49}$$

Define a scalar potential energy $P(q)$ such that $\psi = -\frac{\partial P}{\partial q}$, since the virtual displacement vector δq is unconstrained and its elements δq_j are independent, which leads to $\frac{d}{dt} \frac{\partial K}{\partial \dot{q}} - \frac{\partial K}{\partial q} - \psi = 0$, it can be written as

$$\frac{d}{dt} \frac{\partial L}{\partial \dot{q}} - \frac{\partial L}{\partial q} = 0 \tag{2.50}$$

where $L(q, \dot{q}) = K(q, \dot{q}) - P(q)$ is a Lagrangian function.

Remark 2.2 *Given the generalized co-ordinates, the choice of Lagrangian is not unique for a particular set of equations of motion.*

Remark 2.3 *A necessary and sufficient condition is that F be the gradient of some scalar function P, i. e. , $F = -\frac{\partial P(r)}{\partial q}$, which in turn means that the generalized force is derivable from P by differentiating with respect to q.*

If the generalized force ψ includes an external applied force and a potential field force, suppose there exists a vector τ and a scalar potential function $P(q)$ satisfying $\psi = \tau - \frac{\partial P}{\partial q}$, then, Eq. (2.50) can be written in the form

$$\frac{d}{dt}\frac{\partial L}{\partial \dot{q}} - \frac{\partial L}{\partial q} = \tau \tag{2.51}$$

Eqs. (2.50) and/or (2.51) are called the Lagrangian equations or Lagrange-Euler equations in the robotics literature.

2.4.2 Kinetic Energy

Consider the the velocity of the point in base co-ordinates described by $v_i = \frac{dr_i}{dt} = \sum_{j=1}^{i}[\frac{\partial T_i^0}{\partial q_j}\dot{q}_j]r_i^i = \sum_{j=1}^{n}[\frac{\partial T_i^0}{\partial q_j}\dot{q}_j]r_i^i$ with $\frac{\partial T_i^0}{\partial q_j} = 0, \forall j > i$, we have

$$dK_i = \frac{1}{2}\text{trace}[\sum_{j=1}^{n}\sum_{k=1}^{n}\frac{\partial T_i^0}{\partial q_j}(r_i^i r_i^{iT} dm)\frac{\partial T_i^{0T}}{\partial q_k}\dot{q}_j\dot{q}_k] \tag{2.52}$$

Defining the 4×4 pseudo-inertia matrix for the i-th link as

$$J_i = \int_{linki} r_i^i r_i^{iT} dm \tag{2.53}$$

the total kinetic energy for the i-th link can be expressed as

$$K_i = \int_{linki} dK_i = \frac{1}{2}\text{trace}[\sum_{j=1}^{n}\sum_{k=1}^{n}\frac{\partial T_i^0}{\partial q_j}J_i\frac{\partial T_i^{0T}}{\partial q_k}\dot{q}_j\dot{q}_k] \tag{2.54}$$

Considering the generalized co-ordinates as $r_i^i = \begin{bmatrix} x & y & z & 1 \end{bmatrix}^T$, we can rewrite (2.53) as [111]

$$J_i = \begin{bmatrix} \frac{-I_i^{xx}+I_i^{yy}+I_i^{zz}}{2} & I_i^{xy} & I_i^{xz} & m_i\overline{x}_i \\ I_i^{xy} & \frac{I_i^{xx}-I_i^{yy}+I_i^{zz}}{2} & I_i^{yz} & m_i\overline{y}_i \\ I_i^{xz} & I_i^{yz} & \frac{I_i^{xx}+I_i^{yy}+I_i^{zz}}{2} & m_i\overline{z}_i \\ m_i\overline{x}_i & m_i\overline{y}_i & m_i\overline{z}_i & m_i \end{bmatrix} \tag{2.55}$$

where

$$I_i^{xx} = \int (y^2 + z^2)dm, \ I_i^{yy} = \int (x^2 + z^2)dm, \ I_i^{zz} = \int (x^2 + y^2)dm,$$

$$I_i^{xy} = \int xydm, \ I_i^{xz} = \int xzdm, \ I_i^{yz} = \int yzdm,$$

$$m_i\bar{x}_i = \int xdm, \ m_i\bar{y}_i = \int ydm, \ m_i\bar{z}_i = \int zdm,$$

with m_i as the total mass of the i-th link, and $\bar{r}_i^i = \begin{bmatrix} \bar{x} & \bar{y} & \bar{z} & 1 \end{bmatrix}^T$ as the center of mass vector of the i-th link from the i-th link co-ordinate frame and expressed in the i-th link co-ordinate frame.

Therefore, the total kinetic energy can be written as

$$
\begin{aligned}
K(q,\dot{q}) &= \frac{1}{2}\sum_{i=1}^{n}\text{trace}[\sum_{j=1}^{n}\sum_{k=1}^{n}\frac{\partial T_i^0}{\partial q_j}J_i\frac{\partial T_i^{0T}}{\partial q_k}\dot{q}_j\dot{q}_k] \\
&= \frac{1}{2}\sum_{j=1}^{n}\sum_{k=1}^{n}m_{jk}\dot{q}_j\dot{q}_k = \frac{1}{2}\dot{q}^T M(q)\dot{q}
\end{aligned}
\tag{2.56}
$$

where the jk-th element m_{jk} of the $n \times n$ inertia matrix $M(q)$ is defined as

$$m_{jk}(q) = \sum_{i=1}^{n}\text{trace}[\frac{\partial T_i^0}{\partial q_j}J_i\frac{\partial T_i^{0T}}{\partial q_k}] \tag{2.57}$$

2.4.3   Potential Energy

The total potential energy of the robot is therefore expressed as

$$P(q) = -\sum_{i=1}^{n} P_i \tag{2.58}$$

where P_i is the the potential energy of the ith link with mass m_i and center of gravity $\bar{r}_i^i$ expressed in the co-ordinates of its own frame, the potential energy of the link is given $P_i = -m_i g^T T_i^0 \bar{r}_i^i$, and the gravity vector is expressed in the base co-ordinates as $g = \begin{bmatrix} g_x & g_y & g_z & 0 \end{bmatrix}^T$.

2.4.4   Lagrangian Equations

Consider the kinetic energy $K(q,\dot{q})$ and the potential energy $P(q)$ can be expressed as $K(q,\dot{q}) = \frac{1}{2}\dot{q}^T M(q)\dot{q}$, $P(q) = -\sum_{i=1}^{n} m_i g T_i^0 \bar{r}_i^i$, so the Lagrangian function $L(q,\dot{q}) = K(q,\dot{q}) - P(q)$ is thus given by

$$L(q,\dot{q}) = \frac{1}{2}\dot{q}^T M(q)\dot{q} - P(q) \tag{2.59}$$

We can obtain

$$\frac{\partial L}{\partial \dot{q}_k} = \sum_{j=1}^{n} m_{kj} \dot{q}_j \tag{2.60}$$

$$\frac{d}{dt}\frac{\partial L}{\partial \dot{q}_k} = \sum_{j=1}^{n} m_{kj} \ddot{q}_j + \sum_{j=1}^{n} \frac{d}{dt} m_{kj} \dot{q}_j = \sum_{j=1}^{n} m_{kj} \ddot{q}_j + \sum_{j=1}^{n} \sum_{i=1}^{n} \frac{\partial m_{kj}}{\partial q_i} \dot{q}_i \dot{q}_j \tag{2.61}$$

$$\frac{\partial L}{\partial q_k} = \frac{1}{2} \sum_{i=1}^{n} \sum_{j=1}^{n} \frac{\partial m_{ij}}{\partial q_k} \dot{q}_i \dot{q}_j - \frac{\partial P}{\partial q_k} \tag{2.62}$$

for $k = 1, 2, \ldots, n$.

Considering the symmetry of the inertia matrix, we have

$$\sum_{i=1}^{n} \sum_{j=1}^{n} \frac{\partial m_{kj}}{\partial q_i} \dot{q}_i \dot{q}_j = \frac{1}{2} \sum_{i=1}^{n} \sum_{j=1}^{n} [\frac{\partial m_{kj}}{\partial q_i} \dot{q}_i \dot{q}_j + \frac{\partial m_{ki}}{\partial q_i} \dot{q}_j \dot{q}_i]$$

$$= \frac{1}{2} \sum_{i=1}^{n} \sum_{j=1}^{n} [\frac{\partial m_{kj}}{\partial q_i} + \frac{\partial m_{ki}}{\partial q_j}] \dot{q}_i \dot{q}_j \tag{2.63}$$

and the Lagrange-Euler equations can then be written as

$$\sum_{j=1}^{n} m_{kj} \ddot{q}_j + \sum_{i=1}^{n} \sum_{j=1}^{n} (\frac{\partial m_{kj}}{\partial q_i} - \frac{1}{2}\frac{\partial m_{ij}}{\partial q_k}) \dot{q}_i \dot{q}_j + \frac{\partial P}{\partial q_k}$$

$$= \sum_{j=1}^{n} m_{kj} \ddot{q}_j + \frac{1}{2} \sum_{i=1}^{n} \sum_{j=1}^{n} [\frac{\partial m_{ki}}{\partial q_j} + \frac{\partial m_{kj}}{\partial q_i} - \frac{\partial m_{ij}}{\partial q_k}] \dot{q}_i \dot{q}_j$$

$$= \sum_{j=1}^{n} m_{kj} \ddot{q}_j + \sum_{i=1}^{n} \sum_{j=1}^{n} c_{ijk} \dot{q}_i \dot{q}_j = \tau_k \tag{2.64}$$

where c_{ijk} is the Christoffel symbol (of the first kind) defined as

$$c_{ijk}(q) \triangleq \frac{1}{2}[\frac{\partial m_{kj}(q)}{\partial q_i} + \frac{\partial m_{ki}(q)}{\partial q_j} - \frac{\partial m_{ij}(q)}{\partial q_k}] \tag{2.65}$$

Define $g_k(q) = \frac{\partial P(q)}{\partial q_k}$, then the Lagrange-Euler equations can be written as

$$\sum_{j=1}^{n} m_{kj}(q) \ddot{q}_j + \sum_{i=1}^{n} \sum_{j=1}^{n} c_{ijk}(q) \dot{q}_i \dot{q}_j + g_k(q) = \tau_k \quad k = 1, 2, \ldots, n \tag{2.66}$$

It is common to write the above equations in matrix form

$$M(q)\ddot{q} + C(q, \dot{q})\dot{q} + G(q) = \tau \tag{2.67}$$

where the kj-th element of $C(q, \dot{q})$ defined as

$$c_{kj} = \sum_{i=1}^{n} c_{ijk} \dot{q}_i = \sum_{i=1}^{n} \frac{1}{2} \left[\frac{\partial m_{kj}}{\partial q_i} + \frac{\partial m_{ki}}{\partial q_j} - \frac{m_{ij}}{\partial q_k} \right] \dot{q}_i \qquad (2.68)$$

To facilitate the understanding of the control problems, and to help design controllers for the above systems, it is essential to have a thorough study of the mathematical properties of the system.

2.4.5   Properties of Dynamic Equations

There are some properties summarized for the dynamics of mobile manipulators, which are convenient for controller design.

Property 2.1 *Inertia matrix $M(q)$ is symmetric, i.e. $M(q) = M^T(q)$.*

Property 2.2 *Inertia matrix $M(q)$ is uniformly positive definite, and bounded below and above, i.e., $\exists 0 < \alpha \le \beta < \infty$, such that $\alpha I_n \le M(q) \le \beta I_n, \forall q \in \mathbf{R}^n$, where I_n is the $n \times n$ identity matrix.*

Property 2.3 *The inverse of inertia matrix $M^{-1}(q)$ exists, and is also positive definite and bounded.*

Property 2.4 *Centrifugal and coriolis force $C(q, \dot{q})\dot{q}$ is quadratic in $\dot{q}$.*

Property 2.5 *It may be written in $C(q, \dot{q})\dot{q} = C_1(q)C_2[\dot{q}\dot{q}] = C_3(q)[\dot{q}\dot{q}] + C_4(q)[\dot{q}^2]$, where $[\dot{q}\dot{q}] = [\dot{q}_1\dot{q}_2, \ \dot{q}_1\dot{q}_3, \ \dots, \dot{q}_{n-1}\dot{q}_n]^T$ and $[\dot{q}^2] = [\dot{q}_1^2, \dot{q}_2^2, \ \dots, \dot{q}_n^2]^T$.*

Property 2.6 *Given two n-dimensional vectors x and y, the matrix $C(q, \dot{q})$ defined by Eq. (2.67) implies that $C(q, x)y = C(q, y)x$.*

Property 2.7 *The 2-norm of $C(q, \dot{q})$ satisfies $\|C(q, \dot{q})\| \le k_c(q)\|\dot{q}\|$, where $k_c(q) = \frac{1}{2}\max_{q \in \mathbf{R}^n} \sum_{k=1}^{n} \|C_k(q)\|$. For revolute robots, k_c is a finite constant since the dependence of $C_k(q)$, $k = 1, 2, \ \dots, n$, on q appears only in terms of sine and cosine functions of their entries.*

Property 2.8 *Gravitational force $G(q)$ can be derived from the gravitational potential energy function $P(q)$, i.e. $G(q) = \partial P(q)/\partial q$, and is also bounded, i.e., $\|G(q)\| \le k_{G(q)}$, where $k_{G(q)}$ is a scalar function which may be determined for any given mobile manipulator. For revolute joints, the bound is a constant independent of q whereas for prismatic joints, the bound may depend on q.*

Property 2.9 *If only articulated mobile manipulators are considered, the dependence of $M(q)$, $C(q, \dot{q})$ and $G(q)$ on q will appear only in terms of sine and cosine functions in their entries, so that $M(q)$, $C(q, \dot{q})$ and $G(q)$ have bounds that are independent of q.*

Property 2.10 *By defining each coefficient as a separate parameter, the dynamics can be written in the linear in the parameters (LIPs) form*

$$M(q)\ddot{q} + C(q, \dot{q})\dot{q} + G(q) = Y(q, \dot{q}, \ddot{q})P \tag{2.69}$$

where $Y(q, \dot{q}, \ddot{q})$ is an $n \times r$ matrix of known functions known as the regressor matrix, and P is an r dimensional vector of parameters.

Remark 2.4 *The above equation can also be written as*

$$M(q)\ddot{q}_r + C(q, \dot{q})\dot{q}_r + G(q) = \Phi(q, \dot{q}, \dot{q}_r, \ddot{q}_r)P \tag{2.70}$$

where $\dot{q}_r$ and $\ddot{q}_r$ are the corresponding n-dimensional vectors.

Property 2.11 *The matrix $N(q, \dot{q})$ defined by $N(q, \dot{q}) = \dot{M}(q) - 2C(q, \dot{q})$ is skew-symmetric, i.e., $n_{kj}(q, \dot{q}) = -n_{jk}(q, \dot{q})$, if $C(q, \dot{q})$ is defined using the Christoffel symbols.*

Property 2.12 *Since $M(q)$ and $\dot{M}(q)$ are symmetric matrices, the skew-symmetry of the matrix $\dot{M}(q) - 2C(q, \dot{q})$ can also be seen from the fact that $\dot{M}(q) = C(q, \dot{q}) + C^T(q, \dot{q})$.*

Property 2.13 *The system is passive from τ to $\dot{q}$.*

Property 2.14 *Even though the skew-symmetry property of $N(q, \dot{q})$ is guaranteed if $C(q, \dot{q})$ is defined by the Christoffel symbols, it is always true that $\dot{q}^T[\dot{M}(q) - 2C(q, \dot{q})]\dot{q} = 0$.*

Property 2.15 *The system is feedback linearizable, i.e., there exists a non-linear transformation such that the transformed system is a linear controllable system.*

Remark 2.5 *Let $X = [q^T, \dot{q}^T]^T$, then the system can be fully linearised by using the non-linear control law*

$$\tau = M(q)U + C(q, \dot{q})\dot{q} + G(q) \tag{2.71}$$

as

$$\dot{X} = \begin{bmatrix} 0 & I \\ 0 & 0 \end{bmatrix} X + \begin{bmatrix} 0 \\ I \end{bmatrix} U \qquad (2.72)$$

with $U = \ddot{q}$. *Thus, linear control techniques can be applied to the resulting linearized model.*

2.5   Dynamics in Cartesian Space

It is desirable to express the dynamics of robots in Cartesian space or task space variables rather than in joint space variables q, since the tasks of mobile robotic manipulators are often expressed in Cartesian space. Assume the mobile manipulator is generally redundant, and let $x = [r^T, \theta^T]^T \in \mathbf{R}^n$ with r and θ be the position and orientation in the base frame. According to forward kinematics, x can be expressed as a non-linear function of q as $x = h(q)$.

Redundancy

Let x denote the position and orientation vector of the end-effector, then x is related to $\dot{q}$ the Jacobian matrix $J(q)$ as

$$\dot{x} = J(q)\dot{q} \qquad (2.73)$$

As it is assumed that the manipulators are redundant, considering all the manipulators acting on the object at the same time yields

$$\dot{q} = J^+(q)\dot{x} + (I - J^+(q)J(q))\Gamma_q \qquad (2.74)$$

where $(I - J^+(q)J(q))\Gamma_q$ is a vector in the null space of $J^+(q)$, which describes the redundancy of the robot. By choosing $\Gamma_q = 0$, we have

$$\dot{q} = J^+(q)\dot{x} \qquad (2.75)$$

Differentiating (2.75) with respect to time t leads to

$$\ddot{q} = J^+(q)\ddot{x} + \frac{d}{dt}(J^+(q)J(q))\dot{x} \qquad (2.76)$$

Using equations (2.75) and (2.76), the dynamic model is given by

$$M(q)J^+(q)\ddot{x} + \left(M(q)\frac{d}{dt}(J^+(q)) + C(q, \dot{q})J^+(q) \right)\dot{x} + G = \tau \qquad (2.77)$$

Multiplying both sides of (2.77) by $J^{+T}(q)$, the dynamics of mobile manipulators are given by

$$M_x(q)\ddot{x} + C_x(q,\dot{q})\dot{x} + G_x(q) = \tau_x \qquad (2.78)$$

where

$$
\begin{aligned}
M_x(q) &= J^{+T}(q)M(q)J^+(q) \\
C_x(q,\dot{q}) &= J^{+T}(q)M(q)\frac{d}{dt}(J^+(q)) + J^{+T}(q)C(q,\dot{q})J^+(q) \\
G_x(q) &= J^{+T}(q)G(q) \\
\tau_x &= J^{+T}(q)\tau
\end{aligned}
$$

The dynamics (2.78) have the following structure properties, which can be exploited to facilitate the motion control design.

Property 2.16 *The matrix $M_x(q)$ is symmetric, positive definite, and is bounded from below and above, i.e., $\lambda_{min}I \le M_x(q) \le \lambda_{max}I$, where λ_{min} and $\lambda_{max} \in \mathbf{R}^n$ denote the minimum and maximum eigenvalues of $M_x(q)$.*

Property 2.17 *The matrix $\dot{M}_x - 2C_x$ is skew-symmetric, that is, $x^T(\dot{M}_x - 2C_x)x = 0, \forall x \in \mathbf{R}^n$.*

If $J(q)$ is not square, let the pseudo inverse of $J(q)$ be $J^+(q)$, and F_x be the force in Cartesian space which causes the changes in x. The relationship between F_x and τ is given by

$$F_x = (J^T(q))^+\tau \qquad (2.79)$$

Due to redundancy, the joint torque is not unique for a given F_x and can be formulated as

$$\tau = J^T(q)F_x + [I - J^T(q)(J^T(q))^+]\Gamma_F \qquad (2.80)$$

where I is an $n \times n$ identity matrix, Γ_F is any $n \times 1$ vector, and $[I - J^T(q)(J^T(q))^+]\Gamma_F$ is a vector in the null space of $(J^T(q))^+$, which describes the redundancy of the robot.

By a similar argument,

$$\ddot{q} = J^+(q)(\ddot{x} - \dot{J}(q)\dot{q}) + [I - J^+(q)J(q)]\Gamma_x \qquad (2.81)$$

where Γ_x is any $n \times 1$ vector and $[I - J^+(q)J(q)]\Gamma_x$ is a vector in the null space of $J(q)$, which also describes the redundancy of the robot.

By choosing $\Gamma_x = 0$ and $\Gamma_F = \ddot{q}$, we have the Cartesian dynamics of robots as

$$M(q)J^+(q)(\ddot{x} - \dot{J}(q)\dot{q}) + C(q,\dot{q})\dot{q} + G(q)$$
$$= J^T(q)F_x + [I - J^T(q)J^+(q)]\ddot{q} \qquad (2.82)$$

Nonredundancy Consider a non-redundant mobile manipulator, and assume that the robot manipulator is away from the workspace singularities, and we have

$$\dot{x} = J(q)\dot{q} \qquad (2.83)$$
$$\ddot{x} = J(q)\ddot{q} + \dot{J}(q)\dot{q} \qquad (2.84)$$

where the Jacobian matrix $J(q)$ is defined as $J(q) = \partial h(q)/\partial q$ which is a square matrix, i.e., $|J(q)| \neq 0$ and $J^{-1}(q)$ exists, therefore, we have

$$\tau = J^T(q)F_x \qquad (2.85)$$

Substituting Eqs. (2.83) and (2.84) into Eq. (2.79), yields the Cartesian dynamics of robots as

$$M_x(q)\ddot{x} + C_x(q,\dot{q})\dot{x} + G_x(q) = F_x \qquad (2.86)$$

where

$$M_x(q) = J^{-T}(q)M(q)J^{-1}(q) \qquad (2.87)$$
$$C_x(q,\dot{q}) = J^{-T}(q)[C(q,\dot{q}) - M(q)J^{-1}(q)\dot{J}(q)]J^{-1}(q) \qquad (2.88)$$
$$G_x(q) = J^{-T}(q)G(q) \qquad (2.89)$$

It can be observed that $M_x(q)$, $C_x(q,\dot{q})$, and $G_x(q)$ are functions of q and $\dot{q}$. Thus, strictly speaking, the Cartesian dynamics are not completely given in terms of x, $\dot{x}$, $\ddot{x}$. Most of the properties for joint space dynamics of robots are applicable to Cartesian space dynamics as long as $J(q)$ is non-singular [115, 168]. For example,

Property 2.18 $M_x(q)$ *is symmetric, positive definite and bounded above and below.*

Property 2.19 *The matrix* $N_x(q) = \dot{M}_x(q) - 2C_x(q,\dot{q})$ *is skew-symmetric if* $C(q,\dot{q})$ *is defined using the Christoffel symbols.*

Property 2.20 *The property of linear in the parameters holds, i.e.,*

$$M_x(q)\ddot{x}_r + C_x(q,\dot{q})\dot{x}_r + G_x(q) = Y_x(x,\dot{x},\dot{x}_r,\ddot{x}_r)P_x \qquad (2.90)$$

where P_x is the vector of robot parameters, and $Y_x(x,\dot{x},\ddot{x})$ is the known Cartesian regressor matrix.

Property 2.21 *The system is feedback linearizable, i.e., there exists a nonlinear transformation such that the transformed system is a linear controllable system.*

2.6 Conclusion

In this chapter, we first describe the kinematics and dynamics model for robotic manipulators and wheeled mobile platforms. The Lagrange-Euler equations of motion have been introduced. Based on the Lagrange-Euler formulation, the dynamics for a general n-link mobile robotic manipulator have been presented, which incorporates the dynamic interactions between the mobile platform and the manipulator. The structural properties of robots, which are useful for controller design, have also been briefly summarized. Finally, the dynamic equations for 2-DOF mobile manipulators and 3-DOF mobile manipulators have been derived in a step-by-step manner, which can be found in Chapter 9.

Property 2.2. ...

Property 2.2. ...

2.6 Conclusion

3

Path Planning and Motion Generation

CONTENTS

3.1   Path Planning of Mobile Manipulators

3.1.1   Introduction

Motion planning of mobile manipulators is concerned with obtaining open loop controls that steer the system from an initial configuration to a final one, without violating nonholonomic constraints or collision avoidance constraints. Moving mobile manipulator systems presents many unique problems that are due to the coupling of holonomic manipulators with nonholonomic platforms. The path planning for a mobile manipulator accomplishing a sequence of coordination and manipulation tasks is formulated in [121] as a nonlinear optimization problem with state boundary equality constraints and a general cost function, which was solved using a stochastic algorithm of a simulated annealing. Motion planning of mobile manipulators to execute multiple tasks consisting of a sequence of pre-specified trajectory in a fixed world frame [122] is formulated as a global optimization problem and simultaneously obtains the motion trajectory set and commutation configurations. A general approach based on the calculus of variations was proposed for motion planning for nonholonomic cooperating mobile robots to obtain optimal trajectories and optimal actuator forces/torques in the presence of obstacles [123] such that geometric constraints, kinematic constraints, and dynamic constraints can be easily incorporated into the planning scheme. Navigating a mobile manipulator among obstacles had been studied in [124] by simultaneously considering the obstacle avoidance and the coordination. The developed control allows the system to retain optimal or sub-optimal configurations while the manipulator avoids obstacles using potential functions. In approach, it was assumed that only the manipulator may encounter the obstacle, while in the same study [125], the obstacle avoidance by the entire mobile manipulator system was considered and the proposed nonholonomic motion planner is based on a discontinuous feedback law under the influence of a potential field. Motion planning applicable to handling deformable material by multiple nonholonomic mobile manipulators was described in the obstacles environment [126], which is based on a new class of nonsmooth Lyapunov functions and an extension of the navigation function. The dipolar inverse Lyapunov functions and potential field technique using diffeomorphic transformations were introduced for nonholonomic control. The standard definition of manipulability was generalized to the case of mobile manipulators in [63], and the optimization of criteria inherited from manipulability considerations was given to generate the controls of the system when its end effector motion was imposed. Path planning of nonholonomic mobile platforms with manipulators in the presence of obstacles was developed in [127], which employs smooth and continuous functions such as polynomials and is based on mapping the nonholonomic constraint to a space where it can be satisfied trivially. Motion planning for a mobile manipulator with end-effector along a given path was developed by the randomized generation of configurations that are compatible with the end effector path constraint [128]. A modular fuzzy navigation method and a robust control in unstructured environments were developed for the navigation and control of mobile manipulators by using fuzzy reactive motion planning and robust adaptive control [129]. The probabilistic road map and the fuzzy reactive planner based on elastic band for the vehicle platform to avoid unknown static/dynamic obstacles are also presented [129].

Nonholonomic mobile manipulators, subject to nonholonomic constraints, are fully linearized and input-output decoupled by means of nonlinear dynamic feedback. Based on this result, one can plan smooth trajectories joining in finite time the given initial and final configuration of the robot by polynomial functions. In order to achieve the autonomous obstacle avoidance, the generated path is iteratively deformed to get away from obstacles and satisfy the nonholonomic constraints and collision avoidance constraints, and yields admissible input trajectories that could drive both the manipulator and the platform to a desired configuration. The core idea of the approach is to deform the normal vector along the generated path in order to modify this path, making the collision constraints decrease. Illustrative examples verify the planning methodology in obstacle-free and obstructed environments. The main contributions of this section are listed as follows:

(i) the mobile manipulators are fully linearized and input-output decoupled by means of nonlinear dynamic feedback;

(ii) design smooth and continuous polynomials based on polar form joining the given initial and desired final configuration of the robot in finite time; and

(iii) achieve the autonomous obstacle avoidance by iteratively deforming the generated path such that the mobile manipulator gets away from obstacles and satisfy the nonholonomic constraints and admissible input trajectories to drive both the manipulator and the platform to a desired configuration.

The rest of the chapter is organized as follows. Section 3.1.2 describes the problem formulation considered. Dynamics and kinematics are set up in Section 3.1.3 and the dynamic feedback linearization design is presented. Planning state-to-state trajectories is formulated and solved in 3.1.4.

3.1.2 Preliminaries and Problem Formulation

Consider the general case of mobile manipulators consisting of on-board robotic manipulators mounted on wheeled mobile platforms. Assume that the robot must move on a planar plane, then the motion of the mobile platform on the plane can be described in Fig. 3.1. The coordinate systems are defined as follows: an arbitrary inertial base frame $OXYZ$ is fixed on the motion plane, while $O_v X_v Y_v Z_v$ is a frame attached to the mobile platform. In frame $O_v X_v Y_v Z_v$, the coordinate axis Y_v is along the coaxial line of the two fixed wheels; X_v is vertical with Y_v and passes through the mid point of the line segment connecting the two-fixed wheel centers.

The mobile manipulator configuration is defined by a vector q of n independent coordinates, called generalized coordinates of the mobile manipulator. We can choose: $q = [q_1, q_2, \ldots, q_n]^T = [q_v^T, q_a^T]^T$ and we notice that $n = n_v + n_a$, where n_v and n_a are the dimensions of the generalized spaces associated to the mobile platform and to the robotic manipulator, respectively.

Since the mobile platform is subjected to nonholonomic constraints, the $(n_v - m)$ nonintegrable and independent velocity constraints can be expressed as

$$A(q_v)\dot{q}_v = 0 \tag{3.1}$$

where $A(q_v) \in \mathbf{R}^{(n_v-m)\times n_v}$ is the matrix associated with the constraint.

Consider positions and orientations of the mobile platform's center with respect to target positions in the polar frame denoted with χ, η, respectively, as shown in Fig. 3.1, and the position and orientation of the manipulator's end-effector relative to the mobile platform denoted with $[\rho, \beta, \gamma]^T$.

$$\chi = \sqrt{x_v^2 + y_v^2} \tag{3.2}$$

$$\eta = \text{atan2}(y_v, x_v) \tag{3.3}$$

$$\rho = \sqrt{x_a^2 + y_a^2 + z_a^2} \tag{3.4}$$

$$\beta = \text{atan2}(y_a, x_a) \tag{3.5}$$

$$\gamma = \text{atan2}(z_a, \sqrt{x_a^2 + y_a^2}) \tag{3.6}$$

where (x_v, y_v) is the position of the mobile platform in the fixed frame, and (x_a, y_a, z_a) is the position of the end-effector in $OX_vY_vZ_v$, χ and ρ are the distances to target for the mobile platform and the end-effector, respectively, η is the heading angle of the mobile platform in the fixed frame, β and γ are the pitch angle and yaw angle for the manipulator, respectively. For the mobile platform, the target position is assumed to be $(0, 0, 0)$, and for the manipulator, the target position is assumed to be $\rho_d, \beta_d, \gamma_d$. Our objectives are to design a feasible path to drive the robot to follow it from the start position to the target position.

3.1.3 Dynamics and Kinematics of Mobile Manipulators

Consider an n-DOF redundant manipulator mounted on a nonholonomic mobile platform as shown in Fig. 3.1. The dynamics of a mobile manipulator

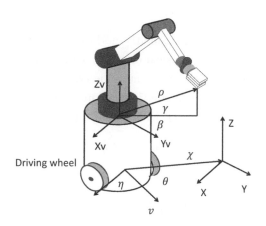

FIGURE 3.1

The mobile manipulator in the spherical coordinate.

consists of the coupled dynamics of the mobile platform and the manipulator:

$$M(q)\ddot{q} + C(q, \dot{q})\dot{q} + G(q) = B(q)\tau + F \tag{3.7}$$

where $M(q) \in \mathbf{R}^{n \times n}$ is the symmetric bounded positive definite inertia matrix, $C(\dot{q}, q)\dot{q} \in \mathbf{R}^n$ denotes the centripetal and Coriolis torques; $G(q) \in \mathbf{R}^n$ is the gravitational torque vector; $\tau \in \mathbf{R}^k$ is the vector of control input; $B(q) \in \mathbf{R}^{n \times k}$ is a full rank input transformation matrix and assumed to be known because it is a function of fixed geometry of the system; $q = [q_v^T, q_a^T]^T \in \mathbf{R}^n$, $q_v \in \mathbf{R}_v^n$ describes the vector of generalized coordinates for the mobile platform; $q_a \in \mathbf{R}_a^n$ is the vector of generalized coordinates for the manipulator; and $n = n_a + n_v$; $F = J^T \lambda \in \mathbf{R}^n$ denotes the vector of constraint forces with $J(q) = [A, 0]$ and $\lambda = [\lambda_n, 0]^T$ for nonholonomic constraints of the system.

Since $A^T(q_v) \in \mathbf{R}^{n_v \times (n_v - m)}$ in (3.1), and the rank of A is $n_v - m$, it is always possible to find an $m + n_a$ rank matrix $R(q) \in \mathbf{R}^{n \times m}$ formed by a set of smooth and linearly independent vector fields spanning the null space of $J(q)$, i.e.,

$$R^T(q)J^T(q) = 0 \tag{3.8}$$

Denote $R(q) = [r_1(q_v), ..., r_m(q_v), r_a(q_a)]$ and define an auxiliary time function

$\dot{z}(t) = [\dot{z}_1(t), ..., \dot{z}_m(t), \dot{z}_a(t)]^T \in \mathbf{R}^m$ such that

$$\dot{q} = R(q)\dot{z}(t) = r_1(q_v)\dot{z}_1(t) + ... + r_m(q_v)\dot{z}_m(t) + r_a(q_v)\dot{z}_a(t) \qquad (3.9)$$

Equation (3.9) is the so-called kinematic model of nonholonomic systems in the literature. Usually, $\dot{z}(t)$ has physical meaning, consisting of the linear velocity v, the angular velocity ω and the joint velocity, i.e., define $\dot{z}(t) = [v \; \omega \; \dot{q}_a]^T$. Equation (3.9) describes the kinematic relationship between the motion vector $q(t)$ and the velocity vector $\dot{z}(t)$. Differentiating (3.9) yields

$$\ddot{q} = \dot{R}(q)\dot{z} + R(q)\ddot{z} \qquad (3.10)$$

From (3.9), $\dot{z}$ can be obtained from q and $\dot{q}$ as

$$\dot{z} = [R^T(q)R(q)]^{-1}R^T(q)\dot{q} \qquad (3.11)$$

The dynamic equation (3.7), which satisfies the nonholonomic constraint (3.1), can be rewritten in terms of the internal state variable $\dot{z}$ as

$$M(q)R(q)\ddot{z} + [M(q)\dot{R}(q) + C(q,\dot{q})R(q)]\dot{z} + G(q) = B(q)\tau + J^T(q)\lambda \qquad (3.12)$$

Substituting (3.9) and (3.10) into (3.7), and then pre-multiplying (3.7) by $R^T(q)$, the constraint matrix $J^T(q)\lambda$ can be eliminated by virtue of (3.8). As a consequence, we have the transformed nonholonomic system

$$\dot{q} = R(q)\dot{z} = r_1(q)\dot{z}_1 + ... + r_m(q)\dot{z}_m + r_a(q)\dot{z}_a \qquad (3.13)$$

$$M_1(q)\ddot{z} + C_1(q,\dot{q})\dot{z} + G_1(q) = B_1(q)\tau \qquad (3.14)$$

where

$$\begin{aligned} M_1(q) &= R^T M(q)R \\ C_1(q,\dot{q}) &= R^T[M(q)\dot{R} + C(q,\dot{q})R] \\ G_1(q) &= R^T G(q) \\ B_1(q) &= R^T B(q) \end{aligned}$$

which is more appropriate for the controller design as the constraint λ has been eliminated from the dynamic equation.

The kinematic equations in Cartesian coordinates corresponding to (3.9) are

$$\begin{bmatrix} x_v \\ y_v \\ \theta \end{bmatrix} = \begin{bmatrix} v\cos\theta \\ v\sin\theta \\ \omega \end{bmatrix} \qquad (3.15)$$

Considering (3.2) and (3.3), the kinematic equations in polar coordinates become

$$
\begin{bmatrix} \dot{\chi} \\ \dot{\phi} \\ \dot{\theta} \end{bmatrix} = \begin{bmatrix} \upsilon \cos(\eta - \theta) \\ \frac{\upsilon}{\chi} \sin(\eta - \theta) \\ \omega \end{bmatrix}
\tag{3.16}
$$

We can modify (3.16) as

$$
T_1(\chi, \eta, \theta) \begin{bmatrix} \dot{\chi} & \dot{\eta} & \dot{\theta} \end{bmatrix}^T = \begin{bmatrix} \upsilon & \omega \end{bmatrix}^T
\tag{3.17}
$$

where $T_1(x_1, x_2, x_3) = \begin{bmatrix} \cos(x_2 - x_3) & -x_1 \sin(x_2 - x_3) & 0 \\ \sin(x_2 - x_3) & x_1 \cos(x_2 - x_3) & 1 \end{bmatrix}$.

Assume that the arm of mobile manipulator is a n_a DOF series-chain multilink redundant spatial manipulator with holonomic constraints (i.e., geometric constraints) as shown in Fig. 3.1. The vector q_a can be further rearranged and partitioned into $q_a = [q_{a1}^T, q_{a2}^T]^T$, $q_{a1}^T = [q_1, \ldots, q_k] \in \mathbf{R}^k$ describes the joint variables associated with β, $q_{a2}^T = [q_{k+1}, \ldots, q_{n_a}] \in \mathbf{R}^{n_a - k}$ denotes the joint variable associated with γ, then

$$
\rho = f(q_a)
\tag{3.18}
$$

$$
\beta = \sum_{i=1}^{k} q_i
\tag{3.19}
$$

$$
\gamma = \sum_{i=k+1}^{n_a} q_i
\tag{3.20}
$$

where $f : q_a \longmapsto \rho$ is any diffeomorphism.

Differentiating (3.4) yields

$$
\begin{aligned}
\dot{\rho} &= \frac{x_a \dot{x}_a + y_a \dot{y}_a + z_a \dot{z}_a}{\sqrt{x_a^2 + y_a^2 + z_a^2}} \\
&= \begin{bmatrix} \frac{x_a}{\rho} & \frac{y_a}{\rho} & \frac{z_a}{\rho} \end{bmatrix} \begin{bmatrix} \dot{x}_a & \dot{y}_a & \dot{z}_a \end{bmatrix}^T \\
&= H J \dot{q}_a
\end{aligned}
\tag{3.21}
$$

where $T_2 = \begin{bmatrix} \cos\beta \sin\gamma & \sin\beta \sin\gamma & \cos\gamma \end{bmatrix}^T$.

Differentiating (3.5) and (3.6) yields

$$
\dot{\beta} = \sum_{i=1}^{k} \dot{q}_i
\tag{3.22}
$$

$$
\dot{\gamma} = \sum_{i=k+1}^{n_a} \dot{q}_i
\tag{3.23}
$$

Integrating (3.21) and (3.22), we can obtain

$$\begin{bmatrix} \dot{\rho} & \dot{\beta} & \dot{\gamma} \end{bmatrix}^T = T_2 \dot{q}_a \tag{3.24}$$

where $T_2 = \begin{bmatrix} HJ_{a1} & HJ_{a2} \\ I_{1\times k} & 0 \\ 0 & I_{1\times n_a - k} \end{bmatrix}$, $q_a = \begin{bmatrix} q_{a1} \\ q_{a2} \end{bmatrix}$, and $I_{1\times k} = \underbrace{[1, \ \dots, 1]}_{k}$.

Considering (3.17) and (3.24), we can obtain

$$\dot{z} = \begin{bmatrix} \upsilon \\ \omega \\ \dot{q}_a \end{bmatrix} = \begin{bmatrix} T_1 & 0 \\ 0 & T_2^{-1} \end{bmatrix} \begin{bmatrix} \dot{\chi} \\ \dot{\eta} \\ \dot{\theta} \\ \dot{\rho} \\ \dot{\beta} \\ \dot{\gamma} \end{bmatrix} \tag{3.25}$$

Let $\mathcal{T} = \begin{bmatrix} T_1 & 0 \\ 0 & T_2^{-1} \end{bmatrix}$ and $\dot{x} = \begin{bmatrix} \dot{\chi} & \dot{\eta} & \dot{\theta} & \dot{\rho} & \dot{\beta} & \dot{\gamma} \end{bmatrix}^T$, considering (3.14), we can obtain

$$\mathcal{M}\ddot{x} + \mathcal{C}\dot{x} + \mathcal{G} = \mathcal{U} \tag{3.26}$$

where

$$
\begin{aligned}
\mathcal{M} &= \mathcal{T}^T M_1(q)\mathcal{T} \\
\mathcal{C} &= \mathcal{T}^T [M_1(q)\dot{\mathcal{T}} + C_1(q,\dot{q})\mathcal{T}] \\
\mathcal{G} &= \mathcal{T}^T G(q) \\
\mathcal{U} &= \mathcal{T}^T B_1(q)\tau
\end{aligned}
$$

3.1.4   Motion Generation

A prerequisite for the successful use of mobile manipulators is the availability of a planning methodology that can generate feasible paths for driving the end effector to the desired coordinates without violating system nonholonomic constraints. However, in many applications, it is required that the platform position and orientation are also specified for several reasons. Such reasons include the particular site geometry or ground morphology, the avoidance of manipulator joint limits or singularities, and the maximization of a systems's manipulability or force output. Moreover, the calculated paths must be computationally inexpensive to compute and should be able to steer the system away from obstacles, which may exist in its workspace. Since nonholonomy is associated with the the system, it is allowed to decouple the systems by input-output linearization. Then we design an admissible path for the mobile manipulator that can drive it from an initial position and orientation to a final desired one. An advantage of this approach is that it is easy to be extended to mobile systems with multiple manipulators on board.

The main idea of exact feedback linearization is to transform a complicated nonlinear control system into a relatively simple, decoupling, and linear one such that the established linear control theory and techniques can be exploited to the control design. Compared with the traditional linearized approaches (e.g., Jacobian equilibrium-based linearization), the exact feedback linearization does not result in the information loss of the dynamics of interest.

Let the state variables be $\mathcal{X} = [x_1^T, x_2^T]^T$ with $x_1 = x$ and $x_2 = \dot{x}$, output variable $y = x_1$, considering (3.26), we can obtain

$$\dot{\mathcal{X}} = \mathcal{F}(x) + \mathcal{H}(x)\mathcal{U}, \quad y = x_1, \quad x_0 = 0 \tag{3.27}$$

where x_0 is the initial state and $\mathcal{F}(x)$, $\mathcal{H}(x)$ and $\mathcal{U}$ are

$$\mathcal{F}(x) \quad = \quad \begin{bmatrix} x_2 \\ \Phi \end{bmatrix} \tag{3.28}$$

$$\mathcal{H}(x) \quad = \quad \begin{bmatrix} 0 \\ \mathcal{M}^{-1} \end{bmatrix} \tag{3.29}$$

where $\Phi = -\mathcal{M}^{-1}(\mathcal{C}+\mathcal{G})x_2$. It is well known that the nonlinear state feedback

$$\mathcal{U} = \mathcal{M}[\mu - \Phi] \tag{3.30}$$

will serve to linear and decouple the input/output map of the system (3.27) such that

$$\ddot{y} = \mu \tag{3.31}$$

where μ is an exogenous input vector.

Planning a feasible motion on the equivalent representation (3.31) can be formulated as an interpolation problem using smooth parametric functions $y(s)$ and with a timing law $s = s(t)$. For simplicity, we directly generate trajectories $y(\lambda)$ as

$$y(\lambda) \quad = \quad \sum_{i=0}^{n} a_i \lambda^i \tag{3.32}$$

with the normalized time $\lambda_i = t/T$.

Determine minimal order polynomial curves which interpolate the given initial configuration $q(0) = [y(0), \dot{y}(0), \ddot{y}(0)]^T$ and the final configuration $q(1) = [y(1), \dot{y}(1), \ddot{y}(1)]^T$.

The general constraint conditions for the object can be expressed as

$$
\begin{aligned}
y(0) &= y_0, \quad y'(0) = 0, \quad y''(0) = 0 & (3.33) \\
y(1) &= y_1, \quad y'(1) = 0, \quad y''(1) = 0 & (3.34)
\end{aligned}
$$

The state trajectory associated to the linearizing output trajectory (3.32) that solves the planning problem is obtained by pure algebraic computations using (3.31). Moreover, the open-loop commands that realize this trajectory are

$$
\mu_i = 20a_5\lambda^3 + 12a_4\lambda^2 + 6a_3\lambda + 2a_2
$$

which represent the nominal inputs of system (3.27), and produce the inputs μ. The polynomial coefficients are detailed by these close-form expressions

$$
\begin{aligned}
a_{i0} &= y_0, \; a_{i1} = y_0', \; a_{i2} = 0, \\
a_{i3} &= 10(y_1 - y_0), \; a_{i4} = -15(y_1 - y_0), \; a_{i5} = 6(y_1 - y_0) \quad (3.35)
\end{aligned}
$$

Give the initial configurations $q(0) = [\chi(0), \eta(0), \kappa_\chi(0), \rho(0), \beta(0), \gamma(0)]^T$, and the final configuration $q(1) = [\chi(1), \eta(1), \kappa_\chi(1), \rho(1), \beta(1), \gamma(1)]^T$, where the scalar curvatures $\kappa_\chi(0)$ and $\kappa_\chi(1)$ are the scalar curvatures of path planning for the mobile platform, determine minimal order polynomial curves which interpolate $P(0)$ and $P(1)$.

The expression for the curvature of curve $y = \chi(\eta)$ in the form of polar polynomial is

$$
\kappa = \frac{\chi^2 + 2\chi' - \chi\chi''}{(\chi^2 + \chi'^2)^{\frac{3}{2}}} \quad (3.36)
$$

where $\chi' = \frac{\partial \chi}{\partial \eta}$ is the first order derivative and $\chi'' = \frac{\partial^2 \chi}{\partial \eta^2}$ is the second order derivative.

The solution proposed for the above interpolating problem is given by

$$\chi(\eta) = \sum_{i=0}^{5} a_{1i}\eta^i \tag{3.37}$$

$$\eta(\lambda) = \sum_{i=0}^{5} a_{2i}\lambda^i \tag{3.38}$$

$$\rho(\lambda) = \sum_{i=0}^{5} a_{3i}\lambda^i \tag{3.39}$$

$$\beta(\lambda) = \sum_{i=0}^{5} a_{4i}\lambda^i \tag{3.40}$$

$$\gamma(\lambda) = \sum_{i=0}^{5} a_{5i}\lambda^i \tag{3.41}$$

The constraints conditions for the mobile platform can be expressed as

$$\chi(0) = \chi_0, \chi'(0) = 0, \kappa_\chi(0) = \kappa_0, \text{ at } \eta(0) = \eta_0, \eta'(0) = 0, \eta''(0) = \varpi$$
$$\chi(1) = \chi_1, \chi'(1) = 0, \kappa_\chi(1) = \kappa_1, \text{ at } \eta(1) = \eta_1, \eta'(1) = 0, \eta''(1) = -\varpi$$

where ϖ is a constant, and for simplification, let $\eta(0) = 0$ and $\eta(1) = \varphi$.

Similarly, the constraints conditions for the end-effector can be expressed as

$$\rho(0) = \rho_0, \rho'(0) = 0, \rho''(0) = 0, \text{ at } \beta(0) = \beta_0, \beta'(0) = 0, \gamma(0) = \gamma_0, \gamma'(0) = 0$$
$$\rho(1) = \rho_1, \rho'(1) = 0, \rho''(1) = 0, \text{ at } \beta(1) = \beta_1, \beta'(1) = 0, \gamma(1) = \gamma_1, \gamma'(1) = 0$$

The state trajectory associated to the linearizing output trajectory (3.37)-(3.41) that solves the planning problem is obtained by pure algebraic computations using (3.31). Moreover, the open-loop commands that realize this trajectory are

$$\mu_1 = \dot{\eta}(20a_{15}\eta^3 + 12a_{14}\eta^2 + 6a_{13}\eta^3 + 2a_{12})$$
$$+ \ddot{\eta}(5a_{15}\eta^4 + 4a_{14}\eta^3 + 3a_{13}\eta^2 + 2a_{12}\eta + a_{11}) \tag{3.42}$$

$$\mu_2 = \begin{cases} 4(\eta_1 - \eta_0)/T & \text{if } (0 < t < \frac{T}{2}) \\ -4(\eta_1 - \eta_0)/T & \text{if } (\frac{T}{2} < t < T)) \end{cases} \tag{3.43}$$

$$\mu_3 = 20a_{35}\lambda^3 + 12a_{34}\lambda^2 + 6a_{33}\lambda^3 + 2a_{32} \tag{3.44}$$

$$\mu_4 = 20a_{45}\lambda^3 + 12a_{44}\lambda^2 + 6a_{43}\lambda^3 + 2a_{42} \tag{3.45}$$

$$\mu_5 = 20a_{55}\lambda^3 + 12a_{54}\lambda^2 + 6a_{53}\lambda^3 + 2a_{52} \tag{3.46}$$

which represent the nominal inputs to system (3.30) and (3.32), and produce the inputs $(a, \omega, u_\rho, u_\beta, u_\gamma)$.

The polynomial coefficients are detailed by these close-form expressions: $a_{10} = \chi_0$, $a_{20} = \eta_0$, $a_{30} = \rho_0$, $a_{40} = \beta_0$, $a_{50} = \gamma_0$, $a_{11} = 0$, $a_{21} = 0$, $a_{31} = \rho_0'$, $a_{41} = \beta_0'$, $a_{51} = \gamma_0'$, $a_{12} = \frac{1}{2}(\chi_0 - \kappa_0 \chi_0^2)$, $a_{22} = \eta_1 - \eta_0$, $a_{32} = 0$, $a_{42} = 0$, $a_{52} = 0$, $a_{13} = \frac{1}{4\varphi}(-8a_{12} + c_1 - \chi_1 - a_{12}\varphi^2) - \frac{1}{4\varphi^3}(5\chi_1 - 3a_{12}\varphi^2 - \chi_0)$, $a_{23} = 0$, $a_{33} = 10(\rho_1 - \rho_0)$, $a_{43} = 10(\beta_1 - \beta_0)$, $a_{53} = 10(\gamma_1 - \gamma_0)$, $a_{14} = \frac{1}{\varphi^4}(5\chi_1 - 3a_{12}\varphi^2 - \chi_0 - 2a_{13}\varphi^3)$, $a_{24} = 0$, $a_{34} = -15(\rho_1 - \rho_0)$, $a_{44} = -15(\beta_1 - \beta_0)$, $a_{54} = -15(\gamma_1 - \gamma_0)$, $a_{15} = \frac{1}{\varphi^5}(\chi_1 - a_{12}\varphi^2 - a_{10} - a_{14}\varphi^4 - a_{13}\varphi^3)$, $a_{25} = 0$, $a_{35} = 6(\rho_1 - \rho_0)$, $a_{45} = 6(\beta_1 - \beta_0)$, $a_{55} = 6(\gamma_1 - \gamma_0)$.

3.2 Path Planning of Coordinated Mobile Manipulators

3.2.1 Introduction

Few previous works have considered motion/trajectory planning for multiple mobile manipulators in the coordination manner for their complex kinematics and dynamics. In [123], general motion planning based on the calculus of variations was proposed for nonholonomic cooperating mobile robots to obtain optimal trajectories and actuator forces/torques in the presence of obstacles, such that geometric constraints such as joint limits, kinematic constraints and dynamic constraints can be easily incorporated into the planning scheme. In [69], obstacle avoidance was proposed for coordinated task involving two mobile manipulators handling a common object. Trajectory planning of the cooperative multiple manipulators is formulated as an optimal control problem in [130], considering dynamic characteristics of mobile manipulators and the grasped object. In [131], motion planning was proposed for cooperative transportation of a large object by multiple mobile robots in three dimensions in space and the motion planner is designed as a local manipulation planner and a global path planner. Motion planning applicable to cooperating nonholonomic manipulators was proposed with guaranteed collision avoidance and convergence properties [132], based on a new class of nonsmooth Lyapunov functions and an extension of the navigation function. In [176], the same leader-follower type coordination motion was proposed for multiple mobile robots engaging in cooperative tasks.

In this section, we propose motion generation for multiple mobile manipulators which carry a common object in coordination manner. We first develop concise dynamics for the coordinated mobile manipulators, which consist of kinematic constraints of mobile platforms and geometrical constraints between the robotic manipulators and the object in the operational space. Then, dynamic feedback linearization is used to decouple and linearize the dynamics of the system with all constraints. Based on this result, smooth trajectories are developed to join given initial and desired final states of the the common object by polynomial functions in finite time, and exponentially stabilizing feedback is given along the planned trajectory. In order to achieve the autonomous obstacle avoidance, the obstacles are deformed to produce the collision avoidance constraints and the generated paths are iteratively deformed to satisfy the collision avoidance constraints and yield admissible input trajectories that drive both the mobile manipulators and the common objects to the desired configurations.

3.2.2   System Description and Assumption

Consider m mobile manipulators holding a common rigid object in a task space with n degrees of freedom shown in Fig. 3.2. $OXYZ$ is the inertial reference frame in which the position and orientation of the mobile manipulator end-effectors and object are referred, $O_oX_oY_oZ_o$ is the object coordinate frame fixed at the center of mass of the object, and $O_{ei}X_{ei}Y_{ei}Z_{ei}$ is the end-effector frame of the ith manipulator located at the grasp point. To facilitate the dynamic formulation, the following assumptions are made

Assumption 3.1 *Each arm of mobile manipulators is non-redundant, i.e., the number of degrees of freedom of each arm is equal to the dimension of task space.*

Assumption 3.2 *All the end-effectors of the mobile manipulators are rigidly attached to the common object so that there is no relative motion between the object and the end-effectors.*

Assumption 3.3 *The object is rigid, that is, the object does not get deformed with the application of forces.*

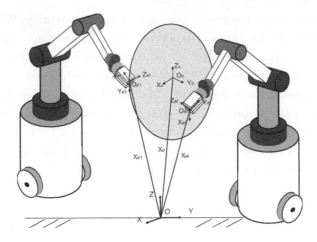

FIGURE 3.2
Coordinated operation of two robots.

3.2.3 Dynamics of System

The dynamics of the ith mobile manipulator in joint space is given by

$$M_i(q_i)\ddot{q}_i + C_i(q_i, \dot{q}_i)\dot{q}_i + G_i(q_i) + d_i = B_i(q_i)\tau_i + J_i^T f_i \qquad (3.47)$$

where $q_i = [q_{i1}, \ldots, q_{in}]^T = [q_{iv}, q_{ia}]^T \in \mathbf{R}^n$ with $q_{iv} \in \mathbf{R}^{n_v}$ describing the generalized coordinates for the mobile platform and $q_{ia} \in \mathbf{R}^{n_a}$ denoting the generalized coordinates of the manipulator, and $n = n_v + n_a$. The symmetric positive definite inertia matrix $M_i(q_i) \in \mathbf{R}^{n \times n}$, the centripetal and Coriolis torques $C_i(\dot{q}_i, q_i) \in \mathbf{R}^{n \times n}$, the gravitational torque vector $G_i(q_i) \in \mathbf{R}^n$, the external disturbances $d_i(t) \in \mathbf{R}^n$ and the control inputs $\tau_i \in \mathbf{R}^p$ could be represented as

$$
M_i(q) = \begin{bmatrix} M_{iv} & M_{iva} \\ M_{iav} & M_{ia} \end{bmatrix}, C_i(\dot{q}_i, q_i) = \begin{bmatrix} C_{iv} & C_{iva} \\ C_{iav} & C_{ia} \end{bmatrix}, G_i(q_i) = \begin{bmatrix} G_{iv} \\ G_{ia} \end{bmatrix}
$$

$$
d(t) = \begin{bmatrix} d_{iv}(t) \\ d_{ia}(t) \end{bmatrix}, \tau = \begin{bmatrix} \tau_{iv} \\ \tau_{ia} \end{bmatrix}, J_i = \begin{bmatrix} A_i & 0 \\ J_{iv} & J_{ia} \end{bmatrix}, f_i = \begin{bmatrix} f_{vi} \\ f_{ei} \end{bmatrix}
$$

where M_{iv}, M_{ia} describe the inertia matrices for the mobile platform and the robotic manipulator, respectively, M_{iva} and M_{iav} are the coupling inertia matrices of the mobile platform and the robotic manipulator, C_{iv}, C_{ia} denote the centripetal and Coriolis torques for the mobile platform and the robotic manipulator, respectively; C_{iva}, and C_{iav} are the coupling centripetal and Coriolis torques of the mobile platform and the robotic manipulator; G_{iv}, G_{ia} are the gravitational torque vectors for the mobile platform and the robotic manipulator, respectively; τ_{iv} and τ_{ia} are the control input vectors for the mobile platform and the robotic manipulator; $d_v(t)$, $d_a(t)$ denote the external disturbances on the mobile platform and the robotic manipulator; J_{iv} and J_{ia} are the Jacobian matrices for the mobile platform and the robotic manipulator; f_{iv} and f_{ia} are the external force on the mobile platform and the robotic manipulator; $B_i(q_i) = \text{diag}[B_{iv}, \ B_{ia}] \in \mathbf{R}^{n \times p}$ is a full rank input transformation matrix for the mobile platform and the robotic manipulator and is assumed to be known because it is a function of fixed geometry of the system; $A_i = [A_{i1}^T(q_{iv}) \ \ldots, \ A_{il}^T(q_{iv})]^T : \mathbf{R}^{n_v} \to \mathbf{R}^{l \times n_v}$ is the kinematic constraint matrix which is assumed to have full rank l; $J_i^T \in \mathbf{R}^{n \times n}$ is Jacobian matrix; and f_{vi} and f_{ei} are the constraint forces corresponding to the nonholonomic and holonomic constraints.

In the chapter, the mobile base is assumed to be completely nonholonomic and the holonomic constraint force is measured by the force sensor mounted on each mobile manipulator's end-effector.

The mobile platform is subjected to nonholonomic constraints and the l non-integrable and independent velocity constraints can be expressed as

$$A(q_{iv})\dot{q}_{iv} = 0 \qquad (3.48)$$

Assume that the annihilator of the co-distribution spanned by the co-vector fields $A_1(q_{iv})$, ..., $A_l(q_{iv})$ is an $(n_v - l)$-dimensional smooth nonsingular distribution Δ on $\mathbf{R}^{n_v}$. This distribution Δ is spanned by a set of $(n_v - l)$ smooth and linearly independent vector fields $H_1(q_{iv})$, ..., $H_{n_v-l}(q_{iv})$, i.e., $\Delta = \text{span}\{H_1(q_{iv}), \ \ldots, \ H_{n_v-l}(q_{iv})\}$. Thus, we have $H^T(q_{iv})A^T(q_{iv}) = 0$, $H(q_{iv}) = [H_1(q_{iv}), \ \ldots, \ H_{n_v-l}(q_{iv})] \in \mathbf{R}^{n_v \times (n_v - l)}$. Note that $H^T H$ is of full rank. Constraint (3.48) implies the existence of vector $\dot{\eta}_i \in \mathbf{R}^{n_v-l}$, such that

$$\dot{q}_{iv} = H(q_{iv})\dot{\eta}_i \qquad (3.49)$$

Considering the nonholonomic constraints (3.48) and (3.49) and their derivative, the dynamics of a mobile manipulator (3.47) can be expressed as

$$M_i^1(\zeta_i)\ddot{\zeta}_i + C^1(\zeta_i, \dot{\zeta}_i)\dot{\zeta}_i + G_i^1 + d_i^1 = u_i + J_{ei}^T f_i \qquad (3.50)$$

where

$$M_i^1 = \begin{bmatrix} H^T M_{iv} H & H^T M_{iva} \\ M_{iav} H & M_{ia} \end{bmatrix}, \zeta_i = \begin{bmatrix} \eta_i \\ q_{ia} \end{bmatrix}, G_i^1 = \begin{bmatrix} H^T G_{iv} \\ G_{ia} \end{bmatrix},$$

$$C_i^1 = \begin{bmatrix} H^T M_{iv} \dot{H} + H^T C_{iv} H & H^T C_{iva} \\ M_{iav} \dot{H} + C_{iav} & C_{ia} \end{bmatrix}, J_{ei}^T = \begin{bmatrix} 0 & 0 \\ J_{iv} & J_{ia} \end{bmatrix}^T$$

$$u_i = B_i^1 \tau_i, B_i^1 = \begin{bmatrix} H^T B_{iv} \\ B_{ia} \end{bmatrix}^T, d_i^1 = \begin{bmatrix} H^T d_{iv} \\ d_{ia} \end{bmatrix}$$

The dynamics of m mobile manipulators from (3.50) can be expressed concisely as

$$M(\zeta)\ddot{\zeta} + C(\zeta, \dot{\zeta})\dot{\zeta} + G + D = u + J_e^T F_e \qquad (3.51)$$

where $M(\zeta) = $ block diag $[M_1^1(\zeta_1), \ldots, M_m^1(\zeta_m)] \in \mathbf{R}^{m(n-l) \times m(n-l)}$; $\zeta = [\zeta_1, \ldots, \zeta_m]^T \in \mathbf{R}^{m(n-l)}$; $u = [B_1^1 \tau_1, \ldots, B_m^1 \tau_m]^T \in \mathbf{R}^{m(n-l)}$; $G = [G_1^1, \ldots, G_m^1]^T \in \mathbf{R}^{m(n-l)}$; $F_e = [f_{e1}, \ldots, f_{em}]^T \in \mathbf{R}^{m(n-l)}$; $C(\zeta, \dot{\zeta}) = $ block diag $[C_1^1(\zeta_1, \dot{\zeta}_1), \ldots, C_m^1(\zeta_m, \dot{\zeta}_m)] \in \mathbf{R}^{m(n-l) \times m(n-l)}$; $D = [d_1^1, \ldots, d_m^1]^T \in \mathbf{R}^{m(n-l)}$; $J_e^T = $ block diag $[J_{e1}^T, \ldots, J_{em}^T] \in \mathbf{R}^{m(n-l) \times m(n-l)}$.

The equation of motion of the object is written by the resultant force vector $F_o \in \mathbf{R}^n$ acting on the center of mass of the object, the symmetric positive definite inertial matrix $M_o(x_o) \in \mathbf{R}^{n \times n}$ of the object, the Corioli and centrifugal matrix $C_o(x_o, \dot{x}_o) \in \mathbf{R}^{n \times n}$, and the gravitational force vector $G_o(x_o) \in \mathbf{R}^n$ as

$$M_o(x_o)\ddot{x}_o + C(x_o, \dot{x}_o)\dot{x}_o + G_o(x_o) = F_o \qquad (3.52)$$

Define $J_o(x_o) \in \mathbf{R}^{mn \times n}$ as $J_o(x_o) = [J_{o1}^T(x_o), \ldots, J_{om}^T(x_o)]^T$ with the Jacobian matrix $J_{oi}(x_o) \in \mathbf{R}^{n \times n}$ from the object frame $O_o X_o Y_o Z_o$ to the ith mobile manipulator's end-effector frame $O_{ei} X_{ei} Y_{ei} Z_{ei}$. Then F_o can be written as

$$F_o = -J_o^T(x_o)F_e \qquad (3.53)$$

Given the resultant force F_o, the end-effector force F_e satisfying (3.53) can be decomposed into two orthogonal components, one contributes to the motion of the object, and another produces the internal force [177]

$$F_e = -(J_o^T(x_o))^+ F_o - F_I \qquad (3.54)$$

where $(J_o^T(x_o))^+ \in \mathbf{R}^{mn \times n}$ is the pseudo-inverse matrix of $J_o^T(x_o)$ and $F_I \in \mathbf{R}^{mn}$ is the internal force vector in the null space of $J_o^T(x_o)$, i.e., satisfying

$$J_o^T(x_o)F_I = 0 \qquad (3.55)$$

Substituting (3.52) into (3.55), we have

$$F_e = -(J_o^T(x_o))^+ (M_o(x_o)\ddot{x}_o + C_o(x_o, \dot{x}_o)\dot{x}_o + G_o(x_o)) - F_I \qquad (3.56)$$

Let $x_{ei} \in \mathbf{R}^n$ denote the position and orientation vector of the i-th end-effector. Then, x_{ei} is related to ζ_i the Jacobian matrix $J_{ei}(\zeta_i)$ as

$$\dot{x}_{ei} = J_{ei}(\zeta_i)\dot{\zeta}_i \qquad (3.57)$$

and the relationship between $\dot{x}_{ie}$ and $\dot{x}_o$ is given by

$$\dot{x}_{ei} = J_{oi}(x_o)\dot{x}_o \qquad (3.58)$$

After combining (3.57) and (3.58), the following relationship between the joint velocity of the ith manipulator and the velocity of the object is obtained

$$J_{ei}(\zeta_i)\dot{\zeta}_i = J_{oi}(x_o)\dot{x}_o \qquad (3.59)$$

As it is assumed that the manipulators work in a nonsingular region, thus the inverse of the Jacobian matrix $J_{ei}(\zeta_i)$ exists. Considering all the manipulators acting on the object at the same time, yields

$$\dot{\zeta} = J_e^{-1}(\zeta)J_o(x_o)\dot{x}_o \qquad (3.60)$$

Differentiating (3.60) with respect to time t leads to

$$\ddot{\zeta} = J_e^{-1}(\zeta)J_o(x_o)\ddot{x}_o + \frac{d}{dt}(J_e^{-1}(\zeta)J_o(x_o))\dot{x}_o \qquad (3.61)$$

Using (3.60) and (3.61) and premultiplying both sides by $J_o^T(x_o)J_e^{-T}(\zeta)$, and using $J_o^T(x_o)F_I = 0$, the dynamics of the multiple mobile manipulators (3.51) system with the object dynamics (3.52) are given by

$$\mathcal{M}(x_o)\ddot{x}_o + \mathcal{C}(x_o, \dot{x}_o)\dot{x}_o + \mathcal{G}(x_o) = \mathcal{U} \qquad (3.62)$$

where

$$
\begin{aligned}
\mathcal{M}(x_o) &= J_o^T(x_o)J_e^{-T}(\zeta)M(\zeta)J_e^{-1}(\zeta)J_o(x_o) + M_o(x_o) \\
\mathcal{C}(x_o, \dot{x}_o) &= J_o^T(x_o)J_e^{-T}(\zeta)M(\zeta)\frac{d}{dt}(J_e^{-1}(\zeta)J_o(x_o)) \\
&\quad + J_o^T(x_o)J_e^{-T}(\zeta)C(\zeta, \dot{\zeta})J_e^{-1}(\zeta)J_o(x_o) + C_o(x_o, \dot{x}_o) \\
\mathcal{G}(x_o) &= J_o^T(x_o)J_e^{-T}(\zeta)G(\zeta) + G_o(x_o) \\
\mathcal{U} &= J_o^T(x_o)J_e^{-T}(\zeta)u
\end{aligned}
$$

The dynamics (3.62) have the following structure properties, which can be exploited to facilitate the motion planning design.

Property 3.1 *The matrix* $\mathcal{M}(x_o)$ *is symmetric, positive definite, and bounded from below and above, i.e.,*$\lambda_{min}I \leq \mathcal{M}(x_o) \leq \lambda_{max}I$, *where* λ_{min} *and* $\lambda_{max} \in \mathbf{R}^n$ *denote the minimum and maximum eigenvalues of* $\mathcal{M}(x_o)$.

Property 3.2 *All Jacobian matrices are uniformly bounded and uniformly continuous if* ζ *and* x_o *are uniformly bounded and continuous, respectively.*

3.2.4   Motion Generation

A prerequisite for the successful use of coordinated multiple mobile manipulators is the availability of a planning methodology that can generate feasible paths for driving the carried object to the desired coordinates without violating nonholonomic constraints. Nonholonomy has been considered in the system dynamics. It allows us to decouple the dynamics by input-output linearization, then we design an admissible path for the object that can drive it from an initial position and orientation to a desired one.

Using the state space vector $x = [x_o^T, \dot{x}_o^T]^T$, and the block partition of the state vector

$$x = \begin{bmatrix} x_1 \\ x_2 \end{bmatrix}, \quad \text{with } x_1 = x_o, \ x_2 = \dot{x}_o$$

we can obtain

$$\dot{x} = \mathcal{F}(x) + \mathcal{H}(x)\mathcal{U}, \quad y = x_1, \quad x(0) = 0 \tag{3.63}$$

where x_0 is the initial state and $\mathcal{F}(x)$, $\mathcal{H}(x)$ and $\mathcal{U}$ are

$$\mathcal{F}(x) = \begin{bmatrix} x_2 \\ \Phi \end{bmatrix} \tag{3.64}$$

$$\mathcal{H}(x) = \begin{bmatrix} 0 \\ \mathcal{M}^{-1} \end{bmatrix} \tag{3.65}$$

where $\Phi = -\mathcal{M}^{-1}(\mathcal{C}+\mathcal{G})x_2$. It is well known that the nonlinear state feedback

$$\mathcal{U} = \mathcal{M}[\mu - \Phi] \tag{3.66}$$

will serve to align and decouple the input/output map of the system (3.62) such that

$$\ddot{y} = \mu \qquad (3.67)$$

where μ is an exogenous input vector.

Planning a feasible motion on the equivalent representation (3.67) can be formulated as an interpolation problem using smooth parametric functions $y(s)$ and with a timing law $s = s(t)$. For simplicity, we directly generate trajectories $y(\lambda)$ as

$$y(\lambda) \;=\; \sum_{i=0}^{n} a_i \lambda^i \qquad (3.68)$$

with the normalized time $\lambda_i = t/T$.

Determine minimal order polynomial curves which interpolate the given configuration $q(0) = [y(0), \dot{y}(0), \ddot{y}(0)]^T$ and the final configuration $q(1) = [y(1), \dot{y}(1), \ddot{y}(1)]^T$.

The general constraint conditions for the object can be expressed as

$$y(0) \;=\; y_0, \quad y'(0) = 0, \quad y''(0) = 0 \qquad (3.69)$$
$$y(1) \;=\; y_1, \quad y'(1) = 0, \quad y''(1) = 0 \qquad (3.70)$$

The state trajectory associated to the linearizing output trajectory (3.68) that solves the planning problem is obtained by pure algebraic computations using (3.67). Moreover, the open-loop commands that realize this trajectory are

$$\mu_i \;=\; 20a_5\lambda^3 + 12a_4\lambda^2 + 6a_3\lambda + 2a_2$$

which represent the nominal inputs to system (3.63), and produce the inputs μ. The polynomial coefficients are detailed by these close-form expressions

$$a_0 \;=\; y_0 \qquad (3.71)$$
$$a_1 \;=\; y_0' \qquad (3.72)$$
$$a_2 \;=\; 0 \qquad (3.73)$$
$$a_3 \;=\; 10(y_1 - y_0) \qquad (3.74)$$
$$a_4 \;=\; -15(y_1 - y_0) \qquad (3.75)$$
$$a_5 \;=\; 6(y_1 - y_0) \qquad (3.76)$$

It is easy to design a linear trajectory tracking control based on the equivalent system (3.67). Given a desired smooth trajectory $y(t)$ for the object, we choose

$$\mu = \ddot{y}_d + K \begin{bmatrix} \dot{y}_d - \dot{y} \\ y_d - y \end{bmatrix} \tag{3.77}$$

where $K = [k_2, k_1]$ indicates the gain matrices to the tracking errors. The actual states $y, \dot{y}$ in (3.77) are computed on-line from the measured joint positions and velocities of the robot by the forward kinematics (3.58), and geometry constraints.

3.2.5 Collision-free Motion Planning

Since the obstacles in the workspace have random shapes, we give the following assumption to modeling the boundary of obstacles.

Assumption 3.4 *Each obstacle in the workspace could be enclosed by a surface*

$$S := \left\{ \sum_{i=1}^{n} \Phi_{obs}^{i}(x) = 0 \right\} \tag{3.78}$$

which consists of a set of N_s surfaces, and the connection of any two surfaces is continuous, Φ_{obs}^{i} is bounded and belong to a class of continuously differentiable manifolds described by $\Phi_{obs}^{i}(x) = A(x_1, x_3) + B(x_2)$ where $x = [x_1, x_2, x_3] \in \mathbf{R}^3$ is the Cartesian position and $A(*)$ is a bounded and uniformly continuous function on $\mathbf{R}^2$, and $B(*)$ has bounded and uniformly continuous derivatives up to the first order.

Remark 3.1 *Assumption 3.4 requires one of the Cartesian coordinates (in this case x_2) to be expressed as a differentiable function of two other coordinates. In some cases, it is possible to rotate the coordinate in such way that x_2 is expressed in terms of other coordinates. Fig. 3.3 demonstrates an example of a surface satisfying the assumption.*

Under Assumption 3.4, the Jacobian matrix of the surface described by $\Phi_{obs}(x) = 0$ becomes

$$J_{\Phi}^{T}(x) = \left(\frac{\partial A}{\partial x} + \frac{\partial B}{\partial x} \right) / \left\| \left(\frac{\partial A}{\partial x} + \frac{\partial B}{\partial x} \right) \right\| \tag{3.79}$$

Letting $S_{\Phi}(x) = [\frac{\partial A}{\partial x_1}, 0, \frac{\partial A}{\partial x_3}]^T$, $C_{\Phi} = [0, \frac{\partial B}{\partial x_2}, 0]^T$, we can obtain

$$J_{\Phi}^{T}(x) = \kappa_{\Phi}(S_{\Phi}(x) + C_{\Phi}) \tag{3.80}$$

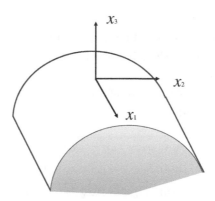

FIGURE 3.3
A demonstration surface.

The scalar valued function $\kappa_\Phi(x)$ is a measure of the curvature of the constraining surface at x. It is easy to know that the constraint Jacobian matrix J has a unit Euclidian norm.

Remark 3.2 *For the obstacle surface, some useful functions are listed as follows:*

    1. $J_\Phi = (\partial \Phi_{obs}(x)/\partial x)/\|(\partial \Phi_{obs}(x)/\partial x)\|$ is the constrained surface normal vector at the point x and $\|J_\phi\| = 1$.

    2. $\kappa_\Phi = 1/\|(\frac{\partial A}{\partial x} + \frac{\partial B}{\partial x})\|$ is smaller than one, at any position x, κ_Φ is a measure of the slope of constraining surface along x_2 axis.

Remark 3.3 *An obstacle can be regarded as an enclosed surface which is composed of a set of surfaces satisfying Assumption 3.4 as Φ_{obs}^1, ..., $\Phi_{obs}^{N_s}$, and each surface in the set has the constants κ_Φ^i and S_Φ^i, $i = 1, ..., N_s$. Therefore, we can design a corresponding path for each surface.*

During the mobile manipulator movement, collision-avoidance (state inequality) constraints resulting from the existence of obstacles in the task space are induced. The general form of these constraints may be written in the following manner:

$$\Phi_c^j(x) \geq 0, \quad j = 1 : N_{obs} \tag{3.81}$$

where Φ_c^j denotes a deform function (expressed in task coordinates) from the analytic description of the jth obstacle, and N_{obs} is the number of the obstacles in the workspace.

Let the center of the obstacle be (x_o, y_o, z_o) in the fixed frame and Φ_{obs} could be represented by

$$\rho_o = \sqrt{(x_{obs} - x_o)^2 + (y_{obs} - y_o)^2 + (z_{obs} - z_o)^2} \tag{3.82}$$

$$\beta_o = \text{atan2}(y_{obs} - y_o, x_{obs} - x_o) \tag{3.83}$$

$$\gamma_o = \text{atan2}(z_{obs} - z_o, \sqrt{(x_{obs} - x_o)^2 + (y_{obs} - y_o)^2}) \tag{3.84}$$

where ρ_o is the boundary $(x_{obs}, y_{obs}, z_{obs})$ distance to the center position of the obstacle, β and γ are the pitch and yaw angles, respectively. Then, we can represent the boundary of obstacle as

$$\Phi_{obs} = \left\{ \rho_o(\beta_o, \gamma_o) \begin{bmatrix} \cos\beta_o \sin\gamma_o \\ \sin\beta_o \sin\gamma_o \\ \cos\gamma_o \end{bmatrix} + \begin{bmatrix} x_o \\ y_o \\ z_o \end{bmatrix} \right\} \tag{3.85}$$

where $\rho_o(\beta_o, \gamma_o)$ is a smooth 2π periodic function. Choose another domain Φ_c as the deform of the domain Φ_{obs} which contains Φ_{obs} as:

$$\Phi_c = \left\{ k_o(\beta_o, \gamma_o)\rho_o(\beta_o, \gamma_o) \begin{bmatrix} \cos\beta_o \sin\gamma_o \\ \sin\beta_o \sin\gamma_o \\ \cos\gamma_o \end{bmatrix} + \begin{bmatrix} x_o \\ y_o \\ z_o \end{bmatrix} \right\} \tag{3.86}$$

where $k_o(\beta_o, \gamma_o) > 1$ is also a smooth periodic function with the periodic 2π.

Define the function Λ, which is monotonously increasing with respect to r, where r is the distance to the center of the obstacles, which could be described as $r = \sqrt{(x - x_o)^2 + (y - y_o)^2 + (z - z_o)^2}$, such that

$$\Lambda(\rho_o, \beta_o, \gamma_o) = 0 \qquad (r = \rho_o(\beta_o, \gamma_o)) \tag{3.87}$$

$$0 < \Lambda(\rho_o, \beta_o, \gamma_o) < \rho_o \quad (\rho_o(\beta_o, \gamma_o) < r < k_o(\beta_o, \gamma_o)\rho_o(\beta_o, \gamma_o)) \tag{3.88}$$

$$\Lambda(\rho_o, \beta_o, \gamma_o) = r \qquad (r \geq k_o(\beta, \gamma)\rho_o(\beta_o, \gamma_o)) \tag{3.89}$$

The function Λ should be chosen in such a way that $\partial\Lambda/\partial r \geq 1$ holds in order to avoid too much distortion in the deformation domain Φ_c. Then we introduce the coordinate transformation Ψ, mapping the modified spherical coordinate $x = (\Lambda, \beta_o, \gamma_o)^T$ to the joint space by

$$\Psi(\zeta) = \begin{bmatrix} \Lambda \\ \beta_o \\ \gamma_o \end{bmatrix} \tag{3.90}$$

where the function Λ is the free parameter to be designed, and new coordinate $\Psi(q)$ does not contain the obstacle domain.

From the above assumption, it is easy to obtain the following properties.

Property 3.3 *For the ith segment on the domain Φ_c, we have $J_c^i = J_\Phi^i$ and $S_c^i = S_\Phi^i / k_o(\beta_o, \gamma_o)$.*

Property 3.4 $S_t(x) = \text{null}(J_c(x))$ *is tangential to the constraints at x.*

Property 3.5 $P_c(x) = I_{n_\Phi} - J^T(JJ^T)^{-1}J$ *is a matrix as the orthogonal projection onto S_t.*

In the obstacles environment, give the start position and target position, if the collision-avoidance constraints the multiple mobile manipulators subjected are active, that is, $\Phi_c(x) = 0$ and $x = \psi(\zeta)$, we have

$$\dot{x} = J_\psi(\zeta)\dot{\zeta} \tag{3.91}$$

$$\frac{\partial \Phi_c(x)}{\partial x} = J_c(x)\dot{x} \tag{3.92}$$

where $J_\psi = \partial\psi/\partial\zeta$.

From (3.91) and (3.92), we can obtain

$$\mathcal{J}\dot{\zeta} = 0 \tag{3.93}$$

where $\mathcal{J} = J_c(x)J_\psi(\zeta)$. The vector $\zeta \in \mathbf{R}^{n_\zeta}$ can always be properly rearranged and partitioned into $\zeta = [\zeta^{1T} \ \zeta^{2T}]^T$, $\zeta^1 = [\zeta_1^1 \ \cdots \ \zeta_{n_\zeta-p}^1] \in \mathbf{R}^{n_\zeta-p}$ describes the constrained motion of coordinated mobile manipulators and $\zeta^2 = [\zeta_1^2 \ \cdots \ \zeta_p^2] \in \mathbf{R}^p$ denotes the remaining variables. There always exists a matrix $\mathcal{T}(\zeta^1) \in \mathbf{R}^{n_\zeta \times (n_\zeta-p)}$ such that $\dot{\zeta} = \mathcal{T}(\zeta^1)\dot{\zeta}^1$ and $\mathcal{T}(\zeta^1)\mathcal{J}^T(\zeta^1) = 0$.

Considering the collision avoidance constraints, (3.51) could be changed to

$$M(\zeta^1)\mathcal{T}(\zeta^1)\ddot{\zeta}^1 + C_2(\zeta^1, \dot{\zeta}^1)\dot{\zeta}^1 + G(\zeta^1) = u + J_e f_e \tag{3.94}$$

where $C_2 = M(\zeta^1)\dot{\mathcal{T}}(\zeta^1) + C(\zeta^1, \dot{\zeta}^1)\mathcal{T}(\zeta^1)$.

Similarly, equation (3.60) is changed to

$$\dot{\zeta}^1 = \mathcal{T}^{-1} J_e^{-1}(\zeta) J_o(x_o) \dot{x}_o \qquad (3.95)$$

therefore, we have

$$\ddot{\zeta}^1 = \mathcal{T}^{-1} J_e^{-1}(\zeta) J_o(x_o) \ddot{x}_o + \frac{d}{dt}(\mathcal{T}^{-1} J_e^{-1}(\zeta) J_o(x_o)) \dot{x}_o \qquad (3.96)$$

Using (3.95) and (3.96), the dynamics of a multiple manipulator system (3.51), coupled with the object's dynamics (3.52), are given by

$$\mathcal{M}_1(x_o) \ddot{x}_o + \mathcal{C}_1(x_o, \dot{x}_o) \dot{x}_o + \mathcal{G}_1(x_o) = \mathcal{U}_1 \qquad (3.97)$$

where

$$
\begin{aligned}
\mathcal{M}_1(x_o) &= J_o^T(x_o) J_e^{-T} \mathcal{T}^{-T}(\zeta) M(\zeta) \mathcal{T}^{-1} J_e^{-1}(\zeta) J_o(x_o) + M_o(x_o) \\
\mathcal{C}_1(x_o, \dot{x}_o) &= J_o^T(x_o) J_e^{-T}(\zeta) M(\zeta) \frac{d}{dt}(\mathcal{T}^{-1} J_e^{-1}(\zeta) J_o(x_o)) \\
&\quad + J_o^T(x_o) \mathcal{T}^{-1} J_e^{-T}(\zeta) C(\zeta, \dot{\zeta}) \mathcal{T}^{-1} J_e^{-1}(\zeta) J_o(x_o) + C_o(x_o, \dot{x}_o) \\
\mathcal{G}_1(x_o) &= J_o^T(x_o) \mathcal{T}^{-1} J_e^{-T}(\zeta) G(\zeta) + G_o(x_o) \\
\mathcal{U}_1 &= J_o^T(x_o) \mathcal{T}^{-1} J_e^{-T}(\zeta) u
\end{aligned}
$$

Using the state space vector $x = [x_o^T, \dot{x}_o^T]^T$, and the block partition of the state vector

$$x = \begin{bmatrix} x_1 \\ x_2 \end{bmatrix}, \text{with } x_1 = x_o, \ x_2 = \dot{x}_o$$

we can obtain

$$\dot{x} = \mathcal{F}_1(x) + \mathcal{H}_1(x)\mathcal{U}_1, \quad y = x_1, \quad x(0) = 0 \qquad (3.98)$$

where x_0 is the initial state and $\mathcal{F}_1(x)$, $\mathcal{H}_1(x)$ and $\mathcal{U}_1$ are

$$\mathcal{F}_1(x) = \begin{bmatrix} x_2 \\ \Phi \end{bmatrix} \qquad (3.99)$$

$$\mathcal{H}_1(x) = \begin{bmatrix} 0 \\ \mathcal{M}_1^{-1} \end{bmatrix} \qquad (3.100)$$

where $\Phi_1 = -\mathcal{M}_1^{-1}(\mathcal{C}_1 + \mathcal{G}_1)x_2$. It is well known that the nonlinear state feedback

$$\mathcal{U}_1 = \mathcal{M}_1[\mu - \Phi_1] \qquad (3.101)$$

will serve to align and decouple the input/output map of the system (3.97) such that

$$\ddot{y} = \mu \tag{3.102}$$

where μ is an exogenous input vector.

Planning a feasible motion on the equivalent representation (3.102) can be formulated as an interpolation problem using using smooth parametric functions $y(s)$ and with a timing law $s = s(t)$. For simplicity, we directly generate trajectories $y(\lambda)$.

$$y(\lambda) = \sum_{i=0}^{n} a_i \lambda^i \tag{3.103}$$

with the normalized time $\lambda = t/T$.

Determine minimal order polynomial curves which interpolate the given configuration $q(0) = [y(0), \dot{y}(0), \ddot{y}(0)]^T$ and the final configuration $q(1) = [y(1), \dot{y}(1), \ddot{y}(1)]^T$.

The general constraint conditions for the object can be expressed as

$$y(0) = y_0, \quad y'(0) = 0, \quad y''(0) = 0 \tag{3.104}$$
$$y(1) = y_1, \quad y'(1) = 0, \quad y''(1) = 0 \tag{3.105}$$

The state trajectory associated to the linearizing output trajectory (3.103) that solves the planning problem is obtained by pure algebraic computations using (3.102). The open-loop command that realizes this trajectory is

$$\mu = 20a_5\lambda^3 + 12a_4\lambda^2 + 6a_3\lambda + 2a_2$$

which represents the nominal inputs to system (3.98), and produces the inputs μ. The polynomial coefficients are detailed by these close-form expressions

$$
\begin{aligned}
a_0 &= y_0 \\
a_1 &= y_0' \\
a_2 &= 0 \\
a_3 &= 10(y_1 - y_0) \\
a_4 &= -15(y_1 - y_0) \\
a_5 &= 6(y_1 - y_0)
\end{aligned}
\tag{3.106}
$$

The design of a linear trajectory tracking control is based on the equivalent system (3.102). Given a desired smooth trajectory $y(t)$ for the object, we choose

$$\mu = \ddot{y}_d + K \begin{bmatrix} \dot{y}_d - \dot{y} \\ y_d - y \end{bmatrix} \tag{3.107}$$

where $K = [k_2, k_1]$ is the gain matrices to the tracking errors. The actual states $y, \dot{y}$ in (3.107) are computed on-line from the measured joint positions and velocities of the robot by the forward kinematics (3.95) and geometry constraints.

If the grasped object is subject to the collision avoidance constraints and the object's path is chosen as $\Phi_c(x_o) = 0$ and $J_c(x_o) = \frac{\partial \Phi_c}{\partial x_o}$, we could obtain

$$J_c(x_o)\dot{x}_o = 0 \tag{3.108}$$

From Property 3.4, let the vector fields $p_{c1}(x_o), \ldots, p_{cr}(x_o)$ form a basis of the null space of S_t at each x_o, and $P_c = [p_{c1}(x_o), \ldots, p_{cr}(x_o)]$. Then, from (3.108), there exists r-vector $v = [v_1, \ldots, v_r]^T$ such that

$$\dot{x}_o = P_c v = p_{c1} v_1 + \cdots + p_{cr} v_r \tag{3.109}$$

Differentiating (3.109) gives

$$\ddot{x}_o = P_c \dot{v} + \dot{P}_c v \tag{3.110}$$

Integrating (3.109) and (3.60), we could obtain

$$\dot{\zeta} = J_e^{-1}(\zeta) J_o(x_o) P_c v \tag{3.111}$$

Considering (3.110) and differentiating (3.111) with respect to time lead to

$$\begin{aligned}
\ddot{\zeta} &= J_e^{-1}(\zeta) J_o(x_o)(P_c \dot{v} + \dot{P}_c v) + \frac{d}{dt}(J_e^{-1}(\zeta) J_o(x_o)) P_c v \\
&= J_e^{-1}(\zeta) J_o(x_o) P_c \dot{v} + (J_e^{-1}(\zeta) J_o(x_o) \dot{P}_c \\
&+ \frac{d}{dt}(J_e^{-1}(\zeta) J_o(x_o)) P_c) v
\end{aligned} \tag{3.112}$$

Using (3.111) and (3.112), after integrating (3.56) into (3.97)and then pre-multiplying both sides by $J_o^T(x_o) J_e^{-T}(\zeta)$ and considering $J_o^T(x_o) F_I = 0$, the dynamics of the multiple mobile manipulators system (3.51) with the object's dynamics (3.52) are given by

$$\mathcal{M}_2 \dot{v} + \mathcal{C}_2 v + \mathcal{G}_2 = \mathcal{U}_2 \tag{3.113}$$

where

$$
\begin{aligned}
\mathcal{M}_2 &= J_o^T(x_o)J_e^{-T}(\zeta)M + M_oJ_o^{-1}(x_o)J_e(\zeta) \\
\mathcal{G}_2 &= J_o^T(x_o)J_e^{-T}(\zeta)G + G_o \\
\mathcal{C}_2 &= J_o^T(x_o)J_e^{-T}(\zeta)C(\zeta,\dot{\zeta}) + M_o\frac{d}{dt}(J_o^{-1}(x_o)J_e(\zeta)) \\
&\quad + C_o(x_o,\dot{x}_o)J_o^{-1}(x_o)J_e(\zeta)) \\
\mathcal{U}_2 &= J_o^T(x_o)J_e^{-T}(\zeta)\tau
\end{aligned}
$$

Using the state space vector $x = [v^T, \dot{v}^T]^T$, and the block partition of the state vector

$$
x = \begin{bmatrix} v_1 \\ v_2 \end{bmatrix}, \quad \text{with } x_1 = v,\ x_2 = \dot{v}
$$

we can obtain

$$
\dot{x} = \mathcal{F}_2(x) + \mathcal{H}_2(x)\mathcal{U}_2, \quad y = x_1, \quad x_0 = 0 \tag{3.114}
$$

where x_0 is the initial state and $\mathcal{F}_2(x)$, $\mathcal{H}_2(x)$ and $\mathcal{U}_2$ are

$$
\mathcal{F}_2(x) = \begin{bmatrix} x_2 \\ \Phi_2 \end{bmatrix} \tag{3.115}
$$

$$
\mathcal{H}_2(x) = \begin{bmatrix} 0 \\ \mathcal{M}_2^{-1} \end{bmatrix} \tag{3.116}
$$

where $\Phi_2 = -\mathcal{M}_2^{-1}(\mathcal{C}_2 + \mathcal{G}_2)x_2$. It is well known that the nonlinear state feedback

$$
\mathcal{U}_2 = \mathcal{M}_2[\mu - \Phi_2] \tag{3.117}
$$

will serve to align and decouple the input/output map of the system (3.113) such that

$$
\dot{y} = \mu \tag{3.118}
$$

where μ is an exogenous input vector.

Planning a feasible motion on the equivalent representation (3.118) can be formulated as an interpolation problem using smooth parametric functions $y(s)$ with a timing law $s = s(t)$. For simplicity, we directly generate trajectories $y(\lambda)$ as

$$
y(\lambda) = \sum_{i=0}^{n} a_i\lambda^i \tag{3.119}
$$

with the normalized time $\lambda = t/T$.

Determine minimal order polynomial curves which interpolate the given configuration $q(0) = [\dot{y}(0), \ddot{y}(0)]^T$ and the final configuration $q(1) = [\dot{y}(1), \ddot{y}(1)]^T$.

The general constraint conditions for the object can be expressed as

$$y'(0) = v_1, \quad y''(0) = 0 \tag{3.120}$$

$$y'(1) = v_2, \quad y''(1) = 0 \tag{3.121}$$

The velocity profile associated to the linearizing output equation (3.119) that solves the planning problem is obtained by pure algebraic computations using (3.118). The open-loop command that realizes this velocity profile is

$$\mu = 3a_3\lambda^2 + 2a_2\lambda + a_1$$

which represents the nominal inputs to system (3.114), and produces the inputs μ. The polynomial coefficients are detailed by these close-form expressions

$$\begin{aligned} a_0 &= v_1 \\ a_1 &= 0 \\ a_2 &= -3(v_2 - v_1) \\ a_3 &= 2(v_2 - v_1) \end{aligned} \tag{3.122}$$

The design of a linear trajectory tracking control is based on the equivalent system (3.118). Given a desired smooth trajectory $y(t)$ for the object, we choose

$$\mu = \ddot{y}_d + k\left[\dot{y}_d - \dot{y}\right] \tag{3.123}$$

where k is positive definite. The actual states $\dot{y}$ in (3.123) are computed on-line from the measured joint positions and velocities of the robot by the forward kinematics (3.111) and geometry constraints.

3.2.6   Simulation Studies

To verify the effectiveness of the proposed control algorithm, we consider two coordinated 2-DOF mobile manipulator systems shown in Fig. 3.4. Each mobile manipulator is subjected to the following constraints:

$$\dot{x}_i \cos \theta_i + \dot{y}_i \sin \theta_i = 0$$

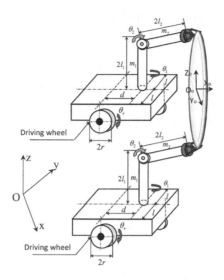

FIGURE 3.4
Cooperating 2-DOF mobile manipulators.

Using Lagrangian approach, we can obtain the standard form with $q_{vi} = [x_i \ y_i \ \theta_i]^T$, $q_{ai} = [\theta_{1i} \ \theta_{2i} \ \theta_{3i}]^T$, $q_i = [q_{vi} \ q_{ai}]^T$, and $A_{vi} = [\cos\theta_i \ \sin\theta_i \ 0.0]^T$. The dynamics of the ith mobile manipulator is given by (3.47) as

$$M_i(q_i)\ddot{q}_i + C_i(q_i, \dot{q}_i)\dot{q}_i + G_i(q_i) = B(q_i)\tau_i + J_{ei}^T f_{ei}$$

The position of end-effector can be given by

$$
\begin{aligned}
x_{ei} &= x_{fi} + 2l_2 \cos\theta_{2i} \cos(\theta_{1i} + \theta_i) \\
y_{ei} &= y_{fi} + 2l_2 \cos\theta_{2i} \sin(\theta_{1i} + \theta_i) \\
z_{ei} &= 2l_1 + 2l_2 \sin\theta_{2i} \\
\alpha_{ei} &= 0.0 \\
\beta_{ei} &= \theta_{1i} + \theta_i \\
\gamma_{ei} &= \theta_{2i}
\end{aligned}
$$

where α_{ei}, β_{ei} and γ_{ei} are the roll angle, the pitch angle and yaw angle for the ith end-effector. The mobile manipulator Jacobian matrix J_i is given by

$$J_i = \begin{bmatrix} J_{i11} & J_{i12} & J_{i13} & J_{i14} & J_{i15} \\ J_{i21} & J_{i22} & J_{i23} & J_{i24} & J_{i25} \\ J_{i31} & J_{i32} & J_{i33} & J_{i34} & J_{i35} \\ J_{i41} & J_{i42} & J_{i43} & J_{i44} & J_{i45} \\ J_{i51} & J_{i52} & J_{i53} & J_{i54} & J_{i55} \end{bmatrix}$$

where

$$\begin{aligned} J_{i11} &= 1, J_{i12} = 0.0, J_{i13} = -2l_2 \cos\theta_{2i} \sin(\theta_{1i} + \theta_i) \\ J_{i14} &= -2l_2 \cos\theta_{2i} \sin(\theta_{1i} + \theta_i), J_{i15} = -2l_2 \sin\theta_{2i} \cos(\theta_{1i} + \theta_i) \\ J_{i21} &= 0.0, J_{i22} = 1, J_{23i} = 2l_2 \cos\theta_{2i} \cos(\theta_{1i} + \theta_i) \\ J_{i24} &= 2l_2 \cos\theta_{2i} \cos(\theta_{1i} + \theta_i), J_{i25} = -2l_2 \sin\theta_{2i} \sin(\theta_{1i} + \theta_i) \\ J_{i31} &= 0.0, J_{i32} = 0.0, J_{i33} = 0.0, J_{i34} = 0.0, J_{i35} = 2l_2 \cos\theta_{2i} \\ J_{i41} &= 0.0, J_{ei42} = 0.0, J_{i43} = 1, J_{i44} = 1.0, J_{i45} = 0.0 \\ J_{i51} &= 0.0, J_{i52} = 0.0, J_{i53} = 0.0, J_{i54} = 0.0, J_{i55} = 1.0 \end{aligned}$$

For the mobile platform, we have

$$\begin{bmatrix} \dot{x}_{fi} \\ \dot{y}_{fi} \\ \dot{\theta}_i \end{bmatrix} = \begin{bmatrix} 1.0 & 0.0 & -d\sin\theta_i \\ 0.0 & 1.0 & d\cos\theta_i \\ 0.0 & 0.0 & 1.0 \end{bmatrix} \begin{bmatrix} \dot{x}_i \\ \dot{y}_i \\ \dot{\theta}_i \end{bmatrix}$$

and

$$\begin{bmatrix} \dot{x}_i \\ \dot{y}_i \\ \dot{\theta}_i \end{bmatrix} = \begin{bmatrix} -\tan\theta_i & 0.0 \\ 1.0 & 0.0 \\ 0.0 & 1.0 \end{bmatrix} \begin{bmatrix} \dot{y}_i \\ \dot{\theta}_i \end{bmatrix}$$

Let the position $x_o = [x_{o1}, x_{o2}, x_{o3}, x_{o4}]^T$ be positions to x axis, y axis, z axis and rotation angle to z axis. The dynamic equation of the object is given by

$$M_o(x_o)\ddot{x}_o + G_o(x_o) = J_{o1}^T(x_o)f_{e1} + J_{o2}^T(x_o)f_{e2} \tag{3.124}$$

where

$$M_o(x_o) = \begin{bmatrix} m_o & 0 & 0 & 0 \\ 0 & m_o & 0 & 0 \\ 0 & 0 & m_o & 0 \\ 0 & 0 & 0 & I_o \end{bmatrix}, \quad G_o(x_o) = \begin{bmatrix} 0 \\ 0 \\ m_o g \\ 0 \end{bmatrix}$$

with m_o and I_o being the mass and inertia of the object, respectively, and

$$J_{o1}(x_o) = \begin{bmatrix} 1 & 0 & 0 & 0 \\ 0 & 1 & 0 & 0 \\ 0 & 0 & 1 & 0 \\ 0 & 0 & 0 & -l_{c1} \end{bmatrix}, \quad J_{o2}(x_o) = \begin{bmatrix} 1 & 0 & 0 & 0 \\ 0 & 1 & 0 & 0 \\ 0 & 0 & 1 & 0 \\ 0 & 0 & 0 & l_{c2} \end{bmatrix} \quad (3.125)$$

with l_{ci} being the length from the ith end-effector to the center of mass of the object.

The parameters in the simulation are set as $d = 6$, $r = 4$, $l = 4$, $l_1 = 10$, and $l_2 = 12$.

(i) The coordinated mobile manipulators move in no-obstacles environments. The polynomial function trajectory is designed as (3.68), the selected initial and final conditions are $x_0 = 2$, $y_0 = 0$, $\dot{x}_0 = 0$, $\dot{y}_T = 0$, $\theta_0 = 60\pi/180$, $\dot{\theta}_0 = 0$, $x_T = 4$, $y_T = 16$, $\dot{x}_T = 0$, $\dot{y}_T = 0$; $\theta_T = -10\pi/180$, $\dot{\theta}_T = 0$. We can obtain a_i from (3.71)-(3.76), the path planning is shown in Fig. 3.5, and the positions are shown in Figs. 3.6 and 3.7. The corresponding trajectory control of the grasping object is shown in Fig. 3.8, where the selected control gains are

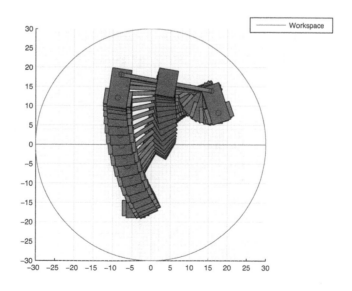

FIGURE 3.5

Cooperating mobile manipulators in no-obstacle environments.

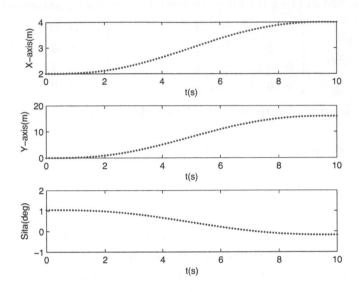

FIGURE 3.6

The positions of the grasping object.

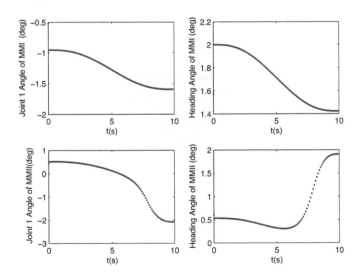

FIGURE 3.7

The joints of the mobile manipulators.

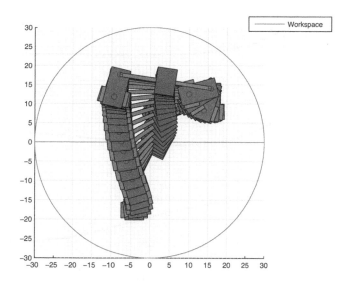

FIGURE 3.8

The trajectory control of the system.

$K_1 = [0.5 \ 0.5]$, $K_2 = [0.8 \ 0.8]$, $K_3 = [0.5 \ 0.5]$. The control error is shown in Fig. 3.9. From this figure, it is found that the planned path is converged.

(ii) The coordinated mobile manipulators move in the environment with obstacles. The obstacle position is $(x - 15)^2 + (y - 5)^2 = 4^2$. We deform the obstacle to obtain the obstacle avoidance constraints as $(x - 15)^2 + (y - 5)^2 = 8^2$. The selected initial and final positions of the grasping object are $x_0 = 2, y_0 = 0, \dot{x}_0 = 0, \dot{y}_T = 0, \theta_0 = 60\pi/180, \dot{\theta}_0 = 0, x_T = 4, y_T = 16, \dot{x}_T = 0, \dot{y}_T = 0; \theta_T = -10\pi/180, \dot{\theta}_T = 0$. In the movement, the left mobile manipulator detects the obstacle, the interconnected system has to satisfy the constraints, the position (x, y) of grasping object is constrained, but the direction θ is a free vector and we design the polynomials function for its trajectory. The path planning of the system is shown in Fig. 3.10, and θ is shown in Fig. 3.11 and the constrained vectors are shown in Fig. 3.12. The tracking control result is shown in Fig. 3.13, and the tracking control error is shown in Fig. 3.14, where the selected control gain is $K = [2 \ 2]$.

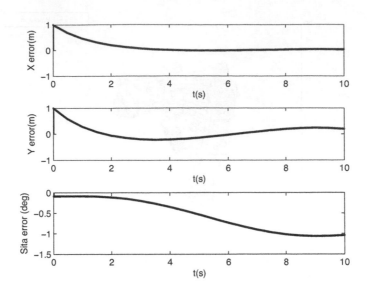

FIGURE 3.9

The trajectory control error of the object.

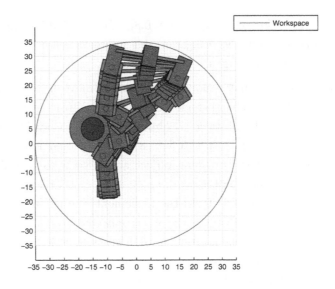

FIGURE 3.10

Cooperating mobile manipulators in obstacle environments.

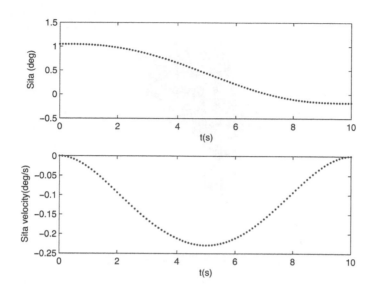

FIGURE 3.11

The free vector of the grasping object.

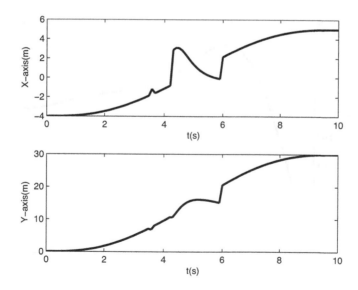

FIGURE 3.12

The constrained vector of the grasping object.

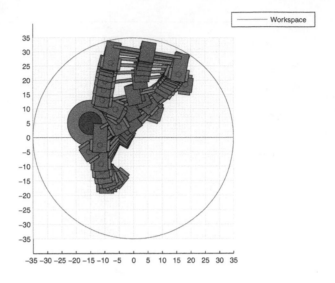

FIGURE 3.13

The trajectory control of the grasping object.

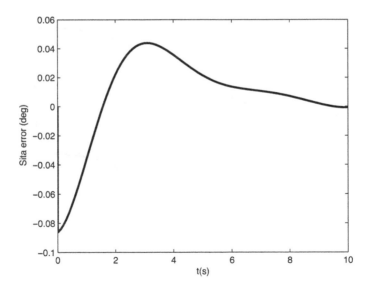

FIGURE 3.14

The tracking control of the free vector of the grasping object.

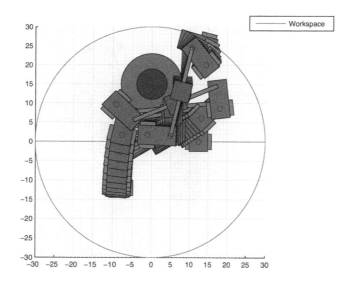

FIGURE 3.15

Cooperating mobile manipulators in obstacle environments.

(iii) The coordinated mobile manipulators move in the environment without obstacles. The obstacle position is $(x-15)^2+(y-5)^2 = 4^2$. We deform the obstacle to obtain the obstacle avoidance constraints as $(x-15)^2+(y-5)^2 = 8^2$. The selected initial positions of the grasping object are $x_0 = 2, y_0 = 0, \dot{x}_0 = 0, \dot{y}_T = 0, \theta_0 = 60\pi/180, \dot{\theta}_0 = 0$. When the grasping object encounters the obstacle, the interconnected system has to satisfy the constraints. The goal is to make the grasped velocity converge to 0. The velocity planning using polynomial function is shown in Fig. 3.15, the tracking control result is shown in Fig. 3.16, and the control error is shown in Fig. 3.17, where the selected gains are $K = [8\ 24]$.

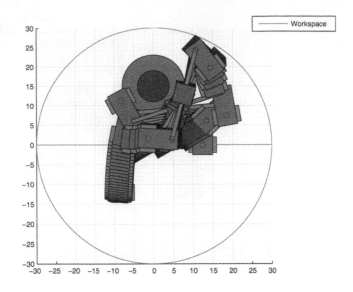

FIGURE 3.16

The tracking control of cooperating mobile manipulators in obstacle environments.

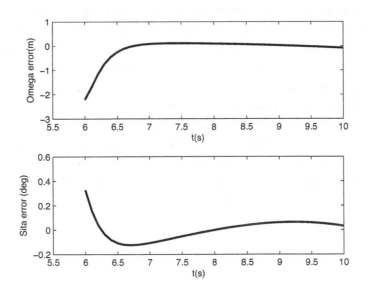

FIGURE 3.17

The tracking control error of cooperating mobile manipulators in obstacle environments.

3.3 Conclusion

In this chapter, a smooth collision-free planning methodology has been developed for a single nonholonomic mobile manipulator without the presence of obstacles and coordinated nonholonomic mobile manipulators carrying a common object in the presence of obstacles, both of which use smooth and continuous polynomials functions based on the input-output feedback linearization to yield admissible input trajectories that drive both the single mobile manipulator and the coordinated mobile manipulators to a final configuration. For the obstacle environments, the obstacles are deformed to produce the obstacle avoidance constraints such that the generated path could satisfy these constraints. Because it requires algebraic manipulations and a single differentiation, implementation of the approach is computationally inexpensive, while it allows direct control over the system. The resulting paths and trajectories are smooth due to the nature of the map and the use of smooth polynomials. Illustrative examples demonstrated the implementation of the methodology in obstacle-free and obstructed spaces.

4

Model-Based Control

CONTENTS

4.1 Introduction

Mobile manipulators process strongly coupled dynamics of mobile platforms and manipulators. If we assume known dynamics, feedback linearization can be utilized to design nonlinear control, which has attracted a great deal of research interest in recent years. The central idea of the approach is to algebraically transform a nonlinear system dynamics into a (fully or partly) linear one, so that linear control techniques can be applied. In [101], input-output feedback linearization was applied to control the mobile platform such that the manipulator can be positioned at the preferred configurations measured by its manipulability. In [61], using nonlinear feedback linearization and decoupling dynamics, force/position control of the end-effector was proposed and applied to nonholonomic cart pushing. In [102], the dynamic interaction effect between the manipulator and the mobile platform of a mobile manipulator was studied, and nonlinear feedback control for the mobile manipulator was developed to compensate for the dynamic interaction. In [62], coordination control of mobile manipulators was presented with two basic task-oriented controls: end-effector task control and joint posture control.

Based on the idea of feedback linearization, model-based control is presented for holonomic constrained nonholonomic mobile manipulators in this

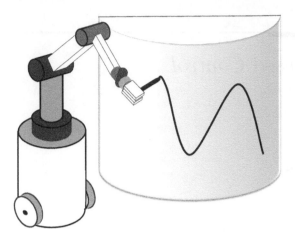

FIGURE 4.1

A holonomic constrained nonholonomic mobile manipulator.

chapter. If accurate knowledge of the dynamic model is available, the model-based control can provide an effective solution to motion/force control. However, the performance of model-based approaches is sensitive to the accuracy of the dynamic model. This is the limitation of model-based control.

The rest of the chapter is organized as follows. The mobile manipulators subject to simultaneous nonholonomic and holonomic constraints are briefly described in Section 4.2. The main results of control design and its stability are presented in Section 4.3. Simulation studies are presented in Section 4.4. Concluding remarks are given in Section 4.5.

4.2   System Description

Consider an n-DOF mobile manipulator mounted on a nonholonomic mobile platform as shown in Fig. 4.1. The constrained dynamics can be described as

$$M(q)\ddot{q} + C(q,\dot{q})\dot{q} + G(q) + d(t) = B(q)\tau + f \qquad (4.1)$$

where $q = [q_1, \ldots, q_n]^T \in \mathbf{R}^n$ denotes the generalized coordinates; $M(q) \in \mathbf{R}^{n \times n}$ is the symmetric bounded positive definite inertia matrix; $C(q, \dot{q})\dot{q} \in \mathbf{R}^n$ denotes the centripetal and Coriolis torques; $G(q) \in \mathbf{R}^n$ is the gravitational torque vector; $d(t)$ denotes the external disturbances; $\tau \in \mathbf{R}^m$ are the control inputs; $B(q) \in \mathbf{R}^{n \times m}$ is a full rank input transformation matrix and is assumed to be known because it is a function of fixed geometry of the system; and $f = [f_n^T, \; f_h^T]^T = J^T\lambda \in \mathbf{R}^n$ with generalized constraint forces f_n and f_h for the nonholonomic and holonomic constraints, respectively, and $\lambda = [\lambda_n, \lambda_h]^T$ denotes the Lagrangian multipliers with both the nonholonomic and holonomic constraints.

The generalized coordinates may be separated as $q = [q_v^T, q_a^T]^T$ where $q_v \in \mathbf{R}^{n_v}$ are the generalized coordinates for the vehicle and $q_a \in \mathbf{R}^{n_a}$ are the coordinates of the arm. Therefore, we can obtain

$$
M(q) = \begin{bmatrix} M_v & M_{va} \\ M_{av} & M_a \end{bmatrix}, C(q, \dot{q}) = \begin{bmatrix} C_v & C_{va} \\ C_{av} & C_a \end{bmatrix}, G(q) = \begin{bmatrix} G_v \\ G_a \end{bmatrix}
$$

$$
B(q) = \begin{bmatrix} B_v & 0 \\ 0 & B_a \end{bmatrix}, \tau = \begin{bmatrix} \tau_v \\ \tau_a \end{bmatrix}, J = \begin{bmatrix} A & 0 \\ J_v & J_a \end{bmatrix}, d(t) = \begin{bmatrix} d_v \\ d_a \end{bmatrix}
$$

The mobile platform is subjected to nonholonomic constraints and the l non-integrable and independent velocity constraints can be expressed as

$$
A(q_v)\dot{q}_v = 0 \tag{4.2}
$$

where $A(q_v) = [A_1^T(q_v), \; \ldots, \; A_l^T(q_v)]^T : \mathbf{R}^{n_v} \rightarrow \mathbf{R}^{l \times n_v}$ is the kinematic constraint matrix which is assumed to have full rank l. The mobile platform is assumed to be completely nonholonomic. The effect of the constraints can be described as a restriction of the dynamics on the manifold Ω_n as

$$
\Omega_n = \{(q_v, \dot{q}_v) | A(q_v)\dot{q}_v = 0\} \tag{4.3}
$$

Then, the generalized constraint forces of the nonholonomic constraints can be given by

$$
f_n = A^T(q_v)\lambda_n \tag{4.4}
$$

Assume that the annihilator of the co-distribution spanned by the covector fields $A_1(q_v), \; \ldots, \; A_l(q_v)$ is an $(n_v - l)$-dimensional smooth nonsingular distribution Δ on $\mathbf{R}^{n_v}$. This distribution Δ is spanned by a set of $(n_v - l)$ smooth and linearly independent vector fields $H_1(q_v), \; \ldots, \; H_{n_v - l}(q_v)$, i.e.,

$\Delta = \text{span}\{H_1(q_v), \ \ldots, \ H_{n_v-l}(q)\}$, which satisfy, in local coordinates, the following relation [110]

$$H^T(q_v)A^T(q_v) = 0 \tag{4.5}$$

where $H(q_v) = [H_1(q_v), \ \ldots, \ H_{n_v-l}(q_v)] \in \mathbf{R}^{n_v \times (n_v-l)}$. Note that $H^T H$ is of full rank. The constraints (4.2) imply the existence of vector $\dot{\eta} \in \mathbf{R}^{n_v-l}$, such that

$$\dot{q}_v = H(q_v)\dot{\eta} \tag{4.6}$$

Considering the nonholonomic constraint (4.2) and its derivative, the dynamics of mobile manipulator can be expressed as

$$
\begin{aligned}
& \begin{bmatrix} H^T M_v H & H^T M_{va} \\ M_{av} H & M_a \end{bmatrix} \begin{bmatrix} \ddot{\eta} \\ \ddot{q}_a \end{bmatrix} + \begin{bmatrix} H^T M_v \dot{H} + H^T C_v H & H^T C_{va} \\ M_{av}\dot{H} + C_{av}H & C_a \end{bmatrix} \\
& \times \begin{bmatrix} \dot{\eta} \\ \dot{q}_a \end{bmatrix} + \begin{bmatrix} H^T G_v \\ G_a \end{bmatrix} + \begin{bmatrix} H^T d_v \\ d_a \end{bmatrix} = \begin{bmatrix} H^T B_v \tau_v \\ B_a \tau_a \end{bmatrix} \\
& + \begin{bmatrix} 0 & 0 \\ J_v & J_a \end{bmatrix}^T \begin{bmatrix} 0 \\ \lambda_h \end{bmatrix} \tag{4.7}
\end{aligned}
$$

Let $\xi = \begin{bmatrix} \eta^T & q_a^T \end{bmatrix}^T$ and assume that the system (4.7) is subjected to k independent holonomic constraints

$$h(\xi) = 0, h(\xi) \in \mathbf{R}^k \tag{4.8}$$

where $h(\xi)$ is of full rank. Define

$$J(\xi) = \partial h/\partial \xi \tag{4.9}$$

then the holonomic constraints could be further written as:

$$J(\xi)\dot{\xi} = 0 \tag{4.10}$$

Then, the holonomic constraint force can be converted to the joint space as

$$f_h = J^T \lambda_h \tag{4.11}$$

The holonomic constraint on the robot's end-effector can be described as restricting only the dynamics on the constraint manifold

$$\Omega_h = \left\{ (\xi, \dot{\xi}) | h(\xi) = 0, J(\xi)\dot{\xi} = 0 \right\} \tag{4.12}$$

Assume that the manipulator consists of series-chain multiple links with holonomic constraints (i.e., geometric constraints). Since the motion is subject to k-dimensional constraint, the configuration space of the holonomic system is left with $n_a - k$ degrees of freedom. From an appropriate manipulation of the constraint $h(\xi) = 0$, the vector q_a can be further rearranged and partitioned into $q_a = [q_a^1, q_a^2]^T$, where $q_a^1 \in \mathbf{R}^{n_a-k}$ describes the constrained motion of the arm and $q_a^2 \in \mathbf{R}^k$ denotes the remaining joint variables. Then,

$$J(\xi) = [\frac{\partial h}{\partial \eta}, \frac{\partial h}{\partial q_a^1}, \frac{\partial h}{\partial q_a^2}] \tag{4.13}$$

From [82], it could be obtained that q is the function of $\zeta = [\eta^T, q_a^{1T}]^T$, that is, $\xi = \varphi(\zeta)$, and we have

$$\dot{\xi} = L(\zeta)\dot{\zeta} \tag{4.14}$$

where $L(\zeta) = \frac{\partial \varphi}{\partial \zeta}$, $\ddot{\xi} = L(\zeta)\ddot{\zeta} + \dot{L}(\zeta)\dot{\zeta}$, and $L(\zeta)$, $J^1(\zeta) = J(\varphi(\zeta))$ satisfies the relationship

$$L^T(\zeta)J^{1T}(\zeta) = 0 \tag{4.15}$$

The dynamics (4.7) can be transformed into the reduced dynamics

$$M^1 L(\zeta)\ddot{\zeta} + C^1\dot{\zeta} + G^1 + d^1(t) = u + J^{1T}\lambda_h \tag{4.16}$$

where $M^1 = \begin{bmatrix} H^T M_v H & H^T M_{va} \\ M_{av} H & M_a \end{bmatrix}$, $G^1 = \begin{bmatrix} H^T G_v \\ G_a \end{bmatrix}$, $d^1(t) = \begin{bmatrix} H^T d_v \\ d_a \end{bmatrix}$,

$C^1 = \begin{bmatrix} H^T M_v \dot{H} & H^T M_{va} \\ M_{av} H & M_a \end{bmatrix} \dot{L}(\zeta) + \begin{bmatrix} H^T M_v \dot{H} + H^T C_v H & H^T C_{va} \\ M_{av} \dot{H} + C_{av} & C_a \end{bmatrix} L(\zeta)$,

$u = B^1 \tau, B^1 = \begin{bmatrix} H^T B_v & 0 \\ 0 & B_a \end{bmatrix}$, $\zeta = \begin{bmatrix} \eta \\ q_a^1 \end{bmatrix}$.

Property 4.1 *The matrices M^1, G^1 are uniformly bounded and uniformly continuous if ζ is uniformly bounded and continuous, respectively. Matrix C^1 is uniformly bounded and uniformly continuous if ζ and $\dot{\zeta}$ are uniformly bounded and continuous, respectively.*

Multiplying L^T on both sides of (4.16), we can obtain

$$M_L \ddot{\zeta} + C_L \dot{\zeta} + G_L + d_L(t) = L^T u \tag{4.17}$$

where

$$M_L = L^T(\zeta)M^1 L, C_L = L^T(\zeta)C^1, G_L = L^T(\zeta)G^1, d_L = L^T(\zeta)d^1(t).$$

The force multiplier λ_h can be obtained by (4.16)

$$\lambda_h = Z(C^1\dot{\zeta} + G^1 + d^1(t) - u) \tag{4.18}$$

where

$$Z = (J^1(M^1)^{-1}J^{1T})^{-1}J^1(M^1)^{-1}$$

Property 4.2 *The matrix M_L is symmetric and positive definite, and we have the following inequalities:*

$$\lambda_{min}(M_L)\|x\|^2 \leq x^T M_L x \leq \lambda_{max}(M_L)\|x\|^2, \ \forall x \in \mathbf{R}^n \tag{4.19}$$

where λ_{min} and λ_{max} denote the minimum and maximum eigenvalues of M_L, respectively [111].

Property 4.3 *The matrix $\dot{M}_L(\zeta) - 2C_L(\zeta, \dot{\zeta})$ is skew-symmetric, and the $C_L(\zeta, \dot{\zeta})$ satisfies $C_L(\zeta, x)y = C_L(\zeta, y)x$ and $C_L(\zeta, z + kx)y = C_L(\zeta, z)y + kC_L(\zeta, x)y, \ \forall x, y,$ and k is scalar.*

Property 4.4 *For holonomic systems, matrices $J^1(\zeta), L(\zeta)$ are uniformly bounded and uniformly continuous, if ζ is uniformly bounded and continuous, respectively.*

Remark 4.1 *The matrix Z is bounded and continuous since M^1 and $J^1(\zeta)$ are bounded and continuous from Property 4.1 and Property 4.4.*

4.3   Model Reference Control

Since the mobile manipulators are subjected to the nonholonomic constraint (4.2) and holonomic constraint (4.8), the state variables q_v, q_a^1, q_a^2 are not independent. By a proper partition of q_a, q_a^2 is uniquely determined by $\zeta = [\eta, q_a^1]^T$. Therefore, it is not necessary to consider the control of q_a^2.

Given a desired motion trajectory $\zeta_d(t) = [\eta_d, \ q_{ad}^1]^T$ and a desired constraint force $f_d(t)$, or, equivalently, a desired multiplier $\lambda_{hd}(t)$, we are to determine a control law such that for any $(\zeta(0), \dot{\zeta}(0)) \in \Omega$, $\zeta, \dot{\zeta}, \lambda_h$ converge to a manifold Ω_d specified as Ω where

$$\Omega_d = \{(\zeta, \dot{\zeta}, \lambda_h) | \zeta = \zeta_d, \dot{\zeta} = \dot{\zeta}_d, \lambda_h = \lambda_{hd}\} \tag{4.20}$$

Assumption 4.1 *The desired trajectory $\zeta_d(t)$ is assumed to be bounded and uniformly continuous, and has bounded and uniformly continuous derivatives up to the second order. The desired Lagrangian multiplier $\lambda_{hd}(t)$ is also bounded and uniformly continuous.*

Consider the following signals

$$
\begin{aligned}
e_\zeta &= \zeta - \zeta_d \\
\dot{\zeta}_r &= \dot{\zeta}_d - K_\zeta e_\zeta \\
r &= \dot{e}_\zeta + K_\zeta e_\zeta \\
e_\lambda &= \lambda_h - \lambda_{hd}
\end{aligned}
$$

where $K_\zeta = \mathrm{diag}[K_{\zeta i}] > 0$.

Consider the control input u in the form:

$$u = u_a - J^{1T} u_b \tag{4.21}$$

Then, (4.17) and (4.18) can be changed to

$$M_L \ddot{\zeta} + C_L \dot{\zeta} + G_L + d_L = L^T u_a \tag{4.22}$$

$$\lambda_h = Z(C^1 \dot{\zeta} + G^1 + d^1(t) - u_a) + u_b \tag{4.23}$$

Under the assumption that the dynamics of mobile manipulators are known without considering external disturbances, consider the following control laws:

$$L^T u_a = -K_p r - K_i \int_0^t r \, ds - \Phi_m \tag{4.24}$$

$$u_b = \chi_m \ddot{\zeta}_d + \lambda_{hd} - K_f e_\lambda \tag{4.25}$$

$$\tag{4.26}$$

where $\Phi_m = C_m \Psi_m$, $\chi_m = ZL^{+T} M_L$, with $C_m = [M_L \ C_L \ G_L]$, $\Psi_m = [\ddot{\zeta}_r \ \dot{\zeta}_r \ 1]^T$, $L^+ = (L^T L)^{-1} L^T$, K_p, K_i and K_f are positive definite.

Theorem 4.1 *Consider the mechanical system without external disturbance described by (4.1), (4.2) and (4.8) with $d(t) = 0$. Using the control law (4.24) and (4.25), the following hold for any $(q(0), \dot{q}(0)) \in \Omega_n \cap \Omega_h$:*

(i) r converges to a set containing the origin as $t \to \infty$;

(ii) e_q and $\dot{e}_q$ converge to 0 as $t \to \infty$; and

(iii) e_λ and τ are bounded for all $t \geq 0$.

Proof (i) The closed-loop system dynamics can be rewritten as

$$M_L \dot{r} = L^T u_a - \mu - C_L r \tag{4.27}$$

where $\mu = M_L \ddot{\zeta}_r + C_L \dot{\zeta}_r + G_L + d_L(t)$.

Substituting (4.24) into (4.27), the closed-loop dynamics are given by

$$M_L \dot{r} = -K_p r - K_i \int_0^t r ds - \Phi_m - \mu - C_L r \tag{4.28}$$

Consider the Lyapunov function candidate

$$V = \frac{1}{2} r^T M_L r + \frac{1}{2} (\int_0^t r ds)^T K_i \int_0^t r ds \tag{4.29}$$

then

$$\dot{V} = r^T (M_L \dot{r} + \frac{1}{2} \dot{M}_L r + K_i \int_0^t r ds) \tag{4.30}$$

Using Property 4.3, the time derivative of V along the trajectory of (4.28) is

$$\begin{aligned} \dot{V} &= -r^T K_p r - r^T \mu - r^T \Phi_m \\ &\leq -r^T K_p r \leq -\lambda_{min}(K_p)\|r\|^2 \leq 0 \end{aligned}$$

Integrating both sides of the above equation gives

$$V(t) - V(0) \leq - \int_0^t r^T K_p r ds \leq 0 \tag{4.31}$$

and, r converges to a small set containing the origin as $t \to \infty$.

(ii) From (4.31), V is bounded, which implies that $r \in L_\infty^{n-k-l}$. We have $\int_0^t r^T K_p r ds \leq V(0) - V(t)$, which leads to $r \in L_2^{n-k-l}$. From $r = \dot{e}_\zeta + K_\zeta e_\zeta$, it can be obtained that $e_\zeta, \dot{e}_\zeta \in L_\infty^{n-k-l}$. As we have established that $e_\zeta, \dot{e}_\zeta \in L_\infty$, from Assumption 4.1, we conclude that $\zeta(t), \dot{\zeta}(t), \dot{\zeta}_r(t), \ddot{\zeta}_r(t) \in L_\infty^{n-k-l}$ and $\dot{q} \in L_\infty^n$.

Therefore, all the signals on the right hand side of (4.28) are bounded, and we can conclude that $\dot{r}$ and therefore $\ddot{\zeta}$ are bounded. Thus, $r \to 0$ as $t \to \infty$ can be obtained. Consequently, we have $e_\zeta \to 0, \dot{e}_\zeta \to 0$ as $t \to \infty$. It follows that $e_q, \dot{e}_q \to 0$ as $t \to \infty$.

(iii) Substituting the control (4.24) and (4.25) into the reduced order dynamics (4.23) yields

$$\begin{aligned} (I + K_f)e_\lambda &= Z(C^1 \dot{\zeta} + G^1 + d^1(t) - u_a) + u_b \\ &= -ZL^{+T} M_L \ddot{\zeta} + \chi_m \ddot{\zeta}_d \end{aligned} \tag{4.32}$$

Since $\ddot{\zeta}$ and Z are bounded, $\zeta \to \zeta_d$, $(-ZL^{+T}M_L\ddot{\zeta} + \chi_m\ddot{\zeta}_d)$ is also bounded and the size of e_λ can be adjusted by choosing the proper gain matrix K_f.

Since r, ζ, $\dot{\zeta}$, ζ_r^1, $\dot{\zeta}_r$, $\ddot{\zeta}_r$, and e_λ are all bounded, it is easy to conclude that τ is bounded from (4.24) and (4.25).

4.4 Simulation Studies

To verify the effectiveness of the model-based control, let us consider the mobile manipulator system shown in Fig. 9.1 in the appendix. The mobile manipulator is subjected to the constraints (2.2), (2.3), and (2.4).

Using the Lagrangian approach, we can obtain the standard form (4.1) with

$$q_v = [x, y, \theta]^T, \; q_a = [\theta_1, \theta_2]^T, \; q = [q_v^T, q_a^T]^T, A = [\cos\theta, \sin\theta, 0]^T$$

Since

$$\begin{bmatrix} \dot{\theta}_l \\ \dot{\theta}_r \\ \dot{\theta}_1 \\ \dot{\theta}_2 \end{bmatrix} = \begin{bmatrix} \frac{1}{2r} & \frac{l}{2r} & 0 & 0 \\ \frac{1}{2r} & -\frac{l}{2r} & 0 & 0 \\ 0 & 0 & 1 & 0 \\ 0 & 0 & 0 & 1 \end{bmatrix} \begin{bmatrix} v \\ \omega \\ \dot{\theta}_1 \\ \dot{\theta}_2 \end{bmatrix}$$

$$\begin{bmatrix} v \\ \omega \\ \dot{\theta}_1 \\ \dot{\theta}_2 \end{bmatrix} = \begin{bmatrix} \cos\theta & \sin\theta & 0 & 0 & 0 \\ 0 & 0 & 1 & 0 & 0 \\ 0 & 0 & 0 & 1 & 0 \\ 0 & 0 & 0 & 0 & 1 \end{bmatrix} \begin{bmatrix} \dot{x} \\ \dot{y} \\ \dot{\theta} \\ \dot{\theta}_1 \\ \dot{\theta}_2 \end{bmatrix},$$

consider the desired trajectories

$$x_d = 1.5\cos(t), \; y_d = 1.5\sin(t), \; \theta_d = 1.0\sin(t), \; \theta_{1d} = \pi/4(1 - \cos(t))$$

and the geometric constraints that the end-effector is subjected to as

$$\Phi = l_1 - l_2\sin(\theta_2) = 0 \; \lambda_{hd} = 10.0N$$

To simplify, let $\frac{1}{2r} = 1$, and $\frac{l}{2r} = 1$ and choose $\zeta = [\zeta_1, \zeta_2, \zeta_3]^T$, and $\dot{\zeta} = [v, \omega, \dot{\theta}_1]^T$. It is easy to have the reduced dynamics model of the mobile manipulator described as in Section 9.1.

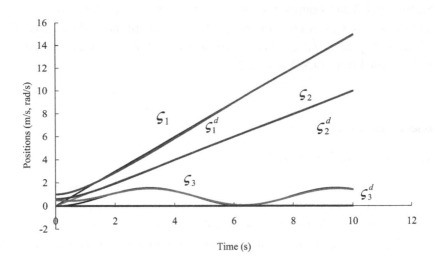

FIGURE 4.2
Position tracking by model-based control.

In the simulation, we assume that the parameters are $p_0 = 6.0 kg \cdot m^2$, $p_1 = 1.0 kg \cdot m^2$, $p_2 = 0.5 kg \cdot m^2$, $p_3 = 1.0 kg \cdot m^2$, $p_4 = 2.0 kg \cdot m^2$, $q_0 = 4.0 kg \cdot m^2$, $q_1 = 1.0 kg \cdot m^2$, $q_2 = 1.0 kg \cdot m^2$, $q_3 = 1.0 kg \cdot m^2$, $q_4 = 0.5 kg \cdot m^2$, $d = 1.0 m$, $r = 0.5 m$, $l = 1.0 m$, $2l_1 = 1.0 m$, $2l_2 = 1.0 m$, $q(0) = [1.0, 0.6, 0.5]^T$, $\dot{q}(0) = [0.0, 0.0, 0.0]^T$ and $\lambda_h(0) = 0$. According to Theorem 4.1, the control gains are selected as $K_p = \text{diag}[10]$, $K_\zeta = \text{diag}[1.0]$, $K_i = 0.0$ and $K_f = 0.5$. The disturbances on the mobile base are set as $1.0 \sin(t)$ and $1.0 \cos(t)$. For the model-based control, we assume that the system model is known beforehand. The trajectory tracking performance of the model-based control is illustrated in Fig. 4.2 and the velocities tracking and the input torques by the adaptive robust control are shown in Figs. 4.3-4.5. The force tracking is shown in Fig. 4.4. Simulation results show that the trajectory and force tracking errors converge to zero, which validates Theorem 4.1.

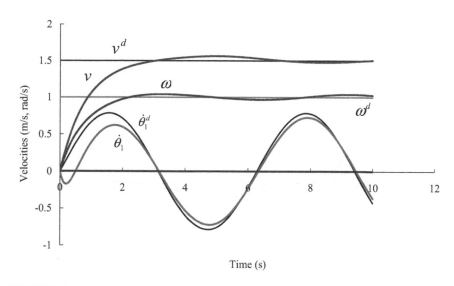

FIGURE 4.3

Velocity tracking by model-based control.

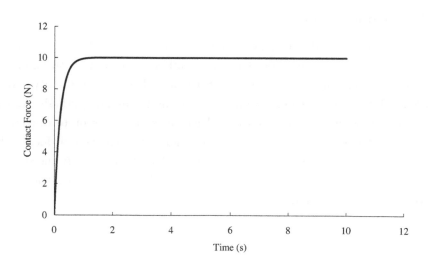

FIGURE 4.4

Contact force.

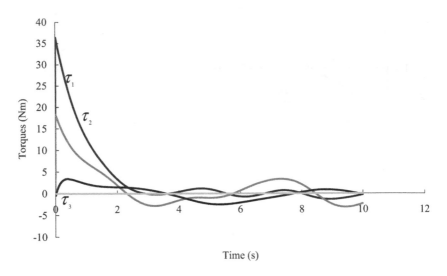

FIGURE 4.5
Input torques by model-based control.

4.5   Conclusion

In this chapter, model reference control has been proposed to control a class
of holonomic constrained noholonomic mobile manipulators with known parameters and ignoring the external disturbances. The system stability and the
boundedness of tracking errors are proved using Lyapunov synthesis. All control strategies have been designed to drive the system motion to converge to
the desired manifold and at the same time guarantee the boundedness of the
constrained force. Simulation studies have verified that not only the states of
the system converge to the desired trajectory, but also the constraint force
converges to the desired force.

5

Adaptive Robust Hybrid Motion/Force Control

CONTENTS

The chapter investigates adaptive hybrid motion/force control for constrained mobile manipulators. The mobile manipulator is subject to nonholonomic constraints, which could be formulated as non-integrable differential equations containing time-derivatives of generalized coordinates (velocity, acceleration, etc.). Due to Brockett's theorem [135], it is well known that nonholonomic systems with restricted mobility cannot be stabilized to a desired configuration nor posture via differentiable, or even continuous, pure state feedback. Moreover, the holonomic constraints also bring the force constraint to the end-effector.

In motion/force control of mobile wheeled manipulators, an important concern is how to deal with unknown perturbations to the nominal model, in the form of parametric and functional uncertainties. In addition, mobile wheeled manipulators are characterized by unmodelled dynamics, and are subject to holonomic and nonholonomic contraints, and time varying external disturbances, which are difficult to model accurately. As a result, traditional model-based control may not be an ideal approach since it works effectively only when the dynamic model is exactly known. The presence of uncertainties and disturbances could disrupt the function of the traditional model-based feedback control and lead to imbalance of the pendulum. How to handle the parametric and functional uncertainties, unmodelled dynamics, and disturbances thus becomes one of the important issues in the motion/force control of mobile manipulators

Therefore, based on the above problems, we divide this chapter into two sections: hybrid force/motion control by state feedback and output feedback in Sections 5.1 and 5.2, in which the robust and adaptive robust designs for the motion/force control and stability analysis, and adaptive output feedback control are presented. The concluding remarks are given in Section 5.4.

5.1   Adaptive Robust Hybrid Motion/Force Control

5.1.1   Introduction

Control of mobile manipulators with uncertainties is essential in many practical applications, especially for the case when the force of the end-effector should be considered. To handle unknown dynamics of mechanical systems, robust and adaptive controls have been extensive investigated for robot manipulators and dynamic nonholonomic systems. Robust controls assume the known boundedness of unknown dynamics of the systems, while adaptive controls could learn the unknown parameters of interest through adaptive tuning laws.

Under the assumption of a good understanding of dynamics of the systems under study, model adaptive controls have been much investigated for dynamic nonholonomic systems. In [103], adaptive control was proposed for trajectory/force control of mobile manipulators subjected to holonomic and nonholonomic constraints with unknown inertia parameters, which ensures the motion of the system to asymptotically converge to the desired trajectory and force. In [105], adaptive state feedback and output feedback control strategies using state scaling and backstepping techniques were proposed for a class of nonholonomic systems in chained form with drift nonlinearity and parametric uncertainties. In [64], the nonholonomic kinematic subsystem was first transformed into a skew-symmetric form, then a virtual adaptive control designed at the actuator level was proposed to compensate for the parametric uncertainties of the kinematic and dynamic subsystems.

In [108], robust adaptive control was proposed for dynamic nonholonomic systems with unknown inertia parameters and disturbances, in which adaptive control techniques were used to compensate for the parametric uncertainties and sliding mode control was used to suppress the bounded disturbances. In [66], adaptive robust force/motion control was presented systematically for holonomic mechanical systems and a large class of nonholonomic mechanical systems in the presence of uncertainties and disturbances.

Because precise dynamics are hard to obtain, adaptive neural network control, a model free-based control, has been extensively studied for different robotic systems, such as robotic manipulators [111, 112, 79], and mobile robots [113]. In [80], adaptive neural network control for a robot manipulator in the task space was proposed, which neither requires the inverse dynamical model nor the time-consuming off-line training process. In [114], the unidirectionality of the contact force of robot manipulators was explicitly included in modeling and the fuzzy control was developed. In [81], adaptive neural fuzzy control for function approximation had been investigated for uncertain nonholonomic mobile robots in the presence of unknown disturbances. In [115], adaptive neural network controls had been developed for the motion control of mobile manipulators subject to kinematic constraints.

In this chapter, we shall consider mobile manipulators with both holonomic and nonholonomic constraints, such as nonholonomic mobile manipulators, and address the force/motion control for holonomic constrained nonholonomic mobile manipulators in the presence of parameter uncertainties and external disturbances.

The stability and the boundedness of tracking errors have been proved using Lyapunov synthesis. The proposed control strategies guarantee that the motion of the system converges to the desired manifold and at the same time guarantee the boundedness of the constrained force. The simulation studies validate not only that the motion of the system converges to the desired trajectory, but also that the constraint force converges to the desired force.

5.1.2 Robust Control

In practice, robust control schemes can handle the uncertainties and external disturbances on the dynamics.

The following assumptions are needed to develop the robust control.

Assumption 5.1 *There exist some finite positive constants $c_{ri} > 0$ $(1 \leq i \leq 4)$, and finite non-negative constant $c_{r5} \geq 0$ such that $\forall \zeta, \dot{\zeta} \in \mathbf{R}^n$ [66]*

$$
\begin{aligned}
||M_L|| &\leq c_{r1} \\
||C_L|| &\leq c_{r2} + c_{r3}||\dot{\zeta}|| \\
||G_L|| &\leq c_{r4} \\
\sup_{t \geq 0} ||d_L|| &\leq c_{r5}
\end{aligned}
$$

Assumption 5.2 *Time varying positive function δ converges to zero as $t \to \infty$ and satisfies*

$$
\lim_{t \to \infty} \int_0^t \delta(\omega)d\omega = \rho < \infty
$$

with finite constant ρ.

There are many choices for δ that satisfy the Assumption 5.2, for example, $\delta = 1/(1+t)^2$.

Consider the following control law for (4.17):

$$
L^T u_a = -K_p r - K_i \int_0^t r ds - \frac{r \Phi_r^2}{||r|| \Phi_r + \delta} \tag{5.1}
$$

$$
u_b = \frac{\chi_r^2}{\chi_r + \delta} \ddot{\zeta}_d + \lambda_{hd} - K_f e_\lambda \tag{5.2}
$$

where

$$
\begin{aligned}
\Phi_r &= C_r^T \Psi_r \\
\chi_r &= c_1 ||Z^* L^{+^T}||
\end{aligned}
$$

with

$$
\begin{aligned}
C_r &= [c_{r1}\ c_{r2}\ c_{r3}\ c_{r4}\ c_{r5}]^T \\
\Psi_r &= [||\ddot{\zeta}_r||\ ||\dot{\zeta}_r||\ ||\dot{\zeta}||||\dot{\zeta}_r||\ 1\ 1]^T \\
Z^* &= (J^1(M^*)^{-1}J^{1T})^{-1}J^1(M^*)^{-1}
\end{aligned}
$$

K_p, K_i, K_f are positive definite and $\|L^T M^* L\| = c_1$.

Theorem 5.1 *Consider the mechanical system described by (4.1), (4.2) and (4.8), using the control law (5.1) and (5.2). The following hold for any $(q(0), \dot{q}(0)) \in \Omega_n \cap \Omega_h$: [(i)]r converges to a set containing the origin as $t \to \infty$; [(ii)]e_q and $\dot{e}_q$ converge to 0 as $t \to \infty$; and [(iii)]e_λ and τ are bounded for all $t \geq 0$.*

Proof (i) Substituting (5.1) into (4.27), the closed-loop dynamics are obtained

$$M_L \dot{r} = -K_p r - K_i \int_0^t r ds - \frac{r \Phi_r^2}{\|r\| \Phi_r + \delta} - \mu - C_L r \qquad (5.3)$$

Considering the Lyapunov function candidate

$$V = \frac{1}{2} r^T M_L r + \frac{1}{2} (\int_0^t r ds)^T K_i \int_0^t r ds \qquad (5.4)$$

from Property 4.2, we have $\frac{1}{2} \lambda_{min}(M_L) r^T r \leq \frac{1}{2} r^T M_L r \leq \frac{1}{2} \lambda_{max}(M_L) r^T r$. By using Property 4.3, the time derivative (4.30) of V along the trajectory of (5.3) is

$$
\begin{aligned}
\dot{V} &= -r^T K_p r - r^T \mu - \frac{\|r\|^2 \Phi_r^2}{\|r\| \Phi_r + \delta} \\
&\leq -r^T K_p r - \frac{\|r\|^2 \Phi_r^2}{\|r\| \Phi_r + \delta} + \|r\| \Phi_r \\
&\leq -r^T K_p r + \delta \\
&\leq -\lambda_{min}(K_p) \|r\|^2 + \delta
\end{aligned}
$$

Since δ is a time varying function converging to zero as $t \to \infty$, and δ is bounded, there exists $t > t_1$, $\delta \leq \rho_1$, when $\|r\| \geq \sqrt{\frac{\rho_1}{\lambda_{min}(K_p)}}$, $\dot{V} \leq 0$. Then r converges to a small set containing the origin as $t \to \infty$.

Integrating both sides of the above equation gives

$$V(t) - V(0) \leq - \int_0^t r^T K_p r ds + \rho < \infty \qquad (5.5)$$

(ii) From (5.5), V is bounded, which implies that $r \in L_\infty^{n-k-l}$. We have $\int_0^t r^T K_p r ds \leq V(0) - V(t) + \rho$, which leads to $r \in L_2^{n-k-l}$. From $r = \dot{e}_\zeta + K_\zeta e_\zeta$, it can be obtained that $e_\zeta, \dot{e}_\zeta \in L_\infty^{n-k-l}$. As we have established $e_\zeta, \dot{e}_\zeta \in L_\infty$, from Assumption 4.1, we conclude that $\zeta(t), \dot{\zeta}(t), \dot{\zeta}_r(t), \ddot{\zeta}_r(t) \in L_\infty^{n-k-l}$ and $\dot{q} \in L_\infty^n$.

Therefore, all the signals on the right hand side of (5.3) are bounded, and we can conclude that $\dot{r}$ and therefore $\ddot{\zeta}$ are bounded. Thus, $r \to 0$ as $t \to \infty$ can be obtained. Consequently, we have $e_\zeta \to 0, \dot{e}_\zeta \to 0$ as $t \to \infty$. It follows that $e_q, \dot{e}_q \to 0$ as $t \to \infty$.

(iii) Substituting the control (5.1) and (5.2) into the reduced order dynamics (4.23) yields

$$
\begin{aligned}
(I + K_f)e_\lambda &= Z(C^1\dot{\zeta} + G^1 + d^1(t) - u_a) + u_b \\
&= -ZL^{+T}M_L\ddot{\zeta} + \frac{\chi^2}{\chi + \delta}\ddot{\zeta}_d
\end{aligned}
\tag{5.6}
$$

Since $\ddot{\zeta}$ and Z are bounded, $\zeta \to \zeta_d$, $-ZL^{+T}M_L\ddot{\zeta} + \frac{\chi^2}{\chi+\delta}\ddot{\zeta}_d$ is also bounded and the size of e_λ can be adjusted by choosing the proper gain matrix K_f.

Since $r, \zeta, \dot{\zeta}, \zeta_r^1, \dot{\zeta}_r, \ddot{\zeta}_r$, and e_λ are all bounded, it is easy to conclude that τ is bounded from (5.1) and (5.2).

5.1.3   Adaptive Robust Control

In developing robust control law (5.1) and (5.2), the vector C_r is assumed to be known. However, in practice, it cannot be obtained easily. Therefore, we can develop an adaptive updating law to estimate the C_r.

Consider the adaptive robust control law for the dynamics (4.17)

$$
L^T u_a = -K_p r - K_i \int_0^t r\,dt - \sum_{i=1}^5 \frac{r\hat{c}_{ri}\Psi_{ri}^2}{\|r\|\Psi_{ri} + \delta_i}
\tag{5.7}
$$

$$
u_b = \frac{\hat{\chi}_r^2}{\hat{\chi}_r + \delta_1}\ddot{\zeta}_d + \lambda_{hd} - K_f e_\lambda
\tag{5.8}
$$

where

$$
\dot{\hat{c}}_{ri} = -\omega_i \hat{c}_{ri} + \sum_{i=1}^5 \frac{\gamma_i \Psi_{ri}^2 \|r\|^2}{\|r\|\Psi_{ri} + \delta_i}, \quad i = 1, \ldots, 5
\tag{5.9}
$$

$$
\hat{\chi}_r = \hat{c}_1 \|\hat{Z}L^{+T}\|
\tag{5.10}
$$

$\hat{Z} = (J^1(\hat{M}^*)^{-1}J^{1T})^{-1}J^1(\hat{M}^*)^{-1}$ with $\|L^T\hat{M}^*L\| = \hat{c}_1$, K_p, K_i, K_f are positive definite; $\gamma_i > 0$; $\delta_i > 0$ and $\omega_i > 0$ satisfy Assumption 5.2:

$$
\int_0^\infty \delta_i(s)\,ds = \rho_{i\delta} < \infty
$$

$$
\int_0^\infty \omega_i(s)\,ds = \rho_{i\omega} < \infty
$$

with the constants $\rho_{i\delta}$ and $\rho_{i\omega}$.

Theorem 5.2 *Considering the mechanical system described by (4.1), (4.2) and (4.8), using the control law (5.7) and (5.8), the following hold for any $(q(0), \dot{q}(0)) \in \Omega_n \cap \Omega_h$:*

    (i) r converges to a set containing the origin as $t \to \infty$;

    (ii) e_q and $\dot{e}_q$ converge to 0 as $t \to \infty$; and

    (iii) e_λ and τ are bounded for all $t \geq 0$.

Proof Substituting (5.7) into (4.27), the closed-loop dynamic can be described as

$$M_L \dot{r} = -K_p r - K_i \int_0^t r ds - \sum_{i=1}^5 \frac{r \hat{c}_{ri} \Psi_{ri}^2}{\|r\| \Psi_{ri} + \delta_i} - \mu - C_L r \qquad (5.11)$$

Consider the Lyapunov function candidate

$$V = \frac{1}{2} r^T M_L r + \frac{1}{2} \left(\int_0^t r ds \right)^T K_i \int_0^t r ds + \frac{1}{2} \tilde{C}_r \Gamma^{-1} \tilde{C}_r^T \qquad (5.12)$$

with $\tilde{C}_r = C_r - \hat{C}_r$.

Its derivative is

$$\dot{V} = r^T \left(M_L \dot{r} + \frac{1}{2} \dot{M}_L r + K_i \int_0^t r dt \right) + \tilde{C}_r \Gamma^{-1} \dot{\tilde{C}}_r^T \qquad (5.13)$$

where $\Gamma = \text{diag}[\gamma_i] > 0$, $i = 1, \ldots, 5$.

From Property 4.2, we have $\frac{1}{2} \lambda_{min}(M_L) r^T r \leq \frac{1}{2} r^T M_L r \leq \frac{1}{2} \lambda_{max}(M_L) r^T r$. By using Property 4.3, the time derivative of V along the trajectory of (5.11) is

$$\begin{aligned}
\dot{V} &= -r^T K_p r - r^T \mu - r^T \sum_{i=1}^5 \frac{r \hat{c}_{ri} \Psi_{ri}^2}{\|r\| \Psi_{ri} + \delta_i} \\
&\quad + \hat{C}_r^T \Omega \Gamma^{-1} \tilde{C}_r - \sum_{i=1}^5 \frac{\|r\|^2 \tilde{c}_{ri} \Psi_{ri}^2}{\|r\| \Psi_{ri} + \delta_i} \\
&\leq -r^T K_p r + \|r\| \Phi_{ri} - \sum_{i=1}^5 \frac{\|r\|^2 c_{ri} \Psi_{ri}^2}{\|r\| \Psi_{ri} + \delta_i} \\
&\quad + \hat{C}_r^T \Omega \Gamma^{-1} \tilde{C}_r \\
&\leq -r^T K_p r + C_r^T \Delta + \hat{C}_r^T \Omega \Gamma^{-1} \tilde{C}_r \\
&= -r^T K_p r + C_r^T \Delta + \hat{C}_r^T \Omega \Gamma^{-1} (C_r - \hat{C}_r) \\
&= -r^T K_p r + C_r^T \Delta - \frac{1}{4} C_r^T \Omega \Gamma^{-1} C_r + \frac{1}{4} C_r^T \Omega \Gamma^{-1} C_r \\
&\quad + \hat{C}_r^T \Omega \Gamma^{-1} C_r - \hat{C}_r^T \Omega \Gamma^{-1} \hat{C}_r
\end{aligned}$$

$$= -r^T K_p r + C_r^T \Delta - (\frac{1}{2}C_r^T - \hat{C}_r^T)\Omega\Gamma^{-1}(\frac{1}{2}C_r - \hat{C}_r)$$

$$+ \frac{1}{4}C_r^T \Omega\Gamma^{-1}C_r$$

$$\leq -r^T K_p r + C_r^T \Delta + \frac{1}{4}C_r^T \Omega\Gamma^{-1}C_r$$

with $\Omega = \text{diag}[\omega_i]$, $\Delta = [\delta_1, \delta_2, \ldots, \delta_5]^T$, $i = 1, \ldots, 5$.

Therefore, $\dot{V} \leq -\lambda_{min}(K_p)\|r\|^2 + C_r^T \Delta + \frac{1}{4}C_r^T\Omega\Gamma^{-1}C_r$. Since $C_r^T \Delta + \frac{1}{4}C_r^T\Omega\Gamma^{-1}C_r$ is bounded, there exists $t > t_2$, $C_r^T\Delta + \frac{1}{4}C_r^T\Omega\Gamma^{-1}C_r \leq \rho_2$, when $\|r\| \geq \sqrt{\frac{\rho_2}{\lambda_{min}(K_p)}}$, $\dot{V} \leq 0$, from above all, r converges to a small set containing the origin as $t \to \infty$.

(ii) Integrating both sides of the above equation gives

$$V(t) - V(0) \leq -\int_0^t r^T K_p r ds$$

$$+ \int_0^t (C_r^T \Delta + \frac{1}{4}C_r^T \Omega\Gamma^{-1}C_r) ds \qquad (5.14)$$

Since C_r and Γ are constant, $\int_0^\infty \Delta ds = \rho_\delta = [\rho_{1\delta}, \ldots, \rho_{5\delta}]^T$, $\int_0^\infty \Omega ds = \rho_\omega = [\rho_{1\omega}, \ldots, \rho_{5\omega}]^T$, we can rewrite (5.14) as

$$V(t) - V(0) \leq -\int_0^t r^T K_p r ds + C_r^T \left(\int_0^t \Delta ds \right)$$

$$+ \frac{1}{4}C_r^T \left(\int_0^t \Omega ds \right) \Gamma^{-1}C_r ds$$

$$\leq -\int_0^t r^T K_p r ds + C_r^T \rho_\delta + C_r^T \rho_\omega \Gamma^{-1}C_r$$

$$< -\lambda_{min}(K_p)\|r\|^2 + C_r^T \rho_\delta + C_r^T \rho_\omega \Gamma^{-1}C_r$$

$$< \infty \qquad (5.15)$$

Thus V is bounded, which implies that $r \in L_\infty^{n-k-l}$. From (5.15), we have

$$\int_0^t r^T K_p r ds \leq V(0) - V(t) + C_r^T \rho_\delta + C_r^T \rho_\omega \Gamma^{-1}C_r \qquad (5.16)$$

which leads to $r \in L_2^{n-k-l}$. From $r = \dot{e}_\zeta + K_\zeta e_\zeta$, it can be obtained that $e_\zeta, \dot{e}_\zeta \in L_\infty^{n-k-l}$. As we have established $e_\zeta, \dot{e}_\zeta \in L_\infty$, from Assumption 4.1, we conclude that $\zeta(t), \dot{\zeta}(t), \dot{\zeta}_r(t), \ddot{\zeta}_r(t) \in L_\infty^{n-k-l}$ and $\dot{q} \in L_\infty^n$.

Therefore, all the signals on the right hand side of (5.11) are bounded, and we can conclude that $\dot{r}$ and therefore $\ddot{\zeta}$ are bounded. Thus, $r \to 0$ as $t \to \infty$ can be obtained. Consequently, we have $e_\zeta \to 0, \dot{e}_\zeta \to 0$ as $t \to \infty$. It follows that $e_q, \dot{e}_q \to 0$ as $t \to \infty$.

(iii) Substituting the control (5.7) and (5.8) into the reduced order dynamics (4.23) yields

$$
\begin{aligned}
(I + K_f)e_\lambda &= Z(C^1\dot{\zeta} + G^1 + d^1(t) - u_a) + u_b \\
&= -ZL^{+T}M_L\ddot{\zeta} + \frac{\hat{\chi}_r^2}{\hat{\chi}_r + \delta_1}\ddot{\zeta}_d
\end{aligned}
\tag{5.17}
$$

Since $\ddot{\zeta}$ and Z are bounded, $\zeta \to \zeta_d$, $-ZL^{+T}M_L\ddot{\zeta} + \frac{\hat{\chi}_r^2}{\hat{\chi}_r + \delta_1}\ddot{\zeta}_d$ is also bounded, the size of e_λ can be adjusted by choosing the proper gain matrix K_f.

Since r, ζ, $\dot{\zeta}$, ζ_r^1, $\dot{\zeta}_r$, $\ddot{\zeta}_r$, and e_λ are all bounded, it is easy to conclude that τ is bounded from (5.7) and (5.8).

5.1.4 Simulation Studies

To verify the effectiveness of the proposed adaptive robust control, let us consider the mobile manipulator system shown in Fig. 9.1. The mobile manipulator is subjected to the following constraints (2.2), (2.3), and (2.4). Using the Lagrangian approach, we can obtain the standard form (4.1) with

$$
q_v = [x, y, \theta]^T, \ q_a = [\theta_1, \theta_2]^T, \ q = [q_v^T, q_a^T]^T, A = [\cos\theta, \sin\theta, 0]^T
$$

We choose $\zeta = [\zeta_1, \zeta_2, \zeta_3, \zeta_4]^T = [y, \theta, \theta_1, \theta_2]^T$, and we have $\dot{\zeta} = [\dot{y}, \dot{\theta}, \dot{\theta}_1, \dot{\zeta}_2]^T$ and we give the desired trajectories as

$$
y_d = 1.5\sin(t), \ \theta_d = 1.0\sin(t), \ \theta_{1d} = \pi/4(1 - \cos(t))
$$

and the geometric constraints that the end-effector is subjected to as

$$
\Phi = l_1 + l_2\sin(\theta_2) = 0, \ \lambda_{hd} = 10.0N
$$

It is easy to have $J = \begin{bmatrix} \cos\theta & \sin\theta & 0 & 0 & 0 \\ 0 & 0 & 0 & 0 & l_2\cos\theta_2 \end{bmatrix}$, $L = \begin{bmatrix} 1 & 0 & 0 \\ 0 & 1 & 0 \\ 0 & 0 & 1 \\ 0 & 0 & 0 \end{bmatrix}$,

$J^1 = \begin{bmatrix} 0 & 0 & 0 & 0 \\ 0 & 0 & 0 & l_2\cos\theta \end{bmatrix}$ and

$$
\begin{bmatrix} \dot{x} \\ \dot{y} \\ \dot{\theta} \\ \dot{\theta}_1 \\ \dot{\theta}_2 \end{bmatrix} = \begin{bmatrix} \tan\theta & 0 & 0 & 0 \\ 1 & 0 & 0 & 0 \\ 0 & 1 & 0 & 0 \\ 0 & 0 & 1 & 0 \\ 0 & 0 & 0 & 1 \end{bmatrix} \begin{bmatrix} \dot{y} \\ \dot{\theta} \\ \dot{\theta}_1 \\ \dot{\theta}_2 \end{bmatrix}
\tag{5.18}
$$

$\frac{1}{2r} = 1$ and $\frac{l}{2r} = 1$, θ_2 is used for the force control and we have

$$\begin{bmatrix} v \\ \dot{\theta} \\ \dot{\theta}_1 \end{bmatrix} = \begin{bmatrix} 1/\cos\theta & 0 & 0 \\ 0 & 1 & 0 \\ 0 & 0 & 1 \end{bmatrix} \begin{bmatrix} \dot{y} \\ \dot{\theta} \\ \dot{\theta}_1 \\ \dot{\theta}_2 \end{bmatrix} \qquad (5.19)$$

The parameters of the dynamics are chosen as $m_p = m_1 = m_2 = 1.0kg$, $I_w = I_p = 1.0kgm^2$, $I_1 = I_2 = 1.0kgm^2$, $I = 0.5kgm^2$, $d = 1.0m$, $r = 0.5m$, $l = 1.0m$, $2l_1 = 1.0m$, $2l_2 = 1.0m$, $q(0) = [0, 4, 0.785, 0.1]^T$, $\dot{q}(0) = [0.0, 0.0, 0.0, 0.0]^T$ and $\lambda_h(0) = 0$. According to Theorem 5.2, the control gains are selected as $K_p = \text{diag}[1.0]$, $K_\zeta = \text{diag}[1.0]$, $K_i = 0.0$ and $K_f = 0.5$. The adaptation gains in control law (5.7) are chosen as $\delta_i = \omega_i = 1/(1+t)^2$, $i = 1, \ldots, 5$, $C_r = [1.0, 1.0, 1.0, 1.0, 1.0]^T$ and $\Gamma = \text{diag}[2.5]$. The disturbances on the mobile base are set as $0.1\sin(t)$ and $0.1\cos(t)$. We conduct the simulations by the adaptive robust control using (5.7) and (5.8) in Theorem 5.2. The trajectory tracking performances of the adaptive robust control are illustrated in Fig. 5.1 and the velocities tracking and the input torques by the adaptive robust control are shown in Figs. 5.2-5.3. The force tracking of the adaptive robust control is shown in Fig. 5.4. The simulation results show that the trajectory and force tracking errors converge to zero, which validates the results of the controls (5.7) and (5.8) in Theorem 5.2.

5.2 Adaptive Robust Output-feedback Control with Actuator Dynamics

5.2.1 Introduction

In previously developed control schemes, the controls are designed at kinematic level with velocity as input or dynamic level with torque as input, but the actuator dynamics are ignored. As demonstrated in [143], actuator dynamics constitute an important component of the complete robot dynamics, especially in the case of high-velocity movement and highly varying loads. Many control methods have therefore been developed to take into account the effects of actuator dynamics (see, for instance, [143, 145, 146]). However, most

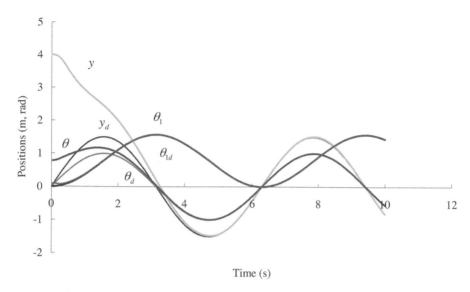

FIGURE 5.1

Position tracking by adaptive robust control.

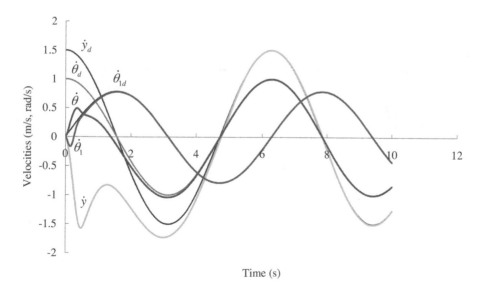

FIGURE 5.2

Velocity tracking by adaptive robust control.

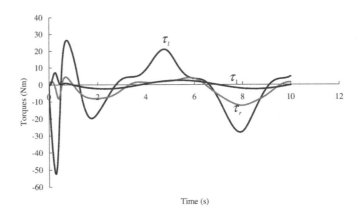

FIGURE 5.3

Input torques by adaptive robust control.

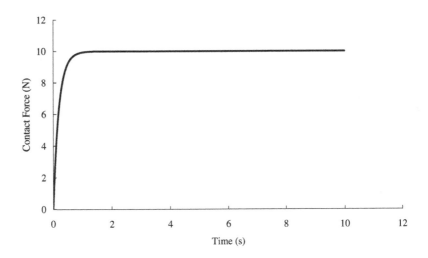

FIGURE 5.4

Contact force.

literature assumes that the actuator velocities are measurable [103], [115], which may deteriorate the control performance of these methods, since velocity measurements are often contaminated by a considerable amount of noise. Therefore, it is desired to achieve good control performance by using only joint position measurement. Moreover, in most research conducted on control of mobile manipulators, joint torques are the control inputs, while in reality the joints are driven by actuators (e.g., DC motors). Therefore, using actuator input voltages as control inputs and designing observer-controller structures for mobile manipulators with only the positions and the driving currents of actuators are more realistic. As such, actuator dynamics are combined with the mobile manipulator's dynamics in this section.

This section addresses adaptive robust output-feedback control of force-motion for mobile manipulators electrically driven by DC motors with both holonomic and nonholonomic constraints in the parameter uncertainties and external disturbances. Simulation results are described in detail that show the effectiveness of the proposed control.

5.2.2 Actuator Dynamics

The joints of the mobile manipulators are assumed to be driven by DC motors. Consider the following notations used to model a DC motor $\nu \in \mathbf{R}^m$ represents the control input voltage vector; I denotes an m-element vector of motor armature current; $K_N \in \mathbf{R}^{m \times m}$ is a positive definite diagonal matrix which characterizes the electromechanical conversion between current and torque; $L_a = \text{diag}\,[L_{a1}, L_{a2}, L_{a3}, ..., L_{am}]$, $R_a = \text{diag}\,[R_{a1}, R_{a2}, R_{a3}, ..., R_{am}]$, $K_e = \text{diag}\,[K_{e1}, K_{e2}, K_{e3}, ..., K_{em}]$, $\omega = [\omega_1, \omega_2, ..., \omega_m]^T$ represent the equivalent armature inductances, resistances, back EMF constants, angular velocities of the driving motors, respectively; $G_r = \text{diag}(g_{ri}) \in \mathbf{R}^{m \times m}$ denotes the gear ratio for m joints; τ_m are the torques exerted by the motor. In order to apply the DC servomotors for actuating an n-DOF mobile manipulator, assume without energy loss a relationship between the ith joint velocity $\dot{\zeta}_i$ and the motor shaft velocity ω_i can be presented as $g_{ri} = \frac{\omega_i}{\dot{\zeta}_i} = \frac{\tau_i}{\tau_{mi}}$, where g_{ri} is the gear ratio of the ith joint, τ_{mi} is the ith motor shaft torque, and τ_i is the ith joint torque. The motor shaft torque is proportional to the motor current as $\tau_m = K_N I$. The back EMF is proportional to the angular velocity of the motor shaft, then one has

$$L_a \frac{dI}{dt} + R_a I + K_e \omega = \nu \qquad (5.20)$$

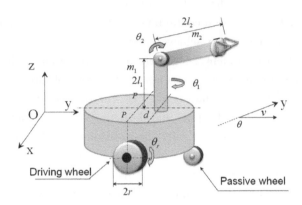

FIGURE 5.5

2-DOF mobile manipulator.

In the actuator dynamics (5.20), the relationship between ω and $\dot{\zeta}$ is dependent on the type of mechanical system and can be generally expressed as

$$\omega = G_r T \dot{\zeta} \tag{5.21}$$

The structure of T depends on the mechanical systems to be controlled. For instance, in the simulation example, a two-wheel differential drive 2-DOF mobile manipulator is used to illustrate the control design. From [147], one has $v = (r\dot{\theta}_l + r\dot{\theta}_r)/2$, $\dot{\theta} = (r\dot{\theta}_r - r\dot{\theta}_l)/2l$, $\dot{\theta}_1 = \dot{\theta}_1$, $\dot{\theta}_2 = \dot{\theta}_2$, where $\dot{\theta}_l$ and $\dot{\theta}_r$ are the angular velocities of the two wheels, respectively and v is the linear velocity of the mobile platform, as shown in Fig. 5.5. Since $\dot{y} = v \cos \theta$, one has $[\; \dot{\theta}_l \;\; \dot{\theta}_r \;\; \dot{\theta}_1 \;\; \dot{\theta}_2 \;]^T = T[\; \dot{y} \;\; \dot{\theta} \;\; \dot{\theta}_1 \;\; \dot{\theta}_2 \;]^T$, $T = [\; T_1 \;\; T_2 \;\; T_3 \;\; T_4 \;]$, $T_1 = [\; \frac{1}{r \cos \theta} \;\; \frac{1}{r \cos \theta} \;\; 0 \;\; 0 \;]^T$, $T_2 = [\; \frac{l}{r} \;\; -\frac{l}{r} \;\; 0 \;\; 0 \;]^T$, $T_3 = [\; 0 \;\; 0 \;\; 1 \;\; 0 \;]^T$, $T_4 = [\; 0 \;\; 0 \;\; 0 \;\; 1 \;]^T$, where r and l are shown in Fig. 5.5.

Eliminating ω from the actuator dynamics (5.20) by substituting (5.21) and considering (4.17), one obtains

$$M_L(\zeta)\ddot{\zeta} + C_L(\zeta, \dot{\zeta})\dot{\zeta} + G_L + d_L(t) = L^T B_1 G_r K_N I \tag{5.22}$$

$$Z(C_2(\zeta, \dot{\zeta})\dot{\zeta} + G_2 + d_2(t) - B_1 G_r K_N I) = \lambda_h \tag{5.23}$$

$$L_a \frac{dI}{dt} + R_a I + K_e G_r T\dot{\zeta} = \nu \tag{5.24}$$

Assumption 5.3 *The unknown actuator parameters L_a, R_a and K_e are bounded and satisfy $\|L_a\| \leq \alpha_1$; $\|R_a\| \leq \alpha_2$; $\|K_e\| \leq \alpha_3$ where $\alpha_\iota, (\iota = 1, 2, 3)$ are finite positive constants.*

Remark 5.1 *In reality, these constants $c_i (1 \leq i \leq 5)$ and $\alpha_\iota (1 \leq \iota \leq 3)$ cannot be obtained easily. Although any fixed large c_i and α_ι can guarantee good performance, it is not recommended in practice as large c_i and α_ι imply, in general, high noise amplification and high cost of control. Therefore, it is necessary to develop a control law which does not require the knowledge of $c_i (1 \leq i \leq 5)$ and $\alpha_\iota (1 \leq \iota \leq 3)$.*

5.2.3 Output-feedback Control Design

Given the desired motion trajectory $\zeta_d(t) = [\eta_d \ q_{rd}^1]^T$ and the desired constraint force $f_d(t)$, or, equivalently, a desired multiplier $\lambda_h(t)$, the trajectory and force tracking controls are to determine control laws such that for any $(\zeta(0), \dot{\zeta}(0)) \in \Omega$, $\zeta, \dot{\zeta}, \lambda_h$ converges to a manifold Ω_d: $\Omega_d = \{(\zeta, \dot{\zeta}, \lambda_h) | \zeta = \zeta_d, \dot{\zeta} = \dot{\zeta}_d, \lambda_h = \lambda_h^d\}$.

The control design consists of two stages: (i) a virtual adaptive control input I_d is designed so that the subsystems (5.22) and (5.23) converge to the desired values, and (ii) the actual control input ν is designed in such a way that $I \to I_d$. In turn, this allows $\zeta - \zeta_d$ and $\lambda_h - \lambda_h^d$ to be stabilized to the origin.

Lemma 5.1 *For $x > 0$ and $\delta \geq 1$, one has $\ln(\cosh(x)) + \delta \geq x$.*

Proof 5.1 *If $x \geq 0$, one has $\int_0^x \frac{2}{e^{2s}+1} ds < \int_0^x \frac{2}{e^{2s}} ds = 1 - e^{-2x} < 1$, Therefore, $\ln(\cosh(x)) + \delta \geq \ln(\cosh(x)) + \int_0^x \frac{2}{e^{2s}+1} ds$ with $\delta \geq 1$. If $f(x) = \ln(\cosh(x)) + \int_0^x \frac{2}{e^{2s}+1} ds - x$, one has $\frac{df(x)}{dx} = \tanh(x) + \frac{2}{e^{2x}+1} - 1 = \frac{e^x - e^{-x}}{e^x + e^{-x}} + \frac{2}{e^{2x}+1} - 1 = 0$. From the mean value theorem, one has $f(x) - f(0) = \frac{df(x)}{dx}(x - 0)$. Since $f(0) = 0$ and $\frac{df(x)}{dx} = 0$ for all x, one has $f(x) = 0$, that is, $\ln(\cosh(x)) + \int_0^x \frac{2}{e^{2s}+1} ds = x$, then, one has $\ln(\cosh(x)) + \delta \geq x$.*

Remark 5.2 *Lemma 5.1 is used to facilitate the control design.*

Definition 5.1 *Consider time varying positive functions $\kappa_\iota(t)$ for $\iota = 1, \ldots, 3$ converging to zero as $t \to \infty$ and satisfying $\lim_{t\to\infty} \int_0^t \kappa_\iota(\omega) d\omega = b_\iota < \infty$ with a finite constant b_ι. There are many choices for $\kappa_\iota(t)$ that satisfy the above condition, for example, $\kappa_\iota = 1/(1 + t)^2$.*

5.2.4 Kinematic and Dynamic Subsystems

Consider the following signals as

$$\dot{\zeta}_r = \dot{\zeta}_d - K_\zeta(\zeta - \zeta_d) = \dot{\zeta}_d - K_\zeta e_\zeta + K_\zeta \tilde{\zeta} \tag{5.25}$$

$$r_1 = \dot{\zeta} - \dot{\zeta}_r = \dot{e}_\zeta + K_\zeta e_\zeta - K_\zeta \tilde{\zeta} \tag{5.26}$$

$$\dot{\zeta}_o = \dot{\hat{\zeta}} - K_\zeta \tilde{\zeta} \tag{5.27}$$

$$r_2 = \dot{\zeta} - \dot{\zeta}_o = \dot{\tilde{\zeta}} + K_\zeta \tilde{\zeta} \tag{5.28}$$

$$r = r_1 + r_2 \tag{5.29}$$

$$e_\lambda = \lambda_h - \lambda_h^d \tag{5.30}$$

where $e_\zeta = \zeta - \zeta_d$, $\tilde{\zeta} = \zeta - \hat{\zeta}$ with $\hat{\zeta}$ denoting the estimate of ζ, and K_ζ is diagonal positive.

The linear observer [148] for velocity estimation adopted

$$\dot{\hat{\zeta}} = z + K_\zeta \tilde{\zeta} + k_d \tilde{\zeta} \tag{5.31}$$

$$\dot{z} = \ddot{\zeta}_r + k_d K_\zeta \tilde{\zeta} \tag{5.32}$$

where k_d is a positive constant.

A decoupling control scheme is introduced to control generalized position and constraint force, separately. Consider the control u in the following form:

$$u = L^{+T} u_a - J^{1T} u_b \tag{5.33}$$

where $u_a = B_{1a} G_{ra} K_{Na} I_a$, $u_b = B_{1b} G_{rb} K_{Nb} I_b$ $B_1 = \text{diag}[B_{1a} \ B_{1b}]$, $G_r = \text{diag}[G_{ra} \ G_{rb}]$, $u_a, B_{1a}, G_{ra}, I_a \in \mathbf{R}^{n-j-k}$, and $u_b, B_{1b}, G_{rb}, I_b \in \mathbf{R}^k$, $L^+ = (L^T L)^{-1} L^T$. Then, equations (4.17) and (4.18) can be rewritten as

$$B_{1a} G_{ra} K_{Na} I_a = M_L(\zeta)\ddot{\zeta} + C_L(\zeta, \dot{\zeta})\dot{\zeta} + G_L + d_L(t) \tag{5.34}$$

$$\lambda_h = Z(C_2(\zeta, \dot{\zeta})\dot{\zeta} + G_2 + d_2(t) - L^{+T} B_{1a} G_{ra} K_{Na} I_a)$$
$$+ B_{1b} G_{rb} K_{Nb} I_b \tag{5.35}$$

Consider the following control laws as

$$I_{ra} = (B_{1a} G_{ra} K_{Na})^{-1}[-K_p(r_1 - r_2) - K_i(\tilde{\zeta} + e_\zeta)$$
$$- \text{sgn}(r)(\ln(\cosh(\hat{\Phi})) + \delta)] \tag{5.36}$$

$$I_{rb} = (B_{1b} G_{rb} K_{Nb})^{-1}[(\ln(\cosh(\hat{\chi})) + \delta)\ddot{\zeta}_d$$
$$+ \lambda_h^d - K_f e_\lambda] \tag{5.37}$$

where $\hat{\Phi} = \hat{C}^T \Psi$, $\hat{C} = [\hat{c}_1 \ \hat{c}_2 \ \hat{c}_3 \ \hat{c}_4 \ \hat{c}_5]^T$, $\dot{\hat{C}} = -\Lambda \hat{C} + \|r\|\Gamma\Psi$, $\hat{\chi} = \hat{c}_1 \|Z^* L^{+T}\|$,

$\Psi = [\|\ddot{\zeta}_r\| \ \|\dot{\zeta}_r\| \ \|\dot{\zeta}_r\|^2 \ 1 \ 1]^T$, K_p, K_i, K_f are diagonal positive; if $r \geq 0$, $\text{sgn}(r) = 1$, otherwise $\text{sgn}(r) = -1$; $\delta \geq 1$ is constant; $\Gamma = \text{diag}[\gamma_i] > 0$; Λ is a diagonal matrix whose each element Λ_i satisfies Definition 5.1, i.e., $\lim_{t \to \infty} \int_0^t \Lambda_i(\omega) d\omega = a_i < \infty$ with the finite constant a_i, $i = 1, \ldots, 5$; $Z^* = (J^1(M^*)^{-1}J^{1T})^{-1}J^1(M^*)^{-1}$ with $\|L^T M^* L\| = \hat{c}_1$, $I_r = [I_{ra}, I_{rb}]^T$ will be defined the following section.

5.2.5   Control Design at the Actuator Level

Until now, ζ tends to ζ_d can be guaranteed, if the actual input control signal of the dynamic system I is of the form I_d which can be realized from the actuator dynamics by the design of the actual control input ν. On the basis of the above statements, it is concluded that if ν is designed in such a way that I tends to I_d then $(\zeta - \zeta_d) \to 0$ and $(\lambda_h - \lambda_h^d) \to 0$.

Define $e_I = \int_0^t (I - I_d)dt$, and $\dot{e}_I = I - I_d$, $I_r = I_d - K_r e_I$, and $s = \dot{e}_I + K_r e_I$ with $K_r = 0$. Substituting I and $\dot{\zeta}$ of (5.24), one obtains

$$L_a \dot{s} + R_a s + K_e G_r T \dot{e}_\zeta = -\phi + \nu \tag{5.38}$$

where $\phi = L_a \dot{I}_r + R_a I_r + K_e G_r T \dot{\zeta}_d$.

Remark 5.3 *Since I is partitioned into I_a and I_b, one has corresponding partitions $e_I = [e_{Ia}, e_{Ib}]^T$, $\dot{e}_I = [\dot{e}_{Ia}, \dot{e}_{Ib}]^T$, $I_r = [I_{ra}, I_{rb}]^T$, $s = [s_a, s_b]^T$, $\phi = [\phi_a, \phi_b]^T$, $T = \text{diag}[T_a, T_b]$, $\nu = [\nu_a, \nu_b]^T$ with the corresponding $\alpha_a = [\alpha_{a1}, \alpha_{a2}, \alpha_{a3}]^T$ and $\alpha_b = [\alpha_{b1}, \alpha_{b2}, \alpha_{b3}]^T$, $\varphi_a = \hat{\alpha}_a^T \mu_a$ and $\varphi_b = \hat{\alpha}_b^T \mu_b$, $\mu_a = [\|\dot{I}_{ad}\| \ \|I_{ad}\| \ \|G_{ra} T_a \dot{\zeta}_d\|]^T$ and $\mu_b = [\|\dot{I}_{bd}\| \ \|I_{bd}\| \ 0]^T$, K_{va} and K_{vb}, $L_a = \text{diag}[L_{aa}, L_{ab}]$, $R_a = \text{diag}[R_{aa}, R_{ab}]$ and $K_e = \text{diag}[K_{ea}, K_{eb}]$.*

Consider the adaptive robust control laws for I_a and I_b, respectively

$$\nu_a = -\text{sgn}(s_a)(\ln(\cosh(\hat{\varphi}_a)) + \delta) - K_{va} s_a \tag{5.39}$$

$$\nu_b = -\text{sgn}(s_b)(\ln(\cosh(\hat{\varphi}_b)) + \delta) - K_{vb} s_b \tag{5.40}$$

where $\hat{\alpha}_a = [\hat{\alpha}_{a1} \ \hat{\alpha}_{a2} \ \hat{\alpha}_{a3}]^T$, $\hat{\alpha}_b = [\hat{\alpha}_{b1} \ \hat{\alpha}_{b2} \ \hat{\alpha}_{b3}]^T$, $\dot{\hat{\alpha}}_a = -\kappa \hat{\alpha}_a + \|s_a\| \beta \mu_a$ and $\dot{\hat{\alpha}}_b = -\kappa \hat{\alpha}_b + \|s_b\| \beta \mu_b$ for I_a and I_b, respectively, and K_{va} and K_{vb} are diagonal positive, $\beta = \text{diag}[\beta_\iota]$, κ is a diagonal matrix whose each element κ_ι satisfies Definition 5.1, i.e., $\lim_{t \to \infty} \int_0^t \kappa_\iota(\omega) d\omega = b_\iota < \infty$ with the finite constant b_ι, $\iota = 1, \ldots, 3$.

5.2.6   Stability Analysis

Theorem 5.3 *Consider the mechanical system described by (4.1), (4.2) and (4.8), using the control laws (5.36), (5.37), (5.39) and (5.40), the following hold for any $(q(0), \dot{q}(0)) \in \Omega_n \cap \Omega_h$: (i) r and s_a converge to a set containing the origin as $t \to \infty$; (ii) e_q and $\dot{e}_q$ converge to 0 as $t \to \infty$; and (iii) s_b converges to a set containing the origin as $t \to \infty$, and e_λ and τ are bounded for all $t \geq 0$.*

Proof (i) By combining (5.22) with (5.26) and considering Property 4.3, the closed-loop system dynamics are given by

$$
\begin{aligned}
M_L(\zeta)\dot{r}_1 \;=\; & B_{1a}G_{ra}K_{Na}I_{ra} + B_{1a}G_{ra}K_{Na}s_a \\
& -\gamma - (C_L(\zeta,\dot{\zeta}) + C_L(\zeta,\dot{\zeta}_r))r_1
\end{aligned}
\tag{5.41}
$$

where $\gamma = M_L(\zeta)\ddot{\zeta}_r + C_L(\zeta,\dot{\zeta}_r)\dot{\zeta}_r + G_L + d_L$. Differentiating (5.31) and considering (5.32), one has $\ddot{\tilde{\zeta}} = \ddot{\zeta}_r + k_d\dot{\tilde{\zeta}} + k_d K_\zeta \tilde{\zeta} + K_\zeta \dot{\tilde{\zeta}}$, which leads to

$$
\dot{r}_2 + k_d r_2 = \dot{r}_1
\tag{5.42}
$$

Substituting (5.36) and (5.28) into (5.41), and considering Property 4.3, the closed-loop dynamic equation is obtained

$$
\begin{aligned}
M_L(\zeta)\dot{r}_1 \;=\; & -C_L(\zeta,\dot{\zeta})r_1 - K_p(r_1 - r_2) - K_i(\tilde{\zeta} + e_\zeta) \\
& - \mathrm{sgn}(r)(\ln(\cosh(\hat{\Phi})) + \delta) - \gamma - C_L(\zeta,\dot{\zeta}_r)r_1 \\
& + B_{1a}G_{ra}K_{Na}s_a
\end{aligned}
\tag{5.43}
$$

Substituting into (5.43), we have

$$
\begin{aligned}
M_L(\zeta)\dot{r}_2 \;=\; & -C_L(\zeta,\dot{\zeta})r_2 - (k_d M_L(\zeta) - K_p)r_2 - K_p r_1 \\
& + C_L(\zeta,\dot{\zeta}_r + r_1)r_2 - C_L(\zeta,r_1)(r_1 + 2\dot{\zeta}_r) \\
& - K_i(\tilde{\zeta} + e_\zeta) - \mathrm{sgn}(r)(\ln(\cosh(\hat{\Phi})) + \delta) \\
& - \gamma + B_{1a}G_{ra}K_{Na}s_a
\end{aligned}
\tag{5.44}
$$

Consider the Lyapunov function candidate

$$
V \;=\; V_1 + V_2
\tag{5.45}
$$

where $V_1 = \frac{1}{2}X^T \mathcal{M} X$ and $V_2 = \frac{1}{2}s_a^T L_{aa}s_a + \frac{1}{2}\tilde{\alpha}_a^T \beta^{-1}\tilde{\alpha}_a$, $X = [r_1\ e_\zeta\ r_2\ \tilde{\zeta}\ \tilde{C}]^T$, $\tilde{C} = C - \hat{C}$, and $\mathcal{M} = \mathrm{diag}[M_L\ K_i\ M_L\ K_i\ \Gamma^{-1}]$, $\tilde{\alpha}_a = \alpha_a - \hat{\alpha}_a$. Differentiating V_1 with respect to time, one has $\dot{V}_1 = r_1^T(M_L\dot{r}_1 + \frac{1}{2}\dot{M}_L r_1) + r_2^T(M_L\dot{r}_2 +$

$\frac{1}{2}\dot{M}_L r_2) + \tilde{C}^T \Gamma^{-1}\dot{\tilde{C}} + e_\zeta^T K_i \dot{e}_\zeta + \tilde{\zeta}^T K_i \dot{\tilde{\zeta}}$. From Property 4.2 and Property 4.3, the time derivative of V_1 along the trajectory of (5.44) is

$$
\begin{aligned}
\dot{V}_1 =\ & -r_1^T K_p r_1 + r_1^T K_p r_2 - r_1^T K_i(\tilde{\zeta} + e_\zeta) \\
& -r_1^T \operatorname{sgn}(r)(\ln(\cosh(\hat{\Phi})) + \delta) - r_1^T \gamma - r_1^T C_L(\zeta, \dot{\zeta}_r) r_1 \\
& +r_1^T B_{1a} G_{ra} K_{Na} s_a - r_2^T (k_d M_L(\zeta) - K_p) r_2 \\
& -r_2^T K_p r_1 + r_2^T C_L(\zeta, \dot{\zeta}_r + r_1) r_2 \\
& -r_2^T C_L(\zeta, r_1)(r_1 + 2\dot{\zeta}_r) - r_2^T K_i(\tilde{\zeta} + e_\zeta) \\
& -r_2^T \operatorname{sgn}(r)(\ln(\cosh(\hat{\Phi})) + \delta) - r_2^T \gamma \\
& +r_2^T B_{1a} G_{ra} K_{Na} s_a + \tilde{C}^T \Gamma^{-1}\dot{\tilde{C}} + e_\zeta^T K_i \dot{e}_\zeta + \tilde{\zeta}^T K_i \dot{\tilde{\zeta}}
\end{aligned}
$$

Considering Lemma 5.1, one has $\ln(\cosh(\hat{\Phi})) + \delta \ge \hat{\Phi}$, and $\|r\|(\ln(\cosh(\hat{\Phi})) + \delta) \ge \|r\|\hat{\Phi}$, and $\dot{e}_\zeta = r_1 - K_\zeta e_\zeta + K_\zeta \tilde{\zeta}$, $\dot{\tilde{\zeta}} = r_2 - K_\zeta \tilde{\zeta}$ from (5.26) and (5.28), one has

$$
\begin{aligned}
\dot{V}_1 =\ & -r_1^T K_p r_1 - r_2^T (k_d M_L(\zeta) - K_p) r_2 \\
& -r_2^T K_i e_\zeta + \tilde{\zeta}^T K_i K_\zeta e_\zeta - r_1^T K_i \tilde{\zeta} - e_\zeta^T K_i K_\zeta e_\zeta - \tilde{\zeta}^T K_i K_\zeta \tilde{\zeta} \\
& -r^T \operatorname{sgn}(r)(\ln(\cosh(\hat{\Phi})) + \delta) - r^T \gamma - r_1^T C_L(\zeta, \dot{\zeta}_r) r_1 \\
& +r_2^T C_L(\zeta, \dot{\zeta}_r + r_1) r_2 - r_2^T C_L(\zeta, r_1)(r_1 + 2\dot{\zeta}_r) \\
& +\tilde{C}^T \Gamma^{-1}\dot{\tilde{C}} + r^T B_{1a} G_{ra} K_{Na} s_a
\end{aligned}
$$

Considering Lemma 5.1, we have $\ln(\cosh(\hat{\Phi})) + \delta \ge \hat{\Phi}$, and $\|r\|(\ln(\cosh(\hat{\Phi})) + \delta) \ge \|r\|\hat{\Phi}$, therefore

$$
\begin{aligned}
\dot{V}_1 \le\ & -r_1^T K_p r_1 - r_2^T (k_d M_L(\zeta) - K_p) r_2 \\
& -e_\zeta^T K_i K_\zeta e_\zeta - \tilde{\zeta}^T K_i K_\zeta \tilde{\zeta} + \|K_i\|\|e_\zeta\|\|r_2\| \\
& +\|K_i K_\zeta\|\|e_\zeta\|\|\tilde{\zeta}\| + \|K_i\|\|\tilde{\zeta}\|\|r_1\| \\
& +\|r\|\|\gamma\| - \|r\|\hat{\Phi} + \tilde{C}^T \Gamma^{-1}\Lambda\hat{C} - \tilde{C}^T \Psi\|r\| \\
& -r_1^T C_L(\zeta, \dot{\zeta}_r) r_1 + r_2^T C_L(\zeta, \dot{\zeta}_r + r_1) r_2 \\
& -r_2^T C_L(\zeta, r_1)(r_1 + 2\dot{\zeta}_r) + r^T B_{1a} G_{ra} K_{Na} s_a
\end{aligned}
$$

Since

$$
\begin{aligned}
\|K_i\|\|e_\zeta\|\|r_2\| &\le \frac{1}{2}\|K_i\|e_\zeta^T e_\zeta + \frac{1}{2}\|K_i\|r_2^T r_2 \\
\|K_i K_\zeta\|\|e_\zeta\|\|\tilde{\zeta}\| &\le \frac{1}{2}\|K_i K_\zeta\|e_\zeta^T e_\zeta + \frac{1}{2}\|K_i K_\zeta\|\tilde{\zeta}^T \tilde{\zeta} \\
\|K_i\|\|\tilde{\zeta}\|\|r_1\| &\le \frac{1}{2}\|K_i\|\tilde{\zeta}^T \tilde{\zeta} + \frac{1}{2}\|K_i\|r_1^T r_1
\end{aligned}
$$

we have

$$
\begin{aligned}
\dot{V}_1 \leq & -r_1^T \left(K_p - \frac{1}{2}\|K_i\|\mathcal{I} \right) r_1 - r_2^T \left((k_d M_L(\zeta) - K_p) - \frac{1}{2}\|K_i\|\mathcal{I} \right) r_2 \\
& -e_\zeta^T \left(K_i K_\zeta - \frac{1}{2}\|K_i K_\zeta\|\mathcal{I} - \frac{1}{2}\|K_i\|\mathcal{I} \right) e_\zeta \\
& -\tilde{\zeta}^T \left(K_i K_\zeta - \frac{1}{2}\|K_i K_\zeta\|\mathcal{I} - \frac{1}{2}\|K_i\|\mathcal{I} \right) \tilde{\zeta} \\
& +\tilde{C}^T \Gamma^{-1}\Lambda\hat{C} + \|C_L(\zeta,\dot{\zeta}_r)\|\|r_1\|^2 + \|C_L(\zeta,\dot{\zeta}_r + r_1)\|\|r_2\|^2 \\
& +\|C_L(\zeta, r_1 + 2\dot{\zeta}_r)\|\|r_1\|\|r_2\| + r^T B_{1a} G_{ra} K_{Na} s_a
\end{aligned}
$$

where $\mathcal{I}$ is the unit matrix.

From Property 5.1, the following relationships are valid:

$$
\|C_L(\zeta,\dot{\zeta}_r)\| \leq \mu_1 \tag{5.46}
$$
$$
\|C_L(\zeta,\dot{\zeta}_r + r_1)\| \leq \mu_2 \tag{5.47}
$$
$$
\|C_L(\zeta, r_1 + 2\dot{\zeta}_r)\| \leq \mu_3 \tag{5.48}
$$

where μ_1, μ_2 and μ_3 are known constants.

Since

$$
\begin{aligned}
\tilde{C}^T \Gamma^{-1}\Lambda\hat{C} &= \hat{C}^T \Gamma^{-1}\Lambda(C - \hat{C}) \\
&= \frac{1}{4}C^T\Gamma^{-1}\Lambda C - \frac{1}{4}C^T\Gamma^{-1}\Lambda C + \hat{C}^T\Gamma^{-1}\Lambda C - \hat{C}^T\Gamma^{-1}\Lambda\hat{C} \\
&= -(\frac{1}{2}C^T - \hat{C}^T)\Gamma^{-1}\Lambda(\frac{1}{2}C - \hat{C}) + \frac{1}{4}C^T\Gamma^{-1}\Lambda C \\
&\leq \frac{1}{4}C^T\Gamma^{-1}\Lambda C \tag{5.49}
\end{aligned}
$$

we have

$$
\begin{aligned}
\dot{V}_1 \leq & -r_1^T \left(K_p - \frac{1}{2}\|K_i\|\mathcal{I} - \mu_1\mathcal{I} - \frac{1}{2}\mu_3\mathcal{I} \right) r_1 \\
& -e_\zeta^T \left(K_i K_\zeta - \frac{1}{2}\|K_i K_\zeta\|\mathcal{I} - \frac{1}{2}\|K_i\|\mathcal{I} \right) e_\zeta \\
& -r_2^T \left(k_d M_L(\zeta) - K_p - \frac{1}{2}\|K_i\|\mathcal{I} - \mu_2\mathcal{I} - \frac{1}{2}\mu_3\mathcal{I} \right) r_2 \\
& -\tilde{\zeta}^T \left(K_i K_\zeta - \frac{1}{2}\|K_i K_\zeta\|\mathcal{I} - \frac{1}{2}\|K_i\|\mathcal{I} \right) \tilde{\zeta} \\
& +\frac{1}{4}C^T\Gamma^{-1}\Lambda C + r^T B_{1a} G_{ra} K_{Na} s_a \tag{5.50}
\end{aligned}
$$

Differentiating $V_2(t)$ with respect to time, considering $n - j - k$ joints, using (5.38) for ν_a, we have

$$
\begin{aligned}
\dot{V}_2 \;=\; & -s_a^T[-\nu_a + (L_{aa}\dot{I}_{ra} + R_{aa}I_{ra} + K_{ea}G_{ra}T_a\dot{\zeta}_d) \\
& + R_{aa}s_a + K_{ea}G_{ra}T_a\dot{e}_\zeta] + \tilde{\alpha}_a^T\beta^{-1}\dot{\hat{\alpha}}_a
\end{aligned}
\tag{5.51}
$$

Substituting ν_a into (5.51) by the control law (5.39) for ν_a, one has

$$
\begin{aligned}
\dot{V}_2 \;=\; & -s_a^T(K_{\nu a} + R_{aa})s_a - s_a^T K_{ea}G_{ra}T_a\dot{e}_\zeta \\
& -s_a^T(L_{aa}\dot{I}_{ra} + R_{aa}I_{ra} + K_{ea}G_{ra}T_a\dot{\zeta}_d) \\
& -s_a^T \operatorname{sgn}(s_a)(\ln\cosh(\hat{\varphi}_a) + \delta) + \tilde{\alpha}^T\beta^{-1}\dot{\hat{\alpha}} \\
\leq\; & -s_a^T(K_{\nu a} + R_{aa})s_a - s_a^T K_{ea}G_{ra}T(r_1 - K_\zeta e_\zeta + K_\zeta\tilde{\zeta}) + \|s_a\|\|\phi_a\| \\
& -s_a^T \operatorname{sgn}(s_a)(\ln\cosh(\hat{\varphi}_a) + \delta) + \tilde{\alpha}_a^T\beta^{-1}\dot{\hat{\alpha}}_a
\end{aligned}
\tag{5.52}
$$

As similar as $\tilde{C}^T\Gamma^{-1}\Lambda\hat{C}$, one has $\tilde{\alpha}_a^T\beta^{-1}\kappa\hat{\alpha}_a \leq \frac{1}{4}\alpha_a^T\beta^{-1}\kappa\alpha_a$. By noting $\|s_a\|(\ln\cosh(\hat{\varphi}_a) + \delta) \geq \|s_a\|\hat{\varphi}_a$, we have

$$
\begin{aligned}
\dot{V}_2 \;\leq\; & -s_a^T(K_{\nu a} + R_{aa})s_a - s_a^T K_{ea}G_{ra}T_a(r_1 - K_\zeta e_\zeta + K_\zeta\tilde{\zeta}) \\
& +\|s_a\|\varphi_a - \|s_a\|\hat{\varphi}_a + \tilde{\alpha}_a^T\beta^{-1}\dot{\hat{\alpha}}_a \\
=\; & -s_a^T(K_{\nu a} + R_{aa})s_a - s_a^T K_{ea}G_{ra}T_a(r_1 - K_\zeta e_\zeta + K_\zeta\tilde{\zeta}) \\
& +\|s_a\|\varphi_a - \|s_a\|\hat{\varphi}_a + \tilde{\alpha}^T\beta^{-1}\kappa\hat{\alpha} - \tilde{\alpha}^T\beta^{-1}\beta\mu_a\|s_a\| \\
\leq\; & -s_a^T(K_{\nu a} + R_{aa})s_a + \frac{1}{4}\alpha_a^T\beta^{-1}\kappa\alpha_a \\
& -s_a^T K_{ea}G_{ra}T_a(r_1 - K_\zeta e_\zeta + K_\zeta\tilde{\zeta})
\end{aligned}
\tag{5.53}
$$

Since the last term in (5.50), we have $r^T B_{1a}G_{ra}K_{Na}s_a \leq \frac{1}{2}\|B_{1a}G_{ra}K_{Na}\|\|r_1\|^2 + \frac{1}{2}\|B_{1a}G_{ra}K_{Na}\|\|r_2\|^2 + \|B_{1a}G_{ra}K_{Na}\|\|s_a\|^2$, and in (5.53), one obtains

$$
\begin{aligned}
-s_a^T K_{ea}G_{ra}T_a(r_1 - K_\zeta e_\zeta + K_\zeta\tilde{\zeta}) \;\leq\; & \frac{3}{2}\|K_{ea}G_{ra}T_a\|\|s_a\|^2 \\
& +\frac{1}{2}\|K_{ea}G_{ra}T_a\|\|r_1\|^2 \\
& +\frac{1}{2}\|K_{ea}G_{ra}K_\zeta\|\|e_\zeta\|^2 \\
& +\frac{1}{2}\|K_{ea}G_{ra}K_\zeta\|\|\tilde{\zeta}\|^2
\end{aligned}
\tag{5.54}
$$

By integrating (5.50) and (5.53) and considering the above two inequalities, $\dot{V}$ can be expressed as

$$
\begin{aligned}
\dot{V} \;\leq\; & -r_1^T\mathcal{A}r_1 - r_2^T\mathcal{B}r_2 - s_a^T\mathcal{C}s_a - e_\zeta^T\mathcal{D}e_\zeta - \tilde{\zeta}^T\mathcal{D}\tilde{\zeta} \\
& +\frac{1}{4}(C^T\Gamma^{-1}\Lambda C + \alpha_a^T\beta^{-1}\kappa\alpha_a)
\end{aligned}
\tag{5.55}
$$

where

$$\mathcal{A} = K_p - \frac{1}{2}\|K_i\|\mathcal{I} - \mu_1\mathcal{I} - \frac{1}{2}\mu_3\mathcal{I} - \frac{1}{2}\|K_{ea}G_{ra}T_a\|\mathcal{I} - \frac{1}{2}\|B_{1a}G_{ra}K_{Na}\|\mathcal{I}$$

$$\mathcal{B} = k_d M_L(\zeta) - K_p - \frac{1}{2}\|K_i\|\mathcal{I} - \mu_2\mathcal{I} - \frac{1}{2}\mu_3\mathcal{I} - \frac{1}{2}\|B_{1a}G_{ra}K_{Na}\|\mathcal{I}$$

$$\mathcal{C} = K_{\nu a} + R_{aa} - \|B_{1a}G_{ra}K_{Na}\|\mathcal{I} - \frac{3}{2}\|K_{ea}G_{ra}T_a\|\mathcal{I}$$

$$\mathcal{D} = K_i K_\zeta - \frac{1}{2}\|K_i K_\zeta\|\mathcal{I} - \frac{1}{2}\|K_i\|\mathcal{I} - \frac{1}{2}\|K_{ea}G_{ra}K_\zeta\|\mathcal{I}$$

From (5.55), we can choose K_p, K_i, $K_{\nu a}$, k_d and K_ζ such that the matrices $\mathcal{A}$, $\mathcal{B}$, $\mathcal{C}$, $\mathcal{D}$ are all positive definite. Then

$$
\begin{aligned}
\dot{V} \leq\ & -\lambda_{min}(\mathcal{A})\|r_1\|^2 - \lambda_{min}(\mathcal{B})\|r_2\|^2 - \lambda_{min}(\mathcal{C})\|s_a\|^2 \\
& -\lambda_{\min}(\mathcal{D})\|e_\zeta\|^2 - \lambda_{\min}(\mathcal{D})\|\tilde{\zeta}\|^2 \\
& +\frac{1}{4}(C^T\Gamma^{-1}\Lambda C + \alpha_a^T\beta^{-1}\kappa\alpha_a) \quad\quad (5.56)
\end{aligned}
$$

noting $\frac{1}{4}(C^T\Gamma^{-1}\Lambda C + \alpha_a^T\beta^{-1}\kappa\alpha_a) \to 0$ as $t \to \infty$ because of Definition 5.1.

Integrating both sides of the above equation gives

$$
\begin{aligned}
V(t) - V(0) \leq\ & -\int_0^t (r_1^T\mathcal{A}r_1 + r_2^T\mathcal{B}r_2 + s_a^T\mathcal{C}s_a + e_\zeta^T\mathcal{D}e_\zeta + \tilde{\zeta}^T\mathcal{D}\tilde{\zeta})d\varpi \\
& +\frac{1}{4}\int_0^t (C^T\Gamma^{-1}\Lambda C + \alpha^T\beta^{-1}\kappa\alpha)d\varpi \\
< & -\int_0^t (r_1^T\mathcal{A}r_1 + r_2^T\mathcal{B}r_2 + s_a^T\mathcal{C}s_a + e_\zeta^T\mathcal{D}e_\zeta \\
& +\tilde{\zeta}^T\mathcal{D}\tilde{\zeta})d\varpi + \frac{1}{4}C^T\mathcal{E}C + \frac{1}{4}\alpha_a^T\mathcal{F}\alpha_a \quad\quad (5.57)
\end{aligned}
$$

with $\mathcal{E} = \text{diag}[a_i/\gamma_i]$, $i = 1, \ldots, 5$ and $\mathcal{F} = \text{diag}[b_\iota/\beta_\iota]$, $\iota = 1, \ldots, 3$.

(ii) From (5.57), one has $V(t) < V(0) + \frac{1}{4}C^T\mathcal{E}C + \frac{1}{4}\alpha_a^T\mathcal{F}\alpha_a$, therefore, V is bounded, which implies that $r_1, r_2, s_a \in L_\infty^{n-j-k}$. From the definitions of r_1 and r_2 and s_a, it can be obtained that $e_\zeta, \dot{e}_\zeta, e_{Ia}, \dot{e}_{Ia} \in L_\infty^{n-j-k}$. As it has been established that $e_\zeta, \dot{e}_\zeta, e_{Ia}, \dot{e}_{Ia} \in L_\infty$, from Assumption 4.1, we conclude that $\zeta(t), \dot{\zeta}(t), \dot{\zeta}_r(t), \ddot{\zeta}_r(t), I_a, \dot{I}_a, I_{ra}, \dot{I}_{ra} \in L_\infty^{n-k}$ and $\dot{q} \in L_\infty^n$.

Therefore, all the signals on the right hand side of (5.44) are bounded and it is concluded that $\dot{r}_1$, $\dot{r}_2$ and $\dot{s}_a$ and therefore $\ddot{\zeta}$ and $\dot{I}_{ra}$ are bounded. Thus, $r, s_a \to 0$ as $t \to \infty$ can be obtained. Consequently, one has $e_\zeta, e_{Ia} \to 0, \dot{e}_\zeta, \dot{e}_{Ia} \to 0$ as $t \to \infty$. It follows that $e_q, \dot{e}_q \to 0$ as $t \to \infty$.

(iii) Substituting the control (5.36) and (5.37) into the reduced order dynamic system model (4.23) yields

$$
\begin{aligned}
(I + K_f)e_\lambda \;=\; & Z(C_2(\zeta, \dot\zeta)\dot\zeta + G_2 + d_2(t) - L^{+^T} B_{1a} G_{rb} K_{Na} I_a) \\
& + B_{1b} G_{rb} K_{Nb} I_{rb} + B_{1b} G_{rb} K_{Nb} s_b \\
& - Z L^{+^T} M_L(\zeta)\ddot\zeta + (\ln(\cosh(\hat\chi)) + \delta)\ddot\zeta_r \\
& + B_{1b} G_{rb} K_{Nb} s_b
\end{aligned}
\tag{5.58}
$$

For the k joints in the force space, $V_1 = 0$, $I_b \in \mathbf{R}^k$, (5.24) could be rewritten as $L_{ab}\frac{dI_b}{dt} + R_{ab}I_b = \nu_b$ and therefore, V_2 could be rewritten as $V_2 = \frac{1}{2}s_b^T L_{ab} s_b + \frac{1}{2}\tilde\alpha_b^T \beta^{-1}\tilde\alpha_b$, and the time derivative of V_2 is as similar as (5.53),

$$
\dot V_2 \;\le\; -s_b^T K_{Nb}(K_{\nu b} + R_{ab})s_b + \frac{1}{4}\alpha_b^T \beta^{-1}\kappa\alpha_b
\tag{5.59}
$$

noting that $\frac{1}{4}\alpha_b^T\beta^{-1}\kappa\alpha_b \to 0$ as $t \to \infty$ because of κ satisfying Definition 5.1.

Integrating both sides of the above equation gives

$$
V_2(t) - V_2(0) < -\int_0^t s_b^T K_{Nb}(K_{\nu b} + R_{ab})s_b d\varpi + \frac{1}{4}\alpha_b^T \mathcal{F}\alpha_b
$$

Since $K_{Nb}(K_{\nu b} + R_{ab}) > 0$ is bounded, $V_2(t) < V_2(0) + \frac{1}{4}\alpha_b^T\mathcal{F}\alpha_b$, therefore, $V_2(t)$ is bounded, $V(t)$ is bounded for the joints in the force space and then $s_b \to 0$ as $t \to \infty$. The proof is completed by noticing that $\ddot\zeta$, Z, K_{Nb} and s_b are bounded. Moreover, $\zeta_r \to \zeta_d$, $\zeta \to \zeta_d$, and the right side of (5.58) are bounded and the size of e_λ can be adjusted by choosing the proper gain matrices K_λ.

Since r, ζ, $\dot\zeta$, ζ_r, $\dot\zeta_r$, $\ddot\zeta_r$, e_λ and s are all bounded, it is easy to conclude that τ is bounded from (5.33).

5.2.7 Simulation Studies

Consider the mobile manipulator shown in Fig. 5.5 with the nonholonomic constraint $\dot x \cos\theta + \dot y \sin\theta = 0$, where $q_v = [x, y, \theta]^T$, $q_r = [\theta_1, \theta_2]^T$, $q = [q_v, q_r]^T$, and $A_v = [\cos\theta, \sin\theta, 0]^T$, $\tau_v = [\tau_{lw}, \tau_{rw}]^T$, $\tau_r = [\tau_1, \tau_2]^T$. The matrices H can be chosen as $H = [H_1, H_2, H_3, H_4, H_5]$, $H_1 = [-\tan\theta, 0.0, 0.0, 0.0]^T$, $H_2 = [1.0, 0.0, 0.0, 0.0]^T$, $H_3 = [0.0, 1.0, 0.0, 0.0]^T$, $H_4 = [0.0, 0.0, 1.0, 0.0]^T$ and $H_5 = [0.0, 0.0, 0.0, 1.0]^T$.

Given the desired trajectories and the geometric constraint, the end-effector is subject to $y_d = 1.5\sin(t)$, $\theta_d = 1.0\sin(t)$, $\theta_{1d} = \pi/4(1 - \cos(t))$, $\Phi = l_1 + l_2\sin(\theta_2) = 0$, and $\lambda_d = 10.0N$.

Remark 5.4 *The existence of sgn-function in the control (5.36), (5.39) and (5.40) may inevitably lead to chattering in control torques. To avoid such a phenomenon, in actual implementation, it can be replaced by the sat function defined as: if $|\sigma| \geq \varsigma$, $\mathrm{sat}(\sigma) = \mathrm{sgn}(\sigma)$, else, $\mathrm{sat}(\sigma) = \varsigma/\sigma$, where $\sigma = r$ or s [23].*

In the simulation, the parameters of the system are assumed to be $m_p = m_1 = m_2 = 1.0kg$, $I_w = I_p = 1.0kgm^2$, $2I_1 = I_2 = 1.0kgm^2$, $I = 0.5kgm^2$, $d = l = r = 1.0m$, $2l_1 = 1.0m$, $2l_2 = 0.6m$, $q(0) = [0, 2.0, 0.6, 0.5]^T m$, $\dot{q}(0) = [0.0, 0.0, 0.0, 0.0]^T m/s$, $K_N = \mathrm{diag}[0.01]Nm/A$, $G_r = \mathrm{diag}[100]$, $L_a = [5.0, \ldots, 5.0]^T mH$, $R_a = [2.5, \ldots, 2.5]^T Ohm$, and $K_e = [0.02, 0.02, 0.02, 0.02]^T Vs/rad$. The disturbances on the mobile base are introduced into the simulation model as $0.1\sin(t)$ and $0.1\cos(t)$. By Theorem 5.3, the observer gain is selected as $k_d = \mathrm{diag}[50]$, the control gains are selected as $K_p = \mathrm{diag}[10.0, 10.0, 10.0]$, $K_\varsigma = \mathrm{diag}[1.0, 1.0, 1.0]$, $K_\nu = \mathrm{diag}[10.0, 10.0, 10.0, 10.0]$, $K_i = \mathrm{diag}[0.5]$ and $K_f = 0.5$, $\hat{C}(0) = [1.0, 1.0, 1.0, 1.0, 1.0]^T$, $K_N = \mathrm{diag}[0.1]$, $G_r = \mathrm{diag}[50]$, $\Lambda = \kappa = \mathrm{diag}[1/(1 + t)^2]$, $\Gamma = \mathrm{diag}[4.8]$, $\beta = \mathrm{diag}[0.2]$, $\hat{a}(0) = [0.001, 1.0, 0.01]^T$, $\varsigma = 0.01$. The simulation results for motion/force are shown in Figs. 5.6-5.8 [168].

The desired currents tracking and input voltages on the motors are shown in Figs. 5.9-5.10 and Figs. 5.12-5.13. The simulation results show the better performance: the trajectory and force tracking errors converge to zero, which validates the effectiveness of the control law in Theorem 5.3. The better tracking performances is largely due to the "adaptive" mechanism. Although the parametric uncertainties and the external disturbances are both introduced into the simulation model, the force/motion control performance of the system, under the proposed control, is not degraded. The simulation results demonstrate the effectiveness of the proposed adaptive control in the presence of unknown nonlinear dynamic system and environments. Different motion/force tracking performance can be achieved by adjusting parameter adaptation gains and control gains.

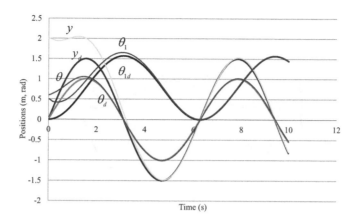

FIGURE 5.6

The positions of the joints.

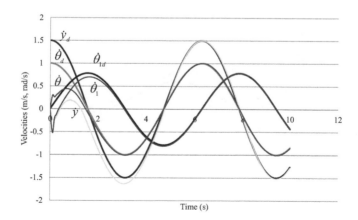

FIGURE 5.7

The velocities of the joints.

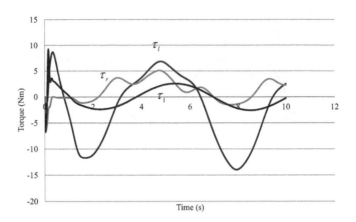

FIGURE 5.8

The torques of the joints.

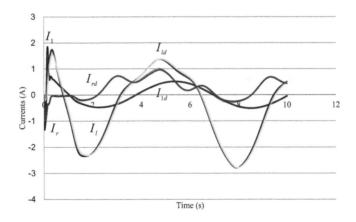

FIGURE 5.9

Tracking the desired currents.

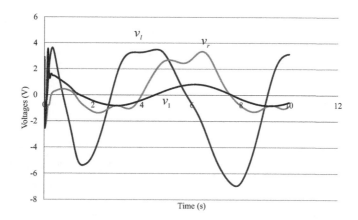

FIGURE 5.10

The input voltages.

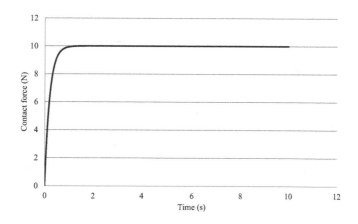

FIGURE 5.11

The constraint force.

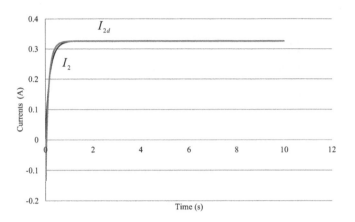

FIGURE 5.12

Tracking the desired current of joint 2.

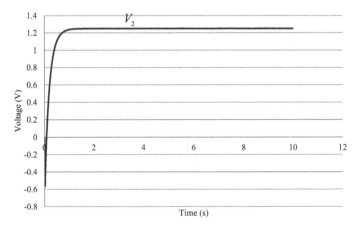

FIGURE 5.13

The input voltage of joint 2.

5.3 Adaptive Robust Hybrid Position/Force Control

5.3.1 Introduction

Fundamentally, the nonholonomic mobile manipulator is a typical nonholonomic mechanical system and possesses a complex and strongly coupled dynamics of the mobile platform and the robotic manipulator. In this section, we shall consider not only position stabilization but also force control under uncertain constraints on the dynamics level. The tracking control of mobile manipulators should include position/force tracking control in the presence of model parameter uncertainty and the constraint uncertainty for the practical applications. Therefore, in this section, we shall consider the case that the end-effector of the mobile manipulator might be in contact with a constrained surface, as in applications involving contour following, grinding, and assembling. Both the positions and constraint forces are required to asymptotically converge to their desired trajectories and constraint forces, respectively, in the presence of uncertainty.

With the assumption of known dynamics, much research has been carried out. In [101], input-output feedback linearization was proposed for the mobile platform such that the manipulator is always positioned at the preferred configurations measured by its manipulability. In [102], the dynamic coupling between the mobile platform and the arm on the tracking performance of a mobile manipulator was studied, and nonlinear feedback control for the mobile manipulator was developed to compensate for the dynamic interaction. Similarly, through nonlinear feedback linearization and decoupling the dynamics in [61], force/position control of the end-effector along the same direction for mobile manipulators was proposed and applied to nonholonomic cart pushing.

Decentralized robust control was developed to overcome the effects of unknown internal interactive forces and parameter uncertainty of the mobile manipulators [7]. Adaptive neural network (NN) control has been proposed for the motion control of a mobile manipulator subject to kinematic constraints [115]. However, for practical applications, not only the motion but also the force of the end-effector of the arm should be considered. Adaptive control was proposed for trajectory/force control of mobile manipulators subjected to holonomic and nonholonomic constraints with unknown inertia parameters [103], which ensures the motion of the system to asymptotically converge to the desired trajectory and force. Most control approaches for mobile manipulators deal with uncertainty in system dynamics and assume that the exact holonomic constraints of the interconnected system is known. However, in practical applications, environmental uncertainties arise in mobile manipulator applications. The motion of the end-effector on the surface together with the contact force are to be controlled, while exact knowledge of the holonomic constraints is not usually available due to the complexity of the constraints or possible change in location and orientation of the constraints. Due to the uncertainty of the system parameters, the holonomic uncertainty in such systems can affect the control stability and performance. If the rigid working environment is partially known, the holonomic uncertainty could be translated into kinematic uncertainty or a constraint Jacobian matrix.

In this section, under holonomic uncertainty, we consider the position and force tracking for general nonholonomic mobile manipulator systems in which the system is transformed into a chained form. Since the general position/force decomposition control is not valid with constraint uncertainties, we develop a low-pass force filter to assure the boundedness of force error and simultaneously estimate the uncertain constraint Jacobian matrix. A main concern in our design is the magnitude of the Jacobian error as it influences the residual force error. By investigating the dynamics error, we apply a reduction procedure and define a mixed tracking error of trajectory and force. Then, robust adaptive force/motion control is presented in which adaptive controls are used to compensate for parametric uncertainties and suppress constraint uncertainties and bounded disturbances. The control guarantees the outputs of the dynamic system to track some bounded hybrid signals which subsequently drive the kinematic system to the desired trajectory/force.

5.3.2 Nonholonomic Constraint in Chained Form

When the system is subjected to nonholonomic constraints, assume the m nonintegrable and independent velocity constraints can be expressed as

$$A(q_v)\dot{q}_v = 0 \tag{5.60}$$

Since $A(q_v) \in \mathbf{R}^{(n_v - m) \times n_v}$, it is always possible to find an m rank matrix $S_v(q_v) \in \mathbf{R}^{n_v \times m}$ formed by a set of smooth and linearly independent vector fields spanning the null space of $A(q_v)$, i.e.,

$$S_v^T(q_v)A^T(q_v) = 0 \tag{5.61}$$

Since $S_v(q_v) = [s_{v1}(q_v), \ldots, s_{vm}(q_v)]$ is formed by a set of smooth and linearly independent vectors spanning the null space of $A(q_v)$, we define an auxiliary function $v_b = [v_{b1}, \ldots, v_{bm}]^T \in \mathbf{R}^m$ such that

$$\dot{q}_v = S_v(q_v)v_b = s_{v1}(q_v)v_{b1} + \cdots + s_{vm}(q_v)v_{bm} \tag{5.62}$$

which is the so-called kinematic model of nonholonomic system. Differentiating (5.62) yields

$$\ddot{q}_v = \dot{S}_v(q_v)v_b + S_v(q_v)\dot{v}_b \tag{5.63}$$

Therefore, the dynamics (4.1), after eliminating the nonholonomic constraint $A^T\lambda_n$, can be reformulated as

$$\dot{q} = S(q)v \tag{5.64}$$
$$M_1(q)\dot{v} + C_1(q,\dot{q})v + G_1(q) + d_1(t) = B_1(q)\tau + f_1 \tag{5.65}$$

where $S = \text{diag}[S_v(q_v), I_a]$, $M_1(q) = S^T(q)M(q)S(q)$, $G_1(q) = S^T(q)G(q)$, $C_1(q,\dot{q}) = S^T(q)[M(q)\dot{S}(q) + C(q,\dot{q})S(q)]$, $d_1(t) = S^T(q)d(t)$, $B_1(q) = S^T(q)B(q)$, $v = [v_b, \dot{q}_a]^T$, and $f_1 = J_1^T\lambda_h$ with $J_1 = [J_v, J_a]$.

5.3.3 Reduced Model and State Transformation

Assume that there exist a coordinate transformation $T_1(q)$ and a state feedback $T_2(q)$

$$\zeta = [\xi, q_a]^T = T_1(q) = [T_{11}(q_v), q_a]^T \tag{5.66}$$

$$v = [v_b, \dot{q}_a]^T = T_2(q)u = [T_{21}(q_v)u_b, u_a]^T \tag{5.67}$$

with $T_2(q) = \text{diag}[T_{21}(q_v), I]$ and $u = [u_b, u_a]^T$, so that the kinematic system (5.64) could be locally or globally converted to the chained form [84, 85]

$$\begin{aligned}
\dot{\xi}_1 &= u_1 \\
\dot{\xi}_i &= u_1\xi_{i+1}(2 \leq i \leq n_v - 1) \\
\dot{\xi}_{n_v} &= u_2 \\
\dot{q}_a &= u_a
\end{aligned} \tag{5.68}$$

where $u_a = [u_{a1}, \ldots, u_{an_a}]^T$.

Remark 5.5 *A necessary and sufficient condition for existence of the transformation of the kinematic system (5.64) with two independent wheels driven* *mobile manipulator into this chained form (single chain) [10]. For the discussion of the other types of mobile platforms (multichain case), the reader may refer to [84].*

Considering the above transformations, the dynamics (5.65) are converted as

$$M_2(\zeta)\dot{u} + C_2(\zeta, \dot{\zeta})u + G_2(\zeta) + d_2 = B_2\tau + f_2 \qquad (5.69)$$

where

$$M_2(\zeta) = T_2^T(q)M_1(q)T_2(q)|_{q=T_1^{-1}(\zeta)} \qquad (5.70)$$

$$C_2(\zeta, \dot{\zeta}) = T_2^T(q)[M_1(q)\dot{T}_2(q) + C_1(q, \dot{q})T_2(q)]|_{q=T_1^{-1}(\zeta)}$$

$$G_2(\zeta) = T_2^T(q)G_1(q)|_{q=T_1^{-1}(\zeta)} \qquad (5.71)$$

$$B_2 = T_2^T(q)B_1(q)|_{q=T_1^{-1}(\zeta)} \qquad (5.72)$$

$$f_2 = T_2^T(q)f_1 = J_2(q)^T\lambda_h = [J_{2v}, J_{2a}]^T\lambda_h|_{q=T_1^{-1}(\zeta)} \qquad (5.73)$$

$$d_2 = T_2^T(q)d_1|_{q=T_1^{-1}(\zeta^1)} \qquad (5.74)$$

Property 5.1 *The matrix M_2 is symmetric and positive definite.*

Property 5.2 *The matrix $\dot{M}_2 - 2C_2$ is skew-symmetric.*

Property 5.3 *There exist some finite non-negative constants $c_i \geq 0$ ($i = 5$) such that $\forall \zeta \in \mathbf{R}^n$, $\forall \dot{\zeta} \in \mathbf{R}^n$, $\|M_2(\zeta)\| \leq c_1$, $\|C_2(\zeta, \dot{\zeta})\| \leq c_2 + c_3\|\dot{\zeta}\|$, $\|G_2(\zeta)\| \leq c_4$, and $\sup_{t \geq 0}\|d_2(t)\| \leq c_5$.*

5.3.4 Uncertain Holonomic Constraints

Assume that the r-rigid-link manipulator in contact with a certain constrained surface $\Phi(q)$ can be represented as $\Phi(\chi(q)) = 0$, where $\Phi(\chi(q))$ is a given scalar function and $\chi(q) \in \mathbf{R}^l$ denotes the position vector of the end-effector in contact with the environment. If the constraint surface is rigid and has a continuous gradient, the contact force in (5.69) can be given as $f_2 = J_2^T(q)\lambda_h$, where λ_h is the magnitude of the contact force. However, when the robot end-effector is rigidly in contact with the uncertain surface, the environmental constraint could be expressed as an algebraic equation of the coordinate χ in the task space. Without loss of generality, the uncertain surface constraint $\bar{\Phi}(\chi(q))$ can be decomposed into a nominal part $\Phi(\chi(q))$ and an unknown constraint error part $\Delta\Phi(\chi(q))$ in an additive manner as follows:

$$\bar{\Phi}(\chi(q)) = \Phi(\chi(q)) + \Delta\Phi(\chi(q)) \qquad (5.75)$$

where $\bar{\Phi}(\chi(q))$ is the constrained surface.

Let $\bar{J}_2$ and J_2 be the Jacobian matrix of $\bar{\Phi}(\chi(q))$ and $\Phi(\chi(q))$ with respect to q, and since $\dot{q}_v = H_v(q_v)T_{21}(q_v)u_b$, $\bar{J}_2(q) = \bar{J}_\chi[J_{2v}H_v(q_v)T_{21}(q_v), J_{2a}]$, $J_2(q) = J_\chi[J_{2v}H_v(q_v)T_{21}(q_v), J_{2a}]$, $\bar{J}_\chi = \bar{\Phi}(\chi(q))/\partial\chi$, $J_\chi = \partial\Phi(\chi(q))/\partial\chi$, $J_{2v} = \partial\chi/\partial q_v$ and $J_{2a} = \partial\chi/\partial q_a$.

Integrating (5.64) and (5.67), we have

$$J_2(q)u + \frac{\partial\Delta\Phi(\chi(q))}{\partial t} = 0 \tag{5.76}$$

Assume that

$$\bar{J}_2(q) = J_2(q) + \Delta J_2(q) \tag{5.77}$$

with $\Delta J_2(q)$ defined later. Since the uncertain constraint errors (5.75) are introduced, integrating (5.77) into (5.69) yields

$$\begin{aligned} M_2(\zeta)\dot{u} \quad + \quad & C_2(\zeta,\dot{\zeta})u + G_2(\zeta) + d_2 = B_2\tau \\ & + (J_2(T_1^{-1}(\zeta)) + \Delta J_2(T_1^{-1}(\zeta)))^T\lambda_h \end{aligned} \tag{5.78}$$

Assumption 5.4 *The Jacobian matrices $\bar{J}_2(q)$ are uniformly bounded and uniformly continuous if q are uniformly bounded and continuous, respectively.*

Assumption 5.5 *The manipulator is operating away from any singularity.*

Remark 5.6 *Under Assumption 5.5, the Jacobian $J_{2a} = \frac{\partial\chi}{\partial q_a}$ is of full row rank l, such that the joint coordinate q_a can be partitioned into $q_a = [q_{a1}, q_{a2}]^T$ where $q_{a1} \in \mathbf{R}^{n_a-l}$ and $q_{a2} \in \mathbf{R}^l$, with $q_{a2} = \Omega(q_{a1})$ with a nonlinear mapping function $\Omega(\cdot)$ from an open set $\mathbf{R}^{n_a-l} \times \mathbf{R} \to \mathbf{R}^l$. The terms $\partial\Omega/\partial q_{a1}$, $\partial^2\Omega/\partial q_{a1}^2$, $\partial\Omega/\partial t$, $\partial^2\Omega/\partial t^2$ exist and are bounded in the workspace.*

Since the dimension of the constraint is l, the configuration space of the manipulator is left with $n_a - l$ degrees of freedom. Based on the full row rank for J_a, the existence of $\Omega(q_{a1})$ [17], [18], it is easy to obtain

$$J_2(q) = J_\chi[J_{2v}H_v(q_v)T_{21}(q_v), J_{2a1}, J_{2a2}] \tag{5.79}$$

Integrating (5.79) into (5.76) and considering (5.79) and letting $\delta_h = \frac{\partial\Delta\Phi(\chi(q))}{\partial t}$, we have

$$u = \begin{bmatrix} u_b \\ \dot{q}_{a1} \\ -J_{2a2}^{-1}[J_{1v}H_v(q_v)T_{21}(q_v)u_b + J_{1a1}\dot{q}_{a1}] \\ -J_{1a2}^{-1}J_\chi^{-1}\delta_h \end{bmatrix} = Lu^1 + \varepsilon \tag{5.80}$$

where

$$L = \begin{bmatrix} L_v & L_a \end{bmatrix}^T = \begin{bmatrix} I_v & 0 \\ 0 & I_{a1} \\ -J_{2a2}^{-1}J_{2v}H_v(q_v)T_{21}(q_v) & -J_{2a2}^{-1}J_{2a1} \end{bmatrix} \quad (5.81)$$

$$u^1 = \begin{bmatrix} u_b & \dot{q}_{a1} \end{bmatrix}^T \quad (5.82)$$

$$\varepsilon = J\delta_h \quad (5.83)$$

$$J = \begin{bmatrix} 0 & 0 & -J_{2a2}^{-1}J_\chi^{-1} \end{bmatrix}^T \quad (5.84)$$

Then, we have

$$L^T J_2^T(q) = 0 \quad (5.85)$$

Differentiating (5.80) and substituting it into (5.78), we have

$$M_3(\zeta)\dot{u}^1 + C_3 u^1 + G_3(\zeta) + d_3 = B_3\tau$$
$$+ (J_2(T_1^{-1}(\zeta)) + \Delta J_2(T_1^{-1}(\zeta)))^T \lambda_h \quad (5.86)$$

where $M_3(\zeta) = M_2(\zeta)L$, $C_3(\zeta, \dot{\zeta}) = M_2(\zeta)\dot{L} + C_2(\zeta, \dot{\zeta})L$, $G_3(\zeta) = G_2(\zeta)$, $B_3 = B_2$, $d_3 = M_2(\zeta)\dot{\varepsilon} + C_2(\zeta, \dot{\zeta})\varepsilon + d_2$.

Assumption 5.6 *The set of the constrained surface reachable by the end-effector of mobile manipulator, defined by*

$$\mathcal{S} := \{\chi : \bar{\Phi}(\chi, \alpha) = 0, \alpha \in \mathbf{R}^{l_1}\} \quad (5.87)$$

is bounded and belongs to a class of continuously differentiable manifolds $\bar{\Phi}(\chi, \alpha) = f(\chi_1, \chi_2, \ldots, \chi_{l_1})\alpha + g(\chi_{l_1+1}, \chi_{l_1+2}, \ldots, \chi_l)\epsilon + \varpi$ *with* $l_1 \leq l \leq n$ *and* $\chi \in \mathbf{R}^n$, *where* $\alpha = [\alpha_1, \ldots, \alpha_{l_1}, 0, \ldots, 0]^T \in \mathbf{R}^l$ *and* $\epsilon = [0, \ldots, 0, 1, \ldots, 1]^T \in \mathbf{R}^l$ *are constant vectors,* $f(*) = [f_1, \ldots, f_{l_1}, 0, \ldots, 0] \in \mathbf{R}^{1 \times l}$ *and* $g(*) = [0, \ldots, 0, g_1, \ldots, g_l] \in \mathbf{R}^{1 \times l}$ *are bounded and uniformly continuous differentiable vectors, and* ϖ *is a constant scalar.*

Remark 5.7 *We can give an example satisfying the Assumption 5.6 as* $z = \sqrt{a^2 - x^2 - y^2}$, $z > h$, *and* a *is a constant scalar,* $0 < h < a$, *as shown in Fig.5.14.*

Considering Assumption 5.6, the uncertainty δ_h could be described as

$$\delta_h = \Delta J_2(T_1^{-1}(\zeta))\frac{d}{dt}(T_1^{-1}(\zeta)) \quad (5.88)$$

$$\Delta J_2^T = J_\zeta^T \rho(S_\Phi(\chi)W + C_\Phi v) \quad (5.89)$$

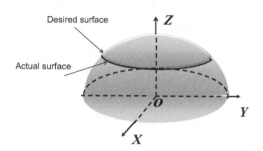

FIGURE 5.14

An example of constraint surface.

where

$$
S_\Phi = [\frac{\partial f}{\partial \chi_1}, \frac{\partial f}{\partial \chi_2}, \ \ldots, \frac{\partial f}{\partial \chi_{l_1}}, 0, \ \ldots, 0]^T \in \mathbf{R}^{l \times l}
$$

$$
C_\Phi = [0, \ \ldots, 0, \frac{\partial g}{\partial \chi_{l_1+1}}, \frac{\partial g}{\partial \chi_{l_1+2}}, \ \ldots, \frac{\partial g}{\partial \chi_l}]^T \in \mathbf{R}^{l \times l}
$$

$$
\rho = 1/\|S_\Phi(\chi)W + C_\Phi v\|
$$

$$
J_\zeta = \frac{\partial \chi}{\partial \zeta}
$$

From Assumption 5.6, the weight vector $W = [W_1, \ \ldots, W_{l_1}, 0, \ \ldots, 0]^T \in \mathbf{R}^l$ is unknown positive, and $v = [0, \ \ldots, 0, 1, \ \ldots, 1]^T \in \mathbf{R}^l$.

Property 5.4 $S_\Phi C_\Phi = \mathbf{0}$ *and* $W^T v = 0$.

Let the estimated Jacobian $\Delta \hat{J}_2$ be

$$
\Delta \hat{J}_2^T = J_\zeta^T \hat{\rho}(S_\Phi \hat{W} + C_\Phi v) \tag{5.90}
$$

with $\hat{\rho} = 1/\|S_\Phi(\chi)\hat{W} + C_\Phi v\|$. Considering Property 5.4, the error in Jacobian matrix $\Delta \tilde{J}_2 = \Delta \hat{J}_2 - \Delta J_2$ can be expressed by

$$
\Delta \tilde{J}_2^T = J_\zeta^T \left(S_\Phi(\hat{\rho}\hat{W} - \rho W) + \tilde{\rho}C_\Phi v \right) \tag{5.91}
$$

where $\tilde{\rho} = \hat{\rho} - \rho$. Considering (5.91), the force error can be expressed as

$$
\begin{aligned}
e_f &= J_\zeta^{+T} J_2^T \lambda_h - F = J_\zeta^{+T}(J_2^T \lambda_h + \Delta \hat{J}_2^T \lambda_h - J_2^T \lambda_h - \Delta J_2^T \lambda_h) \\
&= J_\zeta^{+T} \Delta \tilde{J}_2^T \lambda_h
\end{aligned} \tag{5.92}
$$

where $J_\zeta^{+T} = J_\zeta (J_\zeta^T J_\zeta)^{-1}$ and the force error e_f can be calculated from $J_\zeta^{+T} J_2^T \lambda_h$ and F from the mounted force sensor.

Assumption 5.7 *There exist some finite non-negative known constants $b_{\delta 1}$ and $b_{\delta 2}$, such that, $\forall \chi \in \Omega_\chi$, $\|\delta_h\| \leq b_{\delta 1}\|\frac{d}{dt}T_1^{-1}(\zeta)\|$, $\|\dot{\delta}_h\| \leq b_{\delta 1}\|\frac{d^2}{dt^2}T_1^{-1}(\zeta)\| + b_{\delta 2}\|\frac{d}{dt}T_1^{-1}(\zeta)\|$.*

5.3.5 Adaptive Control

A desired trajectory q_d and a desired constraint force, or, equivalently, a desired multiplier $\lambda_h^d(t)$ should satisfy the constrained equations. Since the desired trajectory q_d should satisfy the equation (5.66), we can have the desired ζ_d. After the transformation for the chained form through $\zeta_d = T_1(q_d)$ and $v_d = T_2(q_d)u_d$, we can obtain ζ_d and u_d and, equivalently, u_d^1. The trajectory and force tracking control can be restated as seeking a strategy for specifying a control law subjected to the uncertain holonomic constraint, such that $\{\lambda_h, \zeta, \dot{\zeta}\} \to \{\lambda_h^d, \zeta_d, \dot{\zeta}_d\}$. For the development of the control law, the following assumption is required.

Assumption 5.8 *The desired reference trajectory $\zeta_d(t)$ is assumed to be bounded and uniformly continuous, and has bounded and uniformly continuous derivatives up to the $(n-1)$th order. The desired Lagrangian multiplier λ_h^d is also bounded and uniformly continuous.*

Assumption 5.9 *The time varying positive functions ϱ_i and h_i converge to zero as $t \to \infty$ and satisfy $\lim_{t \to \infty} \int_0^t \varrho_i(\omega)d\omega = a_i < \infty$, $\lim_{t \to \infty} \int_0^t h_i(\omega)d\omega = b_i < \infty$ with finite constants a_i and b_i, $i = 1, \ldots, 5$.*

There are many choices for ϱ_i and h_i satisfying the Assumption 5.9, for example, $\varrho_i = h_i = 1/(1+t)^2$. Denote the tracking errors as $e = \zeta - \zeta_d$ and

$e_\lambda = \lambda_h - \lambda_h^d$, and define the reference signals

$$u_d^1 = [u_{bd}^1, u_{ad}^1]^T \tag{5.93}$$

$$u_{bd}^1 = \begin{bmatrix} u_{d1} + \eta \\ u_{d2} - s_{n_v-1}u_{d1} - k_{n_v}s_{n_v} \\ + \sum_{i=0}^{n_v-3} \frac{\partial(e_{n_v}-s_{n_v})}{\partial u_{d1}^{(i)}} u_{d1}^{(i+1)} \\ + \sum_{i=2}^{n_v-1} \frac{\partial(e_{n_v}-s_{n_v})}{\partial e_i} e_{i+1} \end{bmatrix}$$

$$u_{ad}^1 = \dot{q}_{a1d} - K_a(q_{a1} - q_{a1d})$$

where

$$s = \begin{bmatrix} e_1 \\ e_2 \\ e_3 + k_2 e_2 u_{d1}^{2p-1} \\ e_4 + s_2 + k_3 s_3 u_{d1}^{2p-1} \\ -\frac{1}{u_{d1}} \sum_{i=0}^{0} \frac{\partial(e_3-s_3)}{\partial u_{d1}^{(i)}} u_{d1}^{(i+1)} - \sum_{i=2}^{2} \frac{\partial(e_3-s_3)}{\partial e_i} e_{i+1} \\ \cdots \\ e_{n_v} + s_{n_v-2} + k_{n_v-1}s_{n_v-1}u_{d1}^{2p-1} \\ -\frac{1}{u_{d1}} \sum_{i=0}^{n_v-4} \frac{\partial(e_{n_v-1}-s_{n_v-1})}{\partial u_{d1}^i} u_{d1}^{(i+1)} \\ -\sum_{i=2}^{n_v-2} \frac{\partial(e_{n_v-1}-s_{n_v-1})}{\partial e_i} e_{i+1} \end{bmatrix} \tag{5.94}$$

$$\dot{\eta} = -k_0\eta - k_1 s_1 - \sum_{i=2}^{n_v-1} s_i \zeta_{i+1} + \sum_{j=3}^{n_v} s_j \sum_{i=2}^{j-1} \frac{\partial(e_j - s_j)}{\partial e_i} \zeta_{i+1}$$

$p = n_v - 2$, $u_{d1}^{(i)}$ is the ith derivative of u_{d1} with respect to t, k_i is a positive constant and $K_a \in \mathbf{R}^{(n_a-l)\times(n_a-l)}$ is diagonal positive.

Let new variables handle the force control

$$\dot{\vartheta} = -K_\vartheta \vartheta - K_\vartheta J_2^T e_\lambda \tag{5.95}$$

where $\vartheta = [0, \vartheta_1]$ with $\vartheta_1 \in \mathbf{R}^{n_a}$, $e_\lambda = \lambda_h - \lambda_h^d$, and $K_\vartheta = \mathrm{diag}[0, k_{\vartheta i}] > 0$. Defining the following auxiliary signals as $\tilde{u}^1 = u^1 - u_d^1 = [\tilde{u}_1, \tilde{u}_2, \tilde{u}_{a1}]^T$, $\mu = \tilde{u}^1$ and $u_r = u_d^1 - K_u \mu$, we have

$$r = \dot{\mu} + K_u \mu \tag{5.96}$$

$$\sigma = Lr + \vartheta \tag{5.97}$$

$$\nu = Lu_r - \vartheta \tag{5.98}$$

where $K_u = \mathrm{diag}[0, K_{u1}] > 0$ and $K_{u1} \in \mathbf{R}^{(n_a-l) \times (n_a-l)}$. From (5.96) and the definitions of $\tilde{u}^1$, u_r and $\dot{\mu}$ above, we have

$$u^1 = u_r + r \tag{5.99}$$

The time derivatives of σ and ν are given by

$$\dot{\sigma} = \dot{L}u^1 + L\dot{u}^1 - \dot{\nu} \tag{5.100}$$

$$\dot{\nu} = \dot{L}u_r + L\dot{u}_r - \dot{\vartheta} \tag{5.101}$$

From (5.97)-(5.99), we have

$$\sigma + \nu = Lu^1 \tag{5.102}$$

From the dynamics (5.86) together with (5.100)-(5.102), we have

$$
\begin{aligned}
M_2(\zeta)\dot{\sigma} \quad &+ C_2(\zeta, \dot{\zeta})\sigma + M_2(\zeta)\dot{\nu} + C_2(\zeta, \dot{\zeta})\nu \\
&+ G_2(\zeta) + d_3(t) = B_2(\zeta)\tau + (J_2^T + \Delta J_2^T)\lambda_h
\end{aligned} \tag{5.103}
$$

Consider the control law given by

$$
\begin{aligned}
B_2\tau = &-\sum_{i=1}^{5} \frac{\hat{c}_i \beta_i^2 \|\sigma\|^2 Y_i^2(\nu, \dot{\nu})}{\beta_i Y_i(\nu, \dot{\nu})\|\sigma\| + \varrho_i} - K_\sigma \sigma - \Lambda \\
&-(J_2^T + \Delta \hat{J}_2^T)\lambda_h^d + (J_2^T + \Delta \hat{J}_2^T)K_\lambda(\lambda_h - \lambda_h^d) - \tau_h
\end{aligned} \tag{5.104}
$$

$$\Lambda = \begin{bmatrix} \Lambda_1 & 0 \end{bmatrix}^T \tag{5.105}$$

$$\Lambda_1 = \begin{bmatrix} k_1 s_1 + \sum_{i=2}^{n_v-1} s_i \zeta_{i+1} \\ -\sum_{j=3}^{n_v} s_j \sum_{i=2}^{j-1} \frac{\partial(e_j - s_j)}{\partial e_i} \zeta_{i+1} \\ s_{n_v} \end{bmatrix}$$

$$
\begin{aligned}
\tau_h = &\frac{\|J_\zeta\|}{2\|C_\Phi v\|}\mathrm{sgn}(\sigma)(\|S_\Phi\|^2 \hat{\omega} + \|C_\Phi v\|^2) \\
&+ \frac{\|J_\zeta\|}{2\|C_\Phi v\|}\mathrm{sgn}(\sigma)\|e_f\|^2 + (K_\lambda + I)\Delta\hat{J}_2^T(\lambda_h - \lambda_h^d)
\end{aligned} \tag{5.106}
$$

$$\mathrm{sgn}(\sigma) = \begin{cases} 1 & \text{if } \sigma \geq 0 \\ -1 & \text{if } \sigma < 0 \end{cases}$$

and the adaptive laws

$$\dot{\hat{W}} = \hat{\rho}\lambda_h S_\Phi^T J_\zeta \sigma \tag{5.107}$$

$$\dot{\hat{\omega}} = \frac{1}{2\|C_\Phi v\|}\|\sigma\|\|J_\zeta\|\|S_\Phi\|^2 \tag{5.108}$$

$$\dot{\hat{c}}_i = -h_i\hat{c}_i + \frac{\gamma_i \beta_i^2 Y_i^2 \|\sigma\|^2}{\beta_i Y_i(\nu, \dot{\nu})\|\sigma\| + \varrho_i} \quad (i = 1, \ldots, 5) \tag{5.109}$$

where

$$
\begin{aligned}
Y_1(\nu, \dot{\nu}) &= \|\dot{\nu}\| + (b_{\delta 1}\|\frac{d}{dt}\mathcal{J}\| + b_{\delta 2}\|\mathcal{J}\|)\|\frac{d}{dt}T_1^{-1}(\zeta)\| \\
&\quad + b_{\delta 1}\|\mathcal{J}\|\|\frac{d^2}{dt^2}T_1^{-1}(\zeta)\| \\
Y_2(\nu, \dot{\nu}) &= \|\nu\| + b_{\delta 1}\|\mathcal{J}\|\|\frac{d}{dt}T_1^{-1}(\zeta)\| \\
Y_3(\nu, \dot{\nu}) &= \|\dot{\zeta}\|(\|\nu\| + b_{\delta 1}\|\mathcal{J}\|\|\frac{d}{dt}T_1^{-1}(\zeta)\|) \\
Y_4(\nu, \dot{\nu}) &= Y_5(\nu, \dot{\nu}) = 1
\end{aligned}
$$

K_σ and K_λ are positive definite matrices, $\beta_i > 0$ and $\gamma_i > 0$ are constant. From the dynamic equation (5.103) together with (5.104), the close-loop system dynamics can be written as

$$
\dot{s}_1 = \eta + \tilde{u}_1 \tag{5.110}
$$

$$
\dot{s}_2 = (\eta + \tilde{u}_1)\zeta_3 + s_3 u_{d1} - k_2 s_2 u_{d1}^{2p} \tag{5.111}
$$

$$
\dot{s}_3 = (\eta + \tilde{u}_1)(\zeta_4 - \frac{\partial(e_3 - s_3)}{\partial e_2}\zeta_3)
$$

$$
\quad + s_4 u_{d1} - s_2 u_{d1} - k_3 s_3 u_{d1}^{2p} \tag{5.112}
$$

$$
\cdots
$$

$$
\dot{s}_{n_v-1} = (\eta + \tilde{u}_1)(\zeta_{n_v} - \sum_{i=2}^{n_v-2} \frac{\partial(e_{n_v-1} - s_{n_v-1})}{\partial e_i}\zeta_{i+1})
$$

$$
\quad + s_{n_v} u_{d1} - s_{n_v-2} u_{d1} - k_{n_v-1} s_{n_v-1} u_{d1}^{2p} \tag{5.113}
$$

$$
\dot{s}_{n_v} = (\eta + \tilde{u}_1) \sum_{i=2}^{n_v-2} \frac{\partial(e_{n_v} - s_{n_v})}{\partial e_i}\zeta_{i+1} - k_{n_v} s_{n_v}
$$

$$
\quad - s_{n_v-1} u_{d1} + \tilde{u}_2 \tag{5.114}
$$

$$
\dot{\eta} = -k_0 \eta - \Lambda_1 \tag{5.115}
$$

$$
\begin{aligned}
M_2 \dot{\sigma} &= -C_2 \sigma - \xi - K_\sigma \sigma - \sum_{i=1}^{5} \frac{\hat{c}_i \beta_i^2 \sigma Y_i^2(\nu, \dot{\nu})}{\beta_i Y_i(\nu, \dot{\nu})\|\sigma\| + \varrho_i} \\
&\quad - (J_2 + \Delta \hat{J}_2)^T (\lambda_h^d - k_\lambda e_\lambda) + (J_2 + \Delta \hat{J}_2)^T \lambda_h \\
&\quad - \tau_h - \Lambda \\
&= -C_2 \sigma - \xi - K_\sigma \sigma - \sum_{i=1}^{5} \frac{\hat{c}_i \beta_i^2 \|\sigma\|^2 Y_i^2(\nu, \dot{\nu})}{\beta_i Y_i(\nu, \dot{\nu})\|\sigma\| + \varrho_i} \\
&\quad - J_2^T (\lambda_h^d - K_\lambda e_\lambda) + J_2^T \lambda_h \\
&\quad - \Delta \tilde{J}_2^T \lambda_h + \Delta \hat{J}_2^T (K_\lambda + I) e_\lambda - \tau_h - \Lambda \tag{5.116}
\end{aligned}
$$

where $\Upsilon = M_2 \dot{\nu} + C_2 \nu + G_2 + d_3$.

Theorem 5.4 *Consider the mechanical system described by (4.1), (5.62) and (5.75). Under Assumption 5.8, the control law (5.104) leads to: (i) $\zeta, \dot{\zeta}, \lambda_h$ converging to $\zeta_d, \dot{\zeta}_d, \lambda_h^d$ at $t \to \infty$; and (ii) all the signals in the closed-loop are bounded for all $t \geq 0$.*

Proof: Consider the Lyapunov function candidate as

$$V(t) = \frac{1}{2}\sum_{i=2}^{n_v} s_i^2 + \frac{1}{2}k_1 s_1^2 + \frac{1}{2}\eta^2 + \frac{1}{2}\sigma^T M_2 \sigma + \sum_{i=1}^{5}\frac{1}{2\gamma_i}\tilde{c}_i^2$$

$$+\frac{1}{2}\tilde{W}^T\tilde{W} + \frac{1}{2}\vartheta^T(I + K_\lambda)K_\vartheta^{-1}\vartheta + \frac{1}{2}(\|W\|^2 - \hat{\omega})^2 \quad (5.117)$$

Where $\tilde{W} = \hat{W} - W$, $\tilde{c}_i = \hat{c}_i - c_i$, we have $\dot{\tilde{W}} = \dot{\hat{W}}$, $\dot{\tilde{c}}_i = \dot{\hat{c}}_i$. Considering Property 5.1 and integrating (5.109) and (5.116) into the derivative of V yields

$$\dot{V} \leq -\sum_{i=2}^{n_v-1} k_i s_i^2 u_{d1}^{2l} - k_{n_v} s_{n_v}^2 - k_0\eta^2 + \tilde{u}^{1T}\Lambda - \sigma^T K_\sigma \sigma$$

$$-\sum_{i=1}^{5}\frac{\hat{c}_i\beta_i^2\|\sigma\|^2 Y_i^2(\nu,\dot{\nu})}{\beta_i Y_i(\nu,\dot{\nu})\|\sigma\| + \varrho_i} + \|\sigma\|\|\Upsilon\|$$

$$-\sigma^T J_2^T(\lambda_h^d - K_\lambda e_\lambda) + \sigma^T J_2^T \lambda_h$$

$$-\sigma^T \Delta \tilde{J}_2^T \lambda_h + \sigma^T \Delta \hat{J}_2^T(K_\lambda + I)e_\lambda - \sigma^T \tau_h - \sigma^T \Lambda$$

$$-\sum_{i=1}^{5}\frac{h_i \hat{c}_i \tilde{c}_i}{\gamma_i} + \sum_{i=1}^{5}\frac{\tilde{c}_i\beta_i^2\|\sigma\|^2 Y_i^2(\nu,\dot{\nu})}{\beta_i Y_i(\nu,\dot{\nu})\|\sigma\| + \varrho_i} + \dot{\hat{W}}^T\tilde{W}$$

$$+\vartheta^T(I + K_\lambda)K_\vartheta^{-1}\dot{\vartheta} - (\|W\|^2 - \hat{\omega})\dot{\hat{\omega}} \quad (5.118)$$

From (5.91) and (5.92), we can obtain

$$e_f = J_\zeta^{+T}\Delta \tilde{J}_2^T \lambda_h$$

$$= J_\zeta^{+T} J_\zeta^T S_\Phi(\hat{\rho}\hat{W} - \rho W)\lambda_h + J_\zeta^+ J_\zeta^T \tilde{\rho}C_\Phi \upsilon \lambda_h \quad (5.119)$$

Considering Property 5.4, we have

$$\|e_f\|^2 = \|J_\zeta^{+T} J_\zeta^T\|^2 \|S_\Phi(\hat{\rho}\hat{W} - \rho W)\|^2\|\lambda_h\|^2$$

$$+\|J_\zeta^{+T} J_\zeta^T\|^2\|C_\Phi \upsilon\|^2\|\tilde{\rho}\lambda_h\| \quad (5.120)$$

Therefore, we can obtain

$$\|\tilde{\rho}\lambda_h\| \leq \|e_f\|/\|C_\Phi \upsilon\| \quad (5.121)$$

Considering (5.91), (5.119) and using the adaptive parameter law (5.107), rewriting the tenth, the eleventh and the sixteenth right-hand terms in (5.118), we have

$$
\begin{aligned}
&-\sigma^T \Delta \tilde{J}_2^T \lambda_h + \dot{\tilde{W}}^T \tilde{W} + \sigma^T \Delta \hat{J}^T (k_\lambda + I) e_\lambda \\
=\ &-\sigma^T J_\zeta^T (S_\Phi \hat{\rho} \tilde{W} + S_\Phi \tilde{\rho} W + \tilde{\rho} C_\Phi v) \lambda_h \\
&+\sigma^T J_\zeta^T \hat{\rho} S_\Phi \lambda_h \tilde{W} + \sigma^T \Delta \hat{J}^T (k_\lambda + I) e_\lambda \\
=\ &-\sigma^T J_\zeta^T \tilde{\rho} (S_\Phi W + C_\Phi v) \lambda_h + \sigma^T \psi
\end{aligned} \tag{5.122}
$$

where $\psi = \Delta \hat{J}_2^T (K_\lambda + I) e_\lambda$. From (5.121) and (5.122), we have

$$
\begin{aligned}
&-\sigma^T J_\zeta^T \lambda_h \tilde{\rho} (S_\Phi W + C_\Phi v) \\
\leq\ &\frac{1}{\|C_\Phi v\|} \|\sigma\| \|J_\zeta\| \|e_f\| \|S_\Phi W + C_\Phi v\| \\
\leq\ &\frac{1}{\|C_\Phi v\|} \|\sigma\| \|J_\zeta\| \|e_f\| (\|S_\Phi\| \|W\| + \|C_\Phi v\|) \\
\leq\ &\frac{\|\sigma\| \|J_\zeta\|}{2 \|C_\Phi v\|} (\|e_f\|^2 + \|S_\Phi\|^2 \|W\|^2 + \|C_\Phi v\|^2)
\end{aligned} \tag{5.123}
$$

by noting Property 5.4. Moreover

$$
\begin{aligned}
&\|\sigma\| \|\Upsilon\| - \sum_{i=1}^{5} \frac{\hat{c}_i \beta_i^2 \|\sigma\|^2 Y_i^2(\nu, \dot{\nu})}{\beta_i Y_i(\nu, \dot{\nu}) \|\sigma\| + \varrho_i} - \sum_{i=1}^{5} \frac{h_i \hat{c}_i \tilde{c}_i}{\gamma_i} \\
&+\sum_{i=1}^{5} \frac{\tilde{c}_i \beta_i^2 \|\sigma\|^2 Y_i^2(\nu, \dot{\nu})}{\beta_i Y_i(\nu, \dot{\nu}) \|\sigma\| + \varrho_i} \\
\leq\ &\|\sigma\| \|\Upsilon\| - \sum_{i=1}^{5} \frac{h_i \hat{c}_i \tilde{c}_i}{\gamma_i} - \sum_{i=1}^{5} \frac{c_i \beta_i^2 \|\sigma\|^2 Y_i^2(\nu, \dot{\nu})}{\beta_i Y_i(\nu, \dot{\nu}) \|\sigma\| + \varrho_i} \\
\leq\ &\sum_{i=1}^{5} (-\frac{h_i \hat{c}_i \tilde{c}_i}{\gamma_i} + c_i \varrho_i) \\
=\ &\sum_{i=1}^{5} (c_i \varrho_i - \frac{h_i}{\gamma_i} (\hat{c}_i - \frac{c_i}{2})^2 + \frac{h_i c_i^2}{4 \gamma_i}) \\
\leq\ &\sum_{i=1}^{5} (\frac{h_i}{4 \gamma_i} c_i^2 + c_i \varrho_i)
\end{aligned} \tag{5.124}
$$

In addition, from (5.95) and (5.97), $\sigma^T = r^T L^T + \vartheta^T$. Thus, we have

$$
\begin{aligned}
&-\sigma^T J_2^T (\lambda_h^d - K_\lambda e_\lambda) + \sigma^T J_2^T \lambda_h + \vartheta^T (I + K_\lambda) K_\vartheta^{-1} \dot{\vartheta} \\
=\ &-\vartheta^T (I + K_\lambda) \vartheta + r^T L^T J_2^T (I + K_\lambda) e_\lambda
\end{aligned} \tag{5.125}
$$

by noting $L^T J_2^T = 0$ from (5.85).

From (5.107), (5.123)–(5.125), it can be shown that

$$\dot{V} \leq -k_0\eta^2 - k_{n_v}s_{n_v} - \sum_{i=2}^{n_v-1} k_i s_i^2 u_{d1}^{2l} + \tilde{u}^{1T}\Lambda - \sigma^T K_\sigma \sigma$$

$$+\sum_{i=1}^{5}(\frac{h_i}{4\gamma_i}c_i^2 + c_i\varrho_i) - \vartheta^T(K_\lambda + I)\vartheta + \frac{\|\sigma\|\|J_\varsigma\|}{2\|C_\Phi v\|}\|e_f\|^2$$

$$+\frac{\|\sigma\|\|J_\varsigma\|}{2\|C_\Phi v\|}(\|S_\Phi\|^2\|W\|^2 + \|C_\Phi v\|^2) + \sigma^T\psi - \sigma^T\tau_h$$

$$-\sigma^T\Lambda - (\|W\|^2 - \hat{\omega})\dot{\hat{\omega}} - \vartheta^T(I + K_\lambda)\vartheta \qquad (5.126)$$

From (5.81), (5.95), and (5.105), considering the fourth and the twelfth right-hand terms in (5.126), we have

$$\tilde{u}^{1T}\Lambda - \sigma^T\Lambda = \tilde{u}^{1T}\Lambda - (r^T L^T\Lambda + \vartheta^T\Lambda)$$

$$= \tilde{u}^{1T}\begin{bmatrix} \Lambda_1 \\ 0 \end{bmatrix} - r^T\begin{bmatrix} L_v^T & L_a^T \end{bmatrix}\begin{bmatrix} \Lambda_1 \\ 0 \end{bmatrix} - \begin{bmatrix} 0 & \vartheta_1 \end{bmatrix}\begin{bmatrix} \Lambda_1 \\ 0 \end{bmatrix}$$

From (5.81) and (5.96), we have $L_v = [I_v,\ 0]$, and $K_u = \text{diag}[0, K_{u1}]$, $r^T\begin{bmatrix} L_v^T & L_a^T \end{bmatrix}\begin{bmatrix} \Lambda_1 & 0 \end{bmatrix}^T = \tilde{u}^{1T}\Lambda_1$ and subsequently we obtain

$$\tilde{u}^{1T}\Lambda - \sigma^T\Lambda = 0 \qquad (5.127)$$

Integrating (5.127), one obtains

$$\dot{V} \leq -k_0\eta^2 - k_{n_v}s_{n_v} - \sum_{i=2}^{n_v-1} k_i s_i^2 u_{d1}^{2l} - \sigma^T K_\sigma \sigma$$

$$+\sum_{i=1}^{5}(\frac{h_i}{4\gamma_i}c_i^2 + c_i\varrho_i) - \vartheta^T(I + K_\lambda)\vartheta$$

$$+\frac{\|\sigma\|\|J_\varsigma\|\|S_\Phi\|^2}{2\|C_\Phi v\|}(\|W\|^2 - \hat{\omega}) - (\|W\|^2 - \hat{\omega})\dot{\hat{\omega}} \qquad (5.128)$$

Considering the parameter $\hat{\omega}$ update law (5.108), $\dot{V} \leq -k_0\eta^2 - k_{n_v}s_{n_v} - \sum_{i=2}^{n_v-1} k_i s_i^2 u_{d1}^{2l} - \sigma^T K_\sigma \sigma + \sum_{i=1}^{5}(\frac{h_i}{4\gamma_i}c_i^2 + c_i\varrho_i) - \vartheta^T(I + K_\lambda)\vartheta$. Noting Assumption 5.9, we have $\sum_{i=1}^{5}(\frac{h_i}{4\gamma_i}c_i^2 + c_i\varrho_i) \to 0$ as $t \to \infty$. Integrating both sides of the above equation gives $V(t) - V(0) < -\int_0^t(k_0\eta^2 + k_{n_v}s_{n_v} + \sum_{i=2}^{n_v-1} k_i s_i^2 u_{d1}^{2l} + \sigma^T K_\sigma \sigma + \vartheta^T(I + K_\lambda)\vartheta)ds + \sum_{i=1}^{5}(\frac{a_i}{4\gamma_i}c_i^2 + c_i b_i) < \infty$. Thus, $V(t) < V(0) + \sum_{i=1}^{5}(\frac{a_i}{4\gamma_i}c_i^2 + c_i b_i)$, therefore $V(t)$ is bounded, which implies that η, s_i, σ, $\hat{c}_i$, $\hat{W}$, ϑ and $\hat{\omega}$ are bounded. From the definition of s_i in (5.94), it can be concluded that $[e_1, e_2,\ \ldots, e_{n_v}]^T$ is bounded, and it follows that η is bounded. Since σ is bounded, we can obtain $r, \tilde{u}^1 \in L_2^{n-l}$

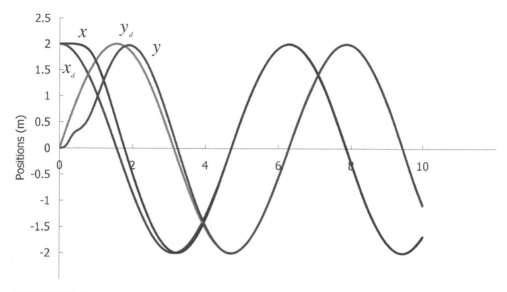

FIGURE 5.15

Tracking x and y.

from (5.97). Therefore, $q_{a1} - q_{a1d}$ and $\dot{q}_{a1} - \dot{q}_{a1d}$ are bounded, which follows that q_{a1} is bounded. Since ϑ is bounded, from (5.95), e_λ is bounded. Therefore, it is concluded that $s_i u_{d1}, s_{n_v}, \eta \in L_2$ and it can be shown that $s_i u_{d1} \to 0, s_{n_v} \to 0, \eta \to 0$ as $t \to 0$, respectively. It is further concluded that $s_i \to 0$ as $t \to 0$. Differentiating $u_{d1}^p \eta$ yields $\frac{d}{dt} u_{d1}^p \eta = -k_1 u_{d1}^p s_1 + l u_{d1}^{p-1} \dot{u}_{d1}^p \eta - k_0 u_{d1}^p \eta - u_{d1}^l \left\{ \sum_{i=2}^{n_v-1} s_i \zeta_{i+1} - \sum_{j=3}^{n_v} s_j \sum_{i=2}^{j-1} \frac{\partial(e_j - s_j)}{\partial e_i} \zeta_{i+1} \right\}$, where the first term is uniformly continuous and the other terms converge to zero. Since $\frac{d}{dt} u_{d1}^p \eta$ converges to zero, s and $\dot{s}$ also tend to zero. It is obvious that $s_i = 0$, yields that $\xi_i \to \xi_{di}$ and $\dot{\xi}_i \to \dot{\xi}_{di}$ as $t \to \infty$. Since $\sigma, \dot{\sigma}, d_3, \Delta\tilde{J}_2, e_\lambda$ and τ_h are all bounded, it can be concluded that τ is bounded from (5.104).

5.3.6 Simulation Studies

Consider a 3-DOF robotic manipulator with two revolute joints and one prismatic joint mounted on a two-wheeled mobile platform shown in Fig. 9.2.

Remark 5.8 *In such case that 3-DOF mobile manipulator consists of two revolute joints and one prismatic joint, $\forall \zeta \in \mathbf{R}^n$, $\forall \dot{\zeta} \in \mathbf{R}^n$, $\|M_2(\zeta)\| \le k_{m1} + k_{m2}\|\zeta_6\|^2$ with k_{m1} and $k_{m2} > 0$, $\|G_2(\zeta)\| \le k_{g1} + k_{g2}\|\zeta_6\|$ with k_{g1}*

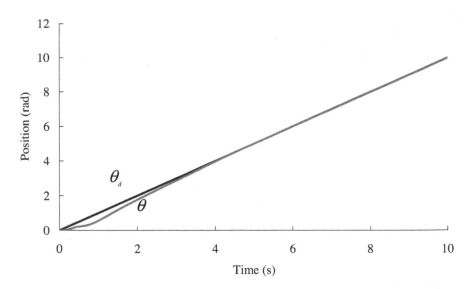

FIGURE 5.16

Tracking θ.

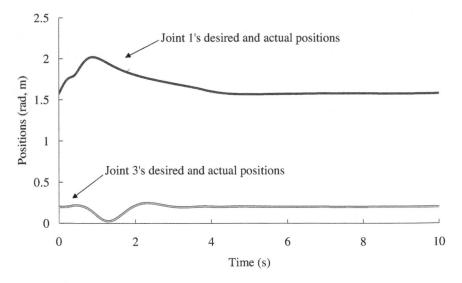

FIGURE 5.17

The position error compensation.

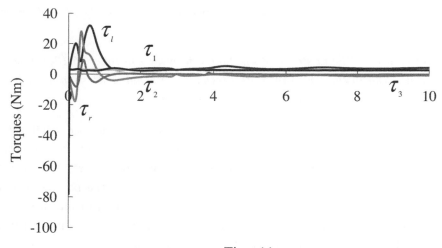

FIGURE 5.18

Torques of joints.

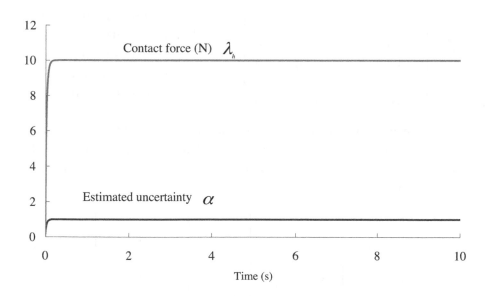

FIGURE 5.19

The contact force (N) and estimated α.

and $k_{g2} > 0$, where $\zeta_6 = \theta_3$, if the boundeness of ζ_6 is known. There still exist some finite non-negative constants $c_i \geq 0$ $(i = 1, \ldots, 4)$; therefore Property 5.3 holds.

Remark 5.9 *The existence of sgn-function in the controller (5.104) may inevitably lead to chattering in control torques. To avoid chattering, a sat function can be used to replace the sgn-function [23].*

Given the desired trajectory $q_d = [x_d, y_d, \theta_d, \theta_{1d}, \theta_{2d}]^T$ with $x_d = 2.0\cos(t)$, $y_d = 2.0\sin(t)$, $\theta_d = t$, $\theta_{1d} = \pi/2$ rad, $\theta_{2d} = \pi/2$ rad and the geometric constraint which the end-effector subjected to: $\Phi = \alpha(x^2 + y^2) + z - c = 0$ with $c = 2.25m$, and $\lambda_d = 10.0N$, the desired value of the parameter α is 1.0, and the joint 3 is the redundant prismatic joint used to compensate the position errors caused by uncertain holonomic constraints. Assume that $\theta_3 \in [0.0m, 0.3m]$. The transformation $T_1(q)$ is defined as: $\zeta_1 = \theta$, $\zeta_2 = x\cos\theta + y\sin\theta$, $\zeta_3 = -x\sin\theta + y\cos\theta$, $\zeta_4 = \theta_1$, $\zeta_5 = \theta_2$, $\zeta_6 = \theta_3$, $u_1 = v_{b2}$, $u_2 = v_{b1} - (x\cos\theta + y\sin\theta)v_{b2}$ and $u_3 = \dot{\zeta}_3$, $u_4 = \dot{\zeta}_4$, $u_5 = \dot{\zeta}_5$. One can obtain the kinematic system in the chained form as: $\dot{\zeta}_1 = u_1$, $\dot{\zeta}_2 = \zeta_3 u_1$, $\dot{\zeta}_3 = u_2$, $\dot{\zeta}_4 = u_3$, $\dot{\zeta}_5 = u_4$, $\dot{\zeta}_6 = u_5$. with

$$
L = \begin{bmatrix}
-\sin\zeta_1 & 0.0 & 0.0 & 0.0 & 0.0 \\
\cos\zeta_1 & 0.0 & 0.0 & 0.0 & 0.0 \\
0.0 & 1.0 & 0.0 & 0.0 & 0.0 \\
0.0 & 0.0 & 1.0 & 0.0 & 0.0 \\
0.0 & 0.0 & 0.0 & 1.0 & 0.0 \\
0.0 & 0.0 & 0.0 & 0.0 & 1.0
\end{bmatrix}
, T_2 = \begin{bmatrix}
-\zeta_1 & 1.0 & 0.0 & 0.0 & 0.0 \\
1.0 & 0.0 & 0.0 & 0.0 & 0.0 \\
0.0 & 0.0 & 1.0 & 0.0 & 0.0 \\
0.0 & 0.0 & 0.0 & 1.0 & 0.0 \\
0.0 & 0.0 & 0.0 & 0.0 & 1.0
\end{bmatrix}
$$

Using the above diffeomorphism transformation, we can obtain $\zeta_{d1} = t$, $\zeta_{d2} = 2.0$, $\zeta_{d3} = 0.0$, $\zeta_{d4} = \pi/2$, $\zeta_{d5} = \pi/2$, $\lambda_h^d = 10.0N$ with $u_{d1} = 1.0$, $u_{d2} = 0.0$, and $u_{d3} = 0.0$, $u_{d4} = 0.0$. If the robust adaptive control (5.104) is used, the tracking error is

$$
\begin{bmatrix}
e_1 \\
e_2 \\
e_3 \\
e_4 \\
e_5 \\
e_\lambda
\end{bmatrix}
=
\begin{bmatrix}
\zeta_1 - \zeta_{d1} \\
\zeta_2 - \zeta_{d2} \\
\zeta_3 - \zeta_{d3} \\
\zeta_4 - \zeta_{d4} \\
\zeta_5 - \zeta_{d5} \\
\lambda_h - \lambda_h^d
\end{bmatrix}
,
\begin{bmatrix}
s_1 \\
s_2 \\
s_3 \\
s_4 \\
s_5 \\
s_6
\end{bmatrix}
=
\begin{bmatrix}
e_1 \\
e_2 \\
e_3 + k_2 e_2 u_{d1} \\
0 \\
0 \\
0
\end{bmatrix}
$$

In the simulation, $m_p = 5.0kg$, $m_1 = 1.0kg$, $m_2 = m_3 = 0.5kg$, $I_p = 2.5kgm^2$, $I_1 = 1.0kgm^2$, $I_2 = 0.5kgm^2$, $I_3 = 0.5 + m_3\theta_3^2 kgm^2$, $d = l = r = 0.5m$, $2l_1 =$

$1.0m$, $2l_2 = 0.3m$. The initial condition selects $x(0) = 2.0m$, $y(0) = 0.0m$, $\theta(0) = \pi/2$ rad, $\theta_1(0) = \pi/2$ rad, $\theta_2(0) = \pi/2$ rad, $\theta_3(0) = 0.1m$, $\lambda(0) = 0.0N$ and $\dot{x}(0) = 0.5m/s$, $\dot{y}(0) = \dot{\theta}(0) = \dot{\theta}_1(0) = \dot{\theta}_2(0) = \dot{\theta}_3(0) = 0.0$, $\alpha(0) = 0.1$, $\hat{c}_i(0) = 1.0, i = 1, \ldots, 5$. In the simulation, the design parameters are set as $b_{\delta 1} = b_{\delta 2} = 1.0$, $k_0 = 30.0$, $k_1 = 200.0$, $k_2 = 1.0$, $k_3 = 1.0$, $K_\vartheta = \text{diag}[0, 0.01]$, $K_\lambda = 0.5$, $\eta(0) = 0$, $K_\sigma = \text{diag}[1.0]$, $K_a = \text{diag}[1.0]$ and the adaptive gains in the adaptive laws are chosen as $\gamma_i = 0.1$, $\beta_i = 1.0$, $h_i = \varrho_i = 1/(1+t)^2$. The disturbances on the mobile base are set to a time-varying form as $0.5\sin(t)$ and $0.5\cos(t)$. The control results are shown in Figs. 5.15-5.19 [166]. Fig. 5.15 shows the trajectory tracking $(q - q_d)$ with the disturbances, and the corresponding torques are shown in Fig. 5.18. Fig. 5.19 shows the contact force tracking $\lambda_h - \lambda_h^d$ and the evolution of α. The simulation results show that the position tracking error converges to zero, the estimated uncertainty converges, and the contact force error converges to the desired contact force in Fig. 5.19.

5.4 Conclusion

In this chapter, adaptive robust control strategies have been presented for a class of holonomic constrained noholonomic mobile manipulators in the presence of uncertainties and disturbances. The system stability and the boundedness of tracking errors are proved using Lyapunov synthesis. All control strategies have been designed to drive the system motion to converge to the desired manifold and at the same time guarantee the boundedness of the constrained force. Simulation studies have verified that not only the states of the system converge to the desired trajectory, but also the constraint force converges to the desired force.

Where state information is unavailable, adaptive robust controls integrating an observer have been presented to control the holonomic constrained noholonomic mobile manipulators in the presence of uncertainties and disturbances and actuator dynamics are considered in the controls. The proposed controls are nonregressor based and require no information on the system dynamics. Simulation studies have verified the effectiveness of the proposed controls.

6

Under-actuated Control

CONTENTS

6.1 Introduction

Mobile under-actuated manipulators are different from under-actuated manipulators, due to simultaneously integrating both kinematic constraints and dynamic constraints. Moreover, they are also more complex than the mobile inverted pendulums [31] or pendulums [32] whose dynamic balances in the vertical plane are achieved due to the existent gravity. The under-actuated joints of mobile under-actuated manipulators may appear anywhere in the manipulators. For these reasons, increasing effort needs to be made towards control design that guarantees stability and robustness for mobile under-actuated manipulators. These systems are intrinsically nonlinear and their dynamics can be described by nonlinear differential equations.

In robotics, nonholonomic constraints formulated as non-integrable differential equations containing time derivatives of generalized coordinates (velocity, acceleration, etc.) are mainly studied. Due to Brockett's theorem [135], it is well known that nonholonomic systems with restricted mobility cannot be stabilized to a desired configuration nor posture via differentiable, or even continuous, pure state feedback. In general, such nonholonomic constraints include: (i) only kinematic constraints which geometrically restrict the direction of mobility, i.e., wheeled mobile robots [163], [105]; (ii) only dynamic

141

constraints due to dynamic balance at passive degrees of freedom where no force or torque is applied, i.e., the manipulators with passive links [106], [107], [109] ; (iii) not only kinematic constraints but also dynamic constraints [30], such as the nonholonomic mobile manipulator with under-actuated joints discussed in this paper. The common features of these systems are governed by under-actuated configuration, i.e., the number of control inputs is fewer than the number of degrees of freedom to be stabilized [142], which makes it difficult to apply the conventional robotics approaches for controlling Euler-Lagrange systems.

For the case (i), these constraints can be represented as a first-order non-integrable differential equation in a Pfaffian form, $A(q)\dot{q} = 0$, where $A(q)$ denotes nonholonomic constraints, and q and $\dot{q}$ are the generalized coordinate vector and velocity vector. The state equation is written as a driftless symmetric affine system with the velocity inputs. For the case (ii), there exists a class of dynamic constraints formulated as a second-order differential equation, for example, $M(q)\ddot{q} + V(q,\dot{q}) = 0$ with the inertia matrix $M(q)$ and the centripetal and Coriolis matrix $V(q,\dot{q})$, which includes the generalized acceleration $\ddot{q}$ and cannot be transformed into a Pfaffian form. These constraints are called second-order nonholonomic constraints. The zero torque at the under-actuated joints results in a second-order nonholonomic constraint, [106], [107], [136], [109]. The under-actuated joints, which can rotate freely, can be indirectly driven by the effect of the dynamic coupling between the active and passive joints. The effective control of under-actuated robotic system could enhance the fault tolerance if the actuator is in failure, for example, robots in the dangerous environments where humans are isolated. The faults can bring risk to the robots, their task, the working environments. On the other hand, for example, in [137], a social robot helps to carry a big or long object in home, office, welfare site, as well as factory. However, the internal force of the carried object would damage the human collaborator. If the under-actuated joints are introduced, it would definitely decrease the internal force and secure the human safety. The control of under-actuated joints for the mobile robots is a challenge. Although the coordination of the multi-manipulators using the passive joints was proposed in [138], [139] to decrease the undesired internal forces, the cooperative robots meet the need with fixed manipulators.

The coupling dynamics between the actuated and the passive joints depend on the dynamic parameters, and are subject to errors if there are uncertainties on the values of these parameters, but in [106], [107], [138], and [139], the precise knowledge of the dynamic models is required in order to achieve satisfactory performance. It was not found how to utilize the dynamic coupling in the presence of the unmodelled dynamics and external unknown disturbances. For the mobile under-actuated manipulators, however, it is impossible to control them with simple control schemes if the dynamics are unknown beforehand, because the dynamics uncertainty causes the unknown coupling between the joints. The mobile under-actuated manipulator is a strong dynamic coupled nonlinear system. There exists strong dynamic coupling between the mobile platform and the manipulator, which is beyond the fixed manipulators. We need to investigate whether the mobile under-actuated manipulators with motion ability could be more flexible than the fixed manipulators with the under-actuated joints. How to utilize this dynamic coupling to control the mobile under-actuated manipulator is one of the important issues and has not been investigated until now. Moreover, mobile under-actuated manipulator control is characterized by unstable balance and unmodelled dynamics, and subject to time-varying external disturbances, which are generally difficult to model accurately. Therefore, the traditional model-based control may not be the ideal approach since it generally works well when the dynamic model is known exactly. The presence of uncertainties and disturbances could disrupt the function of the traditional model-based feedback control and lead to the unstable balance.

Another challenging problem is to control a mobile under-actuated manipulator system whose mobile platform is no longer constrained to the guide rail-like cart-pendulum systems or fixed manipulators, but moves in its terrain while balancing the under-actuated joints. Therefore, the nonholonomic constraint forces between the wheels and the ground should be considered in order to avoid slippage. Most recent works about mechanical under-actuated systems, including mobile inverted pendulum and fixed under-actuated manipulators [107], [139], [31] did not consider nonholonomic constraint forces. The friction forces are assumed beforehand to be enough as needed on the ground, but in the practical applications, this assumption is difficult to satisfy. When the ground friction could not support the system motion, the control performance by previously proposed controllers would be degraded. Therefore, we have to consider the nonholonomic constraint forces in the control design.

Based on our previous works [30] that differ from the known dynamics and available system states in [30], in this paper, we propose adaptive motion control using dynamic coupling and output feedback, and attempt to reduce the workload on modeling dramatically, as well as to compensate for the dynamics uncertainty of the unactuated dynamics. Since the states and the time derivatives of the system output are unavailable, adaptive motion control using a high gain observer to reconstruct the system states is investigated for mobile under-actuated manipulators in the presence of parametric and functional uncertainties in the dynamics. Moreover, considering limited friction forces, we further integrate the force control to ensure the nonholonomic constraint force error within a small deviation from zero.

6.2 System Description

Consider an n-DOF fixed manipulator mounted on a two-wheeled driven mobile platform. The dynamics can be described as

$$M(q)\ddot{q} + V(q,\dot{q})\dot{q} + G(q) + d(t) = B(q)\tau + f \tag{6.1}$$

where

$$M(q) = \begin{bmatrix} M_v & M_{va} & M_{vp} \\ M_{av} & M_a & M_{ap} \\ M_{pv} & M_{pa} & M_p \end{bmatrix}, G(q) = \begin{bmatrix} G_v \\ G_a \\ G_p \end{bmatrix}$$

$$V(q,\dot{q}) = \begin{bmatrix} V_v & V_{va} & V_{vp} \\ V_{av} & V_a & V_{ap} \\ V_{pv} & V_{pa} & V_p \end{bmatrix}, J^T = \begin{bmatrix} J_v^T \\ 0 \\ 0 \end{bmatrix}$$

$$d(t) = \begin{bmatrix} d_v \\ d_a \\ d_p \end{bmatrix}, B(q)\tau = \begin{bmatrix} \tau_v \\ \tau_a \\ 0 \end{bmatrix}, f = J^T\lambda$$

Assumption 6.1 *The mobile under-actuated manipulator is subject to known nonholonomic constraints.*

Remark 6.1 *Since in the actual implementations, the friction between the wheels of the mobile platform and the ground cannot be predicted beforehand, in order to avoid the slipping of the wheels, we have to guarantee the boundedness of the nonholonomic constraint force error with the desired constraint force and make it arbitrarily small within a compact set.*

The mobile platform subject to nonholonomic constraints can be expressed as

$$J_v\dot{q}_v = 0 \tag{6.2}$$

The effect of the constraints can be viewed as a restriction of the dynamics on the manifold Ω_n as $\Omega_n = \{(q_v, \dot{q}_v) | J_v \dot{q}_v = 0\}$.

Assume that the annihilator of the co-distribution spanned by the covector fields $J_1^T(q_v)$, ..., $J_l^T(q_v)$ is an $(n_v - l)$-dimensional smooth nonsingular distribution Δ on R^{n_v}. This distribution Δ is spanned by a set of $(n_v - l)$ smooth and linearly independent vector fields $H_1(q_v)$, ..., $H_{n_v-l}(q_v)$, i.e., $\Delta = \text{span}\{H_1(q_v), \ldots, H_{n_v-l}(q)\}$, which satisfy, in local coordinates, the following relation [110]

$$H^T(q_v)J_v^T(q_v) = 0 \tag{6.3}$$

where $H(q_v) = [H_1(q_v), \ldots, H_{n_v-l}(q_v)] \in R^{n_v \times (n_v-l)}$. Note that $H^T H$ is of full rank. Constraint (6.2) implies the existence of vector $\dot{\eta} \in R^{n_v-l}$, such that

$$\dot{q}_v = H(q_v)\dot{\eta} \tag{6.4}$$

Introducing $\dot{\zeta} = \begin{bmatrix} \dot{\eta}^T & \dot{q}_a^T & \dot{q}_p^T \end{bmatrix}^T$, we have

$$\dot{q} = R(q)\dot{\zeta} \tag{6.5}$$

$$R(q) = \begin{bmatrix} H(q_v) & 0 & 0 \\ 0 & I & 0 \\ 0 & 0 & I \end{bmatrix} \tag{6.6}$$

It is easy to have

$$R^T(q)J^T(q) = 0 \tag{6.7}$$

Differentiating (6.5) yields

$$\ddot{q} = \dot{R}(q)\dot{\zeta} + R(q)\ddot{\zeta} \tag{6.8}$$

From (6.5), $\dot{\zeta}$ can be obtained from q and $\dot{q}$ as

$$\dot{\zeta} = [R^T(q)R(q)]^{-1}R^T(q)\dot{q} \tag{6.9}$$

Then, the dynamic equation (6.1), which satisfying the nonholonomic constraint (6.2), can be rewritten in terms of the internal state variable $\dot{\zeta}$ as

$$M(q)R(q)\ddot{\zeta} + [M(q)\dot{R}(q) + V(q,\dot{q})R(q)]\dot{\zeta}$$
$$+G(q) + d(t) = B(q)\tau + J^T(q)\lambda \tag{6.10}$$

Substituting (6.5) and (6.8) into (6.1), and then pre-multiplying (6.1) by $R^T(q)$, the constraint matrix $J^T(q)\lambda$ can be eliminated by virtue of (6.7). As a consequence, we have the transformed nonholonomic system

$$\mathcal{M}(q)\ddot{\zeta} + \mathcal{V}(q,\dot{q})\dot{\zeta} + \mathcal{G}(q) + \mathcal{D}(t) = R^T\mathcal{U} \qquad (6.11)$$

where

$$
\begin{aligned}
\mathcal{M}(q) &= R^T M(q) R \\
\mathcal{V}(q,\dot{q}) &= R^T[M(q)\dot{R} + V(q,\dot{q})R] \\
\mathcal{G}(q) &= R^T G(q) \\
\mathcal{U} &= B(q)\tau \\
\mathcal{D}(t) &= R^T d(t)
\end{aligned}
$$

which is more appropriate for the controller design as the constraint λ has been eliminated from the dynamic equation.

The force multipliers λ can be obtained by (6.10)

$$\lambda = Z \left([M(q)\dot{R}(q) + V(q,\dot{q})R(q)]\dot{\zeta} + G + d(t) - \mathcal{U} \right) \qquad (6.12)$$

where

$$Z = (JM^{-1}J^T)^{-1}JM^{-1}$$

Consider the control input $\mathcal{U}$ in the form:

$$\mathcal{U} = \tau_\alpha - J^T\tau_\beta \qquad (6.13)$$

then (6.11) and (6.12) can be changed to

$$
\begin{aligned}
R^T\tau_\alpha &= \mathcal{M}\ddot{\zeta} + \mathcal{V}\dot{\zeta} + \mathcal{G} + \mathcal{D}(t) \qquad (6.14)\\
\lambda &= Z\left([M(q)\dot{R}(q) + V(q,\dot{q})R(q)]\dot{\zeta} + G + d(t) \right)\\
&\quad -Z\tau_\alpha + \tau_\beta \qquad (6.15)
\end{aligned}
$$

Exploiting the structure of the dynamic equation (6.11), some properties are listed as follows.

Property 6.1 *Matrix $\mathcal{M}(q)$ is symmetric and positive definite.*

Property 6.2 *Matrix $\dot{\mathcal{M}}(q) - 2\mathcal{V}(q,\dot{q})$ is skew-symmetric.*

Property 6.3 *[140] $\mathcal{M}(q)$, $\mathcal{G}(q)$, $J(q)$ and $R(q)$ are bounded and continuous if q is bounded and uniformly continuous. $\mathcal{V}(q, \dot{q})$ and $\dot{R}(q)$ are bounded if $\dot{q}$ is bounded.*

The internal state variable ζ for mobile manipulators is partitioned in quantities related to the active joints, the passive joints, and the remaining joints as ζ_1, ζ_3 and ζ_2, respectively, such that the dimension of ζ_1 and ζ_3 are equal. According to the above partitions, corresponding to the definition of (6.14), by the investigation of Lagrangian formulation, we can have the partition structure for (6.14) as:

$$\mathcal{M}(\zeta) = \begin{bmatrix} M_{11} & M_{12} & M_{13} \\ M_{21} & M_{22} & M_{23} \\ M_{31} & M_{32} & M_{33} \end{bmatrix}$$

$$\mathcal{V}(\zeta, \dot{\zeta})\dot{\zeta} = \begin{bmatrix} V_1 \\ V_2 \\ V_3 \end{bmatrix} = \begin{bmatrix} V_{11}\dot{\zeta}_1 + V_{12}\dot{\zeta}_2 + V_{13}\dot{\zeta}_3 \\ V_{21}\dot{\zeta}_1 + V_{22}\dot{\zeta}_2 + V_{23}\dot{\zeta}_3 \\ V_{31}\dot{\zeta}_1 + V_{32}\dot{\zeta}_2 + V_{33}\dot{\zeta}_3 \end{bmatrix}$$

$$\mathcal{G} = \begin{bmatrix} G_1 \\ G_2 \\ G_3 \end{bmatrix}, \mathcal{D} = \begin{bmatrix} d_1 \\ d_2 \\ d_3 \end{bmatrix}, R^T \tau_\alpha = \begin{bmatrix} \tau_1 \\ \tau_2 \\ 0 \end{bmatrix} \quad (6.16)$$

In order to make ζ_3 controllable, we assume that matrices M_{13} and M_{31} are not equal to zero and M_{11}^{-1} exists. However, if M_{13} and M_{31} are equal to zero, while M_{12} and M_{21} are not equal to zero, which means that ζ_3 will be coupled with one vector of ζ_2, we only need to exchange ζ_1 with the vector in ζ_2. In this paper, we focus on $M_{13} = M_{31} \neq 0$. Considering the new partition in (6.16), after some simple manipulations, we can obtain three dynamics as

$$M_{11}\ddot{\zeta}_1 = \tau_1 - V_1 - G_1 - d_1 - M_{12}\ddot{\zeta}_2 - M_{13}\ddot{\zeta}_3 \quad (6.17)$$

$$(M_{22} - M_{21}M_{11}^{-1}M_{12})\ddot{\zeta}_2 + (M_{23} - M_{21}M_{11}^{-1}M_{13})\ddot{\zeta}_3$$
$$+V_2 + G_2 + d_2 - M_{21}M_{11}^{-1}V_1 - M_{21}M_{11}^{-1}G_1$$
$$-M_{21}M_{11}^{-1}d_1 = \tau_2 - M_{21}M_{11}^{-1}\tau_1 \quad (6.18)$$

$$(M_{32} - M_{31}M_{11}^{-1}M_{12})\ddot{\zeta}_2 + (M_{33} - M_{31}M_{11}^{-1}M_{13})\ddot{\zeta}_3$$
$$+V_3 + G_3 + d_3 - M_{31}M_{11}^{-1}V_1 - M_{31}M_{11}^{-1}G_1$$
$$-M_{31}M_{11}^{-1}d_1 = -M_{31}M_{11}^{-1}\tau_1 \quad (6.19)$$

Let

$$
\begin{aligned}
\mathcal{A} &= M_{22} - M_{21}M_{11}^{-1}M_{12} \\
\mathcal{L} &= M_{23} - M_{21}M_{11}^{-1}M_{13} \\
\mathcal{C} &= M_{32} - M_{31}M_{11}^{-1}M_{12} \\
\mathcal{J} &= M_{33} - M_{31}M_{11}^{-1}M_{13} \\
\mathcal{E} &= (V_{22} - M_{21}M_{11}^{-1}V_{12})\dot{\zeta}_2 + (V_{23} - M_{21}M_{11}^{-1}V_{13})\dot{\zeta}_3 \\
\mathcal{F} &= (V_{32} - M_{31}M_{11}^{-1}V_{12})\dot{\zeta}_2 + (V_{33} - M_{31}M_{11}^{-1}V_{13})\dot{\zeta}_3 \\
\mathcal{H} &= (V_{21} - M_{21}M_{11}^{-1}V_{11})\dot{\zeta}_1 + G_2 + d_2 - M_{21}M_{11}^{-1}G_1 \\
&\quad - M_{21}M_{11}^{-1}d_1 \\
\mathcal{K} &= (V_{31} - M_{31}M_{11}^{-1}V_{11})\dot{\zeta}_1 + G_3 + d_3 - M_{31}M_{11}^{-1}G_1 \\
&\quad - M_{31}M_{11}^{-1}d_1
\end{aligned}
$$

then we can rewrite (6.17), (6.18) and (6.19) as

$$
M_{11}\ddot{\zeta}_1 = \tau_1 - V_1 - G_1 - d_1 - M_{12}\ddot{\zeta}_2 - M_{13}\ddot{\zeta}_3 \tag{6.20}
$$

$$
\mathcal{A}\ddot{\zeta}_2 + \mathcal{L}\ddot{\zeta}_3 + \mathcal{E} + \mathcal{H} = -M_{21}M_{11}^{-1}\tau_1 + \tau_2 \tag{6.21}
$$

$$
\mathcal{C}\ddot{\zeta}_2 + \mathcal{J}\ddot{\zeta}_3 + \mathcal{F} + \mathcal{K} = -M_{31}M_{11}^{-1}\tau_1 \tag{6.22}
$$

Let $\xi = [\zeta_3^T, \zeta_2^T]^T$, and consider (6.14) and (6.16), the equations (6.21) and (6.22) become

$$
\mathcal{M}_1(\zeta)\ddot{\xi} + \mathcal{V}_1(\zeta, \dot{\zeta})\dot{\xi} + \mathcal{D}_1 = \mathcal{B}_1\Lambda_1\mathcal{U}_1 \tag{6.23}
$$

where

$$
\mathcal{M}_1(\zeta) = \begin{bmatrix} \mathcal{J} & \mathcal{C} \\ \mathcal{L} & \mathcal{A} \end{bmatrix}, \mathcal{D}_1 = \begin{bmatrix} \mathcal{K} \\ \mathcal{H} \end{bmatrix}, \mathcal{B}_1 = \begin{bmatrix} M_{31}M_{11}^{-1} & 0 \\ M_{21}M_{11}^{-1} & I \end{bmatrix}
$$

$$
\mathcal{V}_1(\zeta, \dot{\zeta}) = \begin{bmatrix} V_{33} - M_{31}M_{11}^{-1}V_{13} & V_{32} - M_{31}M_{11}^{-1}V_{12} \\ V_{23} - M_{21}M_{11}^{-1}V_{13} & V_{22} - M_{21}M_{11}^{-1}V_{12} \end{bmatrix}
$$

$$
\Lambda_1 = \begin{bmatrix} -I & 0 \\ 0 & I \end{bmatrix}, \mathcal{U}_1 = \begin{bmatrix} \tau_1 \\ \tau_2 \end{bmatrix}
$$

We then decompose $\mathcal{V}_1 = \hat{\mathcal{V}}_1 + \tilde{\mathcal{V}}_1$ such that

$$
\dot{\mathcal{M}}_1 - 2\tilde{\mathcal{V}}_1 = 0 \tag{6.24}
$$

Property 6.4 *The inertia matrix $\mathcal{M}_1$ is symmetric and positive definite.*

Property 6.5 *The eigenvalues of the inertia matrix $\mathcal{B}_1$ are positive.*

Remark 6.2 *There exist the minimum and maximum eigenvalues $\lambda_{min}(\mathcal{B}_1)$ and $\lambda_{max}(\mathcal{B}_1)$, such that $x^T \lambda_{min}(\mathcal{B}_1)Ix \leq x^T \mathcal{B}_1 x \leq x^T \lambda_{max}(\mathcal{B}_1)Ix$, $\forall x \in R^{(n-l-n_p)}$, with the identity matrix I, and the known positive parameter b satisfying $0 < b \leq \lambda_{min}(\mathcal{B}_1)$, that is, $x^T b I x \leq x^T \lambda_{min}(\mathcal{B}_1)Ix$.*

Assumption 6.2 *[140] The desired trajectories $\zeta_{2d}(t), \zeta_{3d}(t)$ and their time derivatives up to the 3rd order are continuously differentiable and bounded for all $t \geq 0$. The desired Lagrangian multiplier λ_d is also bounded and uniformly continuous.*

The control objective for the motion of the system is to design, if possible, controllers that ensure the tracking errors for the variables ζ_2, ζ_3, and the constraint force vector f, or, equivalently, a multiplier λ, from any $(\zeta_j(0), \dot{\zeta}_j(0), \lambda(0)) \in \Omega$, $\zeta_j, \dot{\zeta}_j$ and λ converge to a manifold Ω_d where

$$\Omega_d = \{(\zeta_j, \dot{\zeta}_j, \lambda) | \ |\zeta_j - \zeta_{jd}| \leq \epsilon_{j1}, |\dot{\zeta}_j - \dot{\zeta}_{jd}| \leq \epsilon_{j2}, \|\lambda - \lambda_d\| \leq \epsilon_\lambda\} \quad (6.25)$$

where $\epsilon_{ji} > 0$, $\epsilon_\lambda > 0$, $i = 1, 2, j = 2, 3$. Ideally, ϵ_{ji} and ϵ_λ should be the threshold of measurable noise. At the same time, all the closed loop signals are to be kept bounded.

In the following, we can analyze and design the control for each subsystem. For clarity, we define the tracking errors and the filtered tracking errors as $e_j = \zeta_j - \zeta_{jd}$, and $r_j = \dot{e}_j + \Lambda_j e_j$ where Λ_j is positive definite, $j = 2, 3$. Then, to study the stability of e_j and $\dot{e}_j$, we only need to study the properties of r_j. In addition, the following computable signals are defined:

$$\dot{\zeta}_{jr} = \dot{\zeta}_{jd} - \Lambda_j e_j \quad (6.26)$$

$$\ddot{\zeta}_{jr} = \ddot{\zeta}_{jd} - \Lambda_j \dot{e}_j \quad (6.27)$$

6.3 High-gain Observer

Since unknown nonlinearities can exist in the control input channel, it may be difficult to measure all the system states in reality. In the case that only the output is available, an observer is needed to estimate the $n - 1$ derivative of the output; therefore, a high-gain observer is employed to estimate the states of a system.

Lemma 6.1 *Suppose the function $\zeta_j(t)$ and its first n derivatives are bounded. Consider the following system: $\epsilon\dot{\vartheta}_{j1} = \vartheta_{j2}$, $\epsilon\dot{\vartheta}_{j2} = \vartheta_{j3}$, $\cdots$, $\epsilon\dot{\vartheta}_{j(n-1)} = \vartheta_{jn}$, $\epsilon\dot{\vartheta}_{jn} = -\kappa_{j1}\vartheta_{jn} - \kappa_{j2}\vartheta_{j(n-1)} - \cdots - \kappa_{j(n-1)}\vartheta_{j2} - \vartheta_{j1} + \zeta_j$, where the parameters κ_{j1} to $\kappa_{j(n-1)}$ are chosen so that the polynomial $s_j^n + \kappa_{j1}s_j^{(n-1)} + \cdots + \kappa_{j(n-1)}s_j + 1$ is Hurwitz. Then, there exist positive constants $h_{j(k+1)}, j = 1, 2, \cdots, m$, $k = 1, 2, \cdots, n-1$ and t^* such that for all $t > t^*$, we have $\frac{\vartheta_{j(k+1)}}{\epsilon^{(k)}} - \zeta_j^{(k)} = -\epsilon\varphi_j^{(k+1)}, \left|\frac{\vartheta_{j(k+1)}}{\epsilon^{(k)}} - \zeta_j^{(k)}\right| \leq \epsilon h_{j(k+1)}$, where ϵ is any small positive constant and $\varphi_j = \vartheta_{jn} + \kappa_{j1}\vartheta_{j(n-1)} + \cdots + \kappa_{j(n-1)}\vartheta_{j1}$ and $\varphi_j^{(k)}$ denote the kth derivative of φ_j and $\left|\varphi_j^{(k)}\right| \leq h_{jk}$.*

Proof: The proof of the Lemma can be found in [141].

Define the signals as $\vartheta_j = [\vartheta_{j1}, \vartheta_{j2}, \vartheta_{j3}]^T$, and let $\hat{\zeta}_j = \vartheta_{j1}$, $\dot{\hat{\zeta}}_j = \frac{\vartheta_{j2}}{\epsilon}$, $\ddot{\hat{\zeta}}_j = \frac{\vartheta_{j3}}{\epsilon^2}$ and $\hat{\mathbf{e}}_j = [\hat{e}_j, \dot{\hat{e}}_j, \ddot{\hat{e}}_j]^T = [\vartheta_{j1} - \zeta_{jd}, \frac{\vartheta_{j2}}{\epsilon} - \dot{\zeta}_{jd}, \frac{\vartheta_{j3}}{\epsilon^2} - \ddot{\zeta}_{jd}]^T$.

Since $\hat{e}_j = \vartheta_{j1} - \zeta_{jd} = \vartheta_{j1} - \zeta_j + \zeta_j - \zeta_{jd} = e_j - \epsilon\varphi_j$, similarly we can deduce the equations as follows:

$$\hat{e}_j = e_j - \epsilon\dot{\varphi}_j \tag{6.28}$$

$$\dot{\hat{e}}_j = \dot{e}_j - \epsilon\ddot{\varphi}_j \tag{6.29}$$

$$\ddot{\hat{e}}_j = \ddot{e}_j - \epsilon\varphi_j^{(3)} \tag{6.30}$$

Then the high-gain observer is designed as

$$\hat{r}_j = r_j - \epsilon\Lambda_j\dot{\varphi}_j - \epsilon\ddot{\varphi}_j \tag{6.31}$$

$$\dot{\hat{r}}_j = \dot{r}_j - \epsilon\Lambda_j\ddot{\varphi}_j - \epsilon\varphi_j^{(3)} \tag{6.32}$$

$$\dot{\hat{\zeta}}_{jr} = \dot{\zeta}_{jr} + \epsilon\Lambda_j\dot{\varphi}_j \tag{6.33}$$

$$\ddot{\hat{\zeta}}_{jr} = \ddot{\zeta}_{jr} + \epsilon\Lambda_j\ddot{\varphi}_j \tag{6.34}$$

Let $\vartheta_{j1} = \zeta_j$, then we can deduce that $\dot{\varphi}_j = 0$, such that we have $\hat{e}_j = e_j$, $\dot{\hat{r}}_j = r_j - \epsilon\ddot{\varphi}_j$ and $\dot{\hat{\zeta}}_{jr} = \dot{\zeta}_{jr}$.

6.4 Adaptive Output Feedback Control

In reality, physical model of the system cannot be exactly known, i.e., there exist model uncertainties, which would cause the dynamics uncertainties. In addition, external disturbances may also affect the performance of the system. In this section, we take both factors into consideration to develop adaptive control schemes to deal with uncertainties as well as external disturbances.

Since $\dot{\xi} = \dot{\xi}_r + r$, $\ddot{\xi} = \ddot{\xi}_r + \dot{r}$ and $\xi = [\zeta_3^T, \zeta_2^T]^T$, the equation (6.23) becomes

$$\mathcal{M}_1 \dot{r} + \tilde{\mathcal{V}}_1 r = -\mathcal{M}_1 \ddot{\xi}_r - \hat{\mathcal{V}}_1 r - \mathcal{V}_1 \dot{\xi}_r - \mathcal{D}_1 + \mathcal{B}_1 \Lambda_1 \mathcal{U}_1 \tag{6.35}$$

where $r = [r_3^T, \ r_2^T]^T$, $\ddot{\xi}_r = [\ddot{\zeta}_{3r}^T, \ \ddot{\zeta}_{2r}^T]^T$.

Let $\mathcal{M}_0$, $\mathcal{V}_0$, $\hat{\mathcal{V}}_0$, $\mathcal{D}_0$, and $\mathcal{B}_0$ be nominal parameter vectors which give the corresponding nominal functions $\mathcal{M}_0 \ddot{\xi}_r + \mathcal{V}_0 \dot{\xi}_r + \hat{\mathcal{V}}_0 r + \mathcal{D}_0$ and $(\mathcal{B}_0)^{-1}$, respectively, and there exist some finite positive constants $c_i > 0$ $(1 \leq i \leq 8)$, such that $\|\mathcal{M}_1 - \mathcal{M}_0\| \leq c_1$, $\|\mathcal{V}_1 - \mathcal{V}_0\| \leq c_2 + c_3\|\dot{\zeta}\|$, $\|\hat{\mathcal{V}}_1 - \hat{\mathcal{V}}_0\| \leq c_4 + c_5\|\dot{\zeta}\|$, $\|\mathcal{D} - \mathcal{D}_0\| \leq c_6 + c_7\|\dot{\zeta}\|$, $\|\mathcal{B}_1 - \mathcal{B}_0\| \leq c_8$.

Considering (6.31)–(6.34), the equation (6.35) becomes

$$\mathcal{M}_1 \dot{\hat{r}} + \tilde{\mathcal{V}}_1 \hat{r} = \mu + \mathcal{B}_1 \Lambda_1 \mathcal{U}_1 \tag{6.36}$$

where $\mu = -\mathcal{M}_1 \ddot{\hat{\xi}}_r - \mathcal{V}_1 \dot{\hat{\xi}}_r - \hat{\mathcal{V}}_1 \hat{r} - \epsilon \mathcal{M}_1 \varphi^{(3)} - \mathcal{V}_1 \epsilon \ddot{\varphi} - \mathcal{D}_1$, $\Lambda = \text{diag}[\Lambda_j]$, $\ddot{\varphi} = [\ddot{\varphi}_3^T, \ddot{\varphi}_2^T]^T$, $\varphi^{(3)} = [\varphi_3^{(3)T}, \varphi_2^{(3)T}]^T$.

Since $|\frac{\vartheta_{j(k+1)}}{\epsilon^{(k)}} - \zeta_j^{(k)}| \leq \epsilon h_{j(k+1)}$ holds in Lemma 6.1, we have $|\frac{\vartheta_{j(k+1)}}{\epsilon^{(k)}}| - |\zeta_j^{(k)}| \leq \epsilon h_{j(k+1)}$, and $|\zeta_j^{(k)}| \leq |\frac{\vartheta_{j(k+1)}}{\epsilon^{(k)}}| - \epsilon h_{j(k+1)}$ and it is easy to have the following assumption:

Assumption 6.3 *There exist some finite positive constants $\alpha_i > 0$ $(1 \leq i \leq 11)$ such that $\forall \zeta \in R^{n-l}$, $\dot{\zeta} \in R^{n-l}$, $\|\mathcal{M}_1 - \mathcal{M}_0\| \leq \alpha_1$, $\|\epsilon \mathcal{M}_1 \varphi^{(3)}\| \leq \alpha_2$, $\|\hat{\mathcal{V}}_1 - \hat{\mathcal{V}}_0\| \leq \alpha_3 + \alpha_4\|\dot{\zeta}\|$, $\|\mathcal{V}_1 - \mathcal{V}_0\| \leq \alpha_5 + \alpha_6\|\dot{\zeta}\|$, $\|\mathcal{V}_1 \epsilon \ddot{\varphi}\| \leq \alpha_7 + \alpha_8\|\dot{\zeta}\|$, $\|\mathcal{D}_1 - \mathcal{D}_0\| \leq \alpha_9 + \alpha_{10}\|\dot{\zeta}\|$, $\|\mathcal{B}_1 - \mathcal{B}_0\| \leq \alpha_{11}$.*

Remark 6.3 *Although we assume some finite positive constants $\alpha_i > 0$ $(1 \leq i \leq 11)$ in Assumption 6.3, these constants are unknown beforehand; therefore, we need to propose the following adaptive laws to approximate them.*

Assumption 6.4 *A time-varying positive function ϖ converges to zero as $t \to \infty$ and satisfies $\lim_{t \to \infty} \int_0^t \varpi(s)ds = \rho < \infty$ with finite constant ρ.*

Consider the following output feedback control design:

$$\Lambda_1 \mathcal{U}_1 = -\mathcal{B}_0^{-1} K_P \hat{r} + \mathcal{B}_0^{-1} \mu_0 - \frac{1}{b} \sum_{i=1}^{11} \frac{\hat{r} \hat{\alpha}_i \Upsilon_i^2}{\Upsilon_i \|\hat{r}\| + \delta_i} \tag{6.37}$$

with $\mu_0 = \mathcal{M}_0 \ddot{\xi}_r + \mathcal{V}_0 \dot{\xi}_r + \hat{\mathcal{V}}_0 \hat{r} + \mathcal{D}_0$, which is adaptively tuned according to

$$\dot{\hat{\alpha}}_i = -\sigma_i \hat{\alpha}_i + \frac{\gamma_i \Upsilon_i^2 \|\hat{r}\|^2}{\|\hat{r}\| \Upsilon_i + \delta_i}, \quad i = 1, \ldots, 11 \tag{6.38}$$

where K_P is positive definite, $\gamma_i > 0$, $\delta_i > 0$ and $\sigma_i > 0$ ($1 \leq i \leq 11$) satisfies Assumption 6.4: $\lim_{t \to \infty} \delta_i(t) = 0$, $\int_0^\infty \delta_i(s)ds = \rho_{i\delta} < \infty$, $\lim_{t \to \infty} \sigma_i(t) = 0$, $\int_0^\infty \sigma_i(s)ds = \rho_{i\sigma} < \infty$ with the constants $\rho_{i\delta}$ and $\rho_{i\sigma}$. Let

$$\hat{\alpha} = [\hat{\alpha}_1, \ldots, \hat{\alpha}_{11}]^T$$
$$\Upsilon = [\|\ddot{\xi}_r\|, 1, \|\hat{r}\|, \|\hat{r}\|\|\dot{\zeta}\|, \|\dot{\xi}_r\|, \|\dot{\xi}_r\|\|\dot{\zeta}\|, 1, \|\dot{\zeta}\|, 1, \|\dot{\zeta}\|, \Upsilon_{11}]^T$$

where $\Upsilon_{11} = \| - \mathcal{B}_0^{-1} K_P \hat{r} + \mathcal{B}_0^{-1} \mu_0 \|$, and $\Phi = \alpha^T \Upsilon$.

Consider the following Lyapunov function candidate for the ζ_2 and ζ_3-subsystem

$$\mathbb{V}_1 = \frac{1}{2} \hat{r}^T \mathcal{M}_1 \hat{r} + \frac{1}{2} \tilde{\alpha}^T \Gamma^{-1} \tilde{\alpha} \tag{6.39}$$

where $\Gamma = \text{diag}[\gamma_1, \ldots, \gamma_{11}]$, and $\tilde{\alpha} = \alpha - \hat{\alpha}$. Its time derivative is given by

$$\dot{\mathbb{V}}_1 = \frac{1}{2} \hat{r}^T (\dot{\mathcal{M}}_1 \hat{r} + 2\mathcal{M}_1 \dot{\hat{r}}) + \tilde{\alpha}^T \Gamma^{-1} \dot{\tilde{\alpha}} \tag{6.40}$$

Considering Property 6.4, and (6.24), and substituting (6.36) into (6.40), and integrating (6.37), we have

$$\begin{aligned}
\dot{\mathbb{V}}_1 &= \hat{r}^T (\mathcal{B}_1 \Lambda_1 \mathcal{U}_1 + \mu) + \tilde{\alpha}^T \Gamma^{-1} \dot{\tilde{\alpha}} \\
&= \hat{r}^T [(\mathcal{B}_1 - \mathcal{B}_0)\Lambda_1 \mathcal{U}_1 + \mathcal{B}_0 \Lambda_1 \mathcal{U}_1 + \mu] + \tilde{\alpha}^T \Gamma^{-1} \dot{\tilde{\alpha}} \\
&\leq -\hat{r}^T K_P \hat{r} - \hat{r}^T \mathcal{B}_1 \frac{1}{b} \sum_{i=1}^{11} \frac{\hat{r} \hat{\alpha}_i \Upsilon_i^2}{\|\hat{r}\| \Upsilon_i + \delta_i} \\
&\quad + \|\hat{r}\| \|\mathcal{M}_0 - \mathcal{M}_1\| \|\ddot{\xi}_r\| + \|\hat{r}\| \|\mathcal{V}_0 - \mathcal{V}_1\| \|\dot{\xi}_r\| \\
&\quad + \|\hat{r}\| \|\hat{\mathcal{V}}_0 - \hat{\mathcal{V}}_1\| \|\hat{r}\| + \|\hat{r}\| \epsilon \mathcal{M}_1 \varphi^{(3)}\| \\
&\quad + \|\hat{r}\| \|\mathcal{V}_1 \epsilon \ddot{\varphi}\| + \|\hat{r}\| \|\mathcal{D}_0 - \mathcal{D}_1\| \\
&\quad + \|\hat{r}\| \|(\mathcal{B}_1 - \mathcal{B}_0)\| \| - \mathcal{B}_0^{-1} K_P \hat{r} + \mathcal{B}_0^{-1} \mu_0\| \\
&\quad + \tilde{\alpha}^T \Gamma^{-1} \dot{\tilde{\alpha}} \\
&\leq -\hat{r}^T K_P \hat{r} - \sum_{i=1}^{11} \frac{\|\hat{r}\|^2 \hat{\alpha}_i \Upsilon_i^2}{\|\hat{r}\| \Upsilon_i + \delta_i} + \|\hat{r}\| \alpha^T \Upsilon \\
&\quad + \hat{\alpha}^T \Sigma \Gamma^{-1} \tilde{\alpha} - \sum_{i=1}^{11} \frac{\|\hat{r}\|^2 \tilde{\alpha}_i \Upsilon_i^2}{\|\hat{r}\| \Upsilon_i + \delta_i} \\
&\leq -\hat{r}^T K_P \hat{r} + \alpha^T \Delta + \hat{\alpha}^T \Sigma \Gamma^{-1} (\alpha - \hat{\alpha}) \\
&\leq -\hat{r}^T K_P \hat{r} + \alpha^T \Delta + \frac{1}{4} \alpha^T \Sigma \Gamma^{-1} \alpha \tag{6.41}
\end{aligned}$$

with $\Sigma = \text{diag}[\sigma_1, \ldots, \sigma_{11}]$, $\Delta = [\delta_1, \ldots, \delta_{11}]^T$. Therefore,

$$\dot{\mathbb{V}}_1 \leq -\lambda_{min}(K_P)\|\hat{r}\|^2 + \alpha^T\Delta + \frac{1}{4}\alpha^T\Sigma\Gamma^{-1}\alpha \tag{6.42}$$

Since $\alpha^T\Delta + \frac{1}{4}\alpha^T\Sigma\Gamma^{-1}\alpha$ is bounded, and $t > t_1$, $\alpha^T\Delta + \frac{1}{4}\alpha^T\Sigma\Gamma^{-1}\alpha \leq \rho_1$ with the finite constant ρ_1, we have,

$$\dot{\mathbb{V}}_1 \leq -\lambda_{min}(K_P)\|\hat{r}\|^2 + \rho_1 \tag{6.43}$$

which leads to

$$\begin{aligned}\|\hat{r}\|^2 &\leq \exp(-\lambda_{min}(K_P)t)\|\hat{r}(0)\|^2 \\ &+ \frac{\rho_1}{\lambda_{min}(K_P)}(1 - \exp(-\lambda_{min}(K_P)t))\end{aligned} \tag{6.44}$$

or

$$\|\hat{r}\| \leq \sqrt{\Theta} \tag{6.45}$$

where $\Theta = \exp(\lambda_{min}(K_P)t)\|\hat{r}(0)\|^2 + \frac{\rho_1}{\lambda_{min}(K_P)}(1-\exp(-\lambda_{min}(K_P)t))$. Therefore, we can get the conclusion that $\hat{r}$ converges to a small bounded set.

When $\|\hat{r}\| \geq \sqrt{\frac{\rho_1}{\lambda_{min}(K_P)}}$, then $\dot{\mathbb{V}}_1 \leq 0$, from above and $\hat{r}$ converges to a small set Ω_1 containing the origin as $t \to \infty$,

$$\Omega_1 : \|\hat{r}\| \leq \sqrt{\frac{\rho_1}{\lambda_{min}(K_P)}} \tag{6.46}$$

By Assumption 6.4, because $\rho_1 \to 0$ as $t \to \infty$, from (6.46), we have $\hat{r} \to 0$, $\dot{\mathbb{V}}_1 \to 0$ with $t \to \infty$, then $\mathbb{V}_1$ is bounded and hence $\lim_{t\to\infty} \hat{r} = 0$. From (6.31), where ϵ is any small positive constant and $\ddot{\varphi}, \dot{\varphi}$ are bounded, $\hat{r} \to 0$ as $t \to \infty$.

Integrating both sides of (6.43) gives

$$\begin{aligned}\mathbb{V}_1(t) - \mathbb{V}_1(0) &\leq -\int_0^t \lambda_{min}(K_P)\|\hat{r}\|^2 ds \\ &+ \int_0^t (\alpha^T\Delta + \frac{1}{4}\alpha^T\Sigma\Gamma^{-1}\alpha)ds\end{aligned} \tag{6.47}$$

Since α and Γ are constant, $\int_0^\infty \Delta ds = \rho_\delta = [\rho_{1\delta}, \ldots, \rho_{11\delta}]^T$, $\int_0^\infty \Sigma ds = \rho_\sigma = \text{diag}[\rho_{1\sigma}, \ldots, \rho_{11\sigma}]$, then, we can rewrite (6.47) as $\mathbb{V}_1(t) - \mathbb{V}_1(0) \leq -\int_0^t \lambda_{min}(K_P)\|\hat{r}\|^2 ds + \alpha^T\rho_\delta + \frac{1}{4}\alpha^T\rho_\sigma\Gamma^{-1}\alpha < \infty$. Thus $\mathbb{V}_1$ is bounded, which implies that $\hat{r} \in L_\infty$. From (6.47), we have $\int_0^t \lambda_{min}(K_P)\|\hat{r}\|^2 ds \leq \mathbb{V}_1(0) - \mathbb{V}_1(t) + \alpha^T\rho_\delta + \frac{1}{4}\alpha^T\rho_\sigma\Gamma^{-1}\alpha$, which leads to $\hat{r} \in L_2$. From $\hat{r}_j = \dot{\hat{e}}_j + \Lambda_j\hat{e}_j$, it can be obtained that $\hat{e}_j, \dot{\hat{e}}_j \in L_\infty$. As we have established $\hat{e}_j, \dot{\hat{e}}_j \in L_\infty$, from Assumption 6.2, we conclude that $\hat{\zeta}_j, \dot{\hat{\zeta}}_j, \dot{\hat{\xi}}_r, \ddot{\hat{\xi}}_r \in L_\infty$. Therefore, all the signals on the right hand side of (6.35), if (6.37) is integrated, are bounded, and we can conclude that $\dot{r}$ and therefore $\ddot{\hat{\zeta}}_j$ are bounded. Consequently, it follows that $\hat{e}_j \to 0, \dot{\hat{e}}_j \to 0$ as $t \to \infty$. Since $\hat{r}, \hat{\zeta}_j, \dot{\hat{\zeta}}_j, \hat{\zeta}_{jr}, \dot{\hat{\zeta}}_{jr}, \ddot{\hat{\zeta}}_{jr}$ are all bounded it is easy to conclude that $\mathcal{U}_1$ is bounded from (6.37).

Since (6.31), the sliding error r exponentially converges to a small value and is bounded by

$$\|r\| \leq \sqrt{\frac{\rho_1}{\lambda_{min}(K_P)}} + \epsilon\|\ddot{\varphi}\| + \epsilon\Lambda\|\dot{\varphi}\| \tag{6.48}$$

where ϵ is any small positive constant and $\ddot{\varphi}$ is bounded. It is easy to know $\dot{e}$ and e converge to zero.

Finally, for system (6.17)–(6.19) under control laws (6.37), apparently, the ζ_1-subsystem (6.17) can be rewritten as

$$\dot{\varphi} = f(\nu, \varphi, \mathcal{U}) \tag{6.49}$$

where $\varphi = [\zeta_1^T, \dot{\zeta}_1^T]^T$, $\nu = [r^T, \dot{r}^T]^T$, $\mathcal{U} = [\tau_1^T, \tau_2^T]^T$.

Assumption 6.5 *From (6.18) and (6.19), the reference signal satisfies Assumption 6.2, and the following functions are Lipschitz in $\gamma = [\zeta_2, \zeta_3, \dot{\zeta}_2, \dot{\zeta}_3]^T$, i.e., there exist Lipschitz positive constants L_γ and L_f such that*

$$\|V_1 + G_1 + d_1\| \leq L_{1\gamma}\|\gamma\| + L_{1f} \tag{6.50}$$

$$\|\|\mathcal{F} + \mathcal{K}\| \leq L_{2\gamma}\|\gamma\| + L_{2f} \tag{6.51}$$

moreover, from the stability analysis of ζ_2 and ζ_3 subsystems, γ converges to a small neighborhood of $\gamma_d = [\zeta_{2d}, \zeta_{3d}, \dot{\zeta}_{2d}, \dot{\zeta}_{3d}]^T$.

Remark 6.4 *Under the stability of ζ_2 and ζ_3 subsystems, and considering (6.25), let $\|\gamma - \gamma_d\| \leq \varsigma_1$. It is easy to obtain $\|\gamma\| \leq \|\gamma_d\| + \varsigma_1$, and similarly, let $\mu = [\ddot{\zeta}_2, \ddot{\zeta}_3]^T$, and $\mu_d = [\ddot{\zeta}_{2d}, \ddot{\zeta}_{3d}]^T$, $\|\mu\| \leq \|\mu_d\| + \varsigma_2$, where ς_1 and ς_2 are small bounded errors.*

Lemma 6.2 *The ζ_1-subsystem (6.17), if ζ_2-subsystem and ζ_3-subsystem are stable, is asymptotically stable, too.*

Proof: From (6.17), (6.18), and (6.19), we choose the following function as

$$\mathbb{V}_2 = \mathbb{V}_1 + \ln(\cosh(\dot{\zeta}_1)) \tag{6.52}$$

Differentiating (6.52) along (6.17) gives

$$
\begin{aligned}
\dot{V}_2 &= \dot{V}_1 + \tanh(\dot{\zeta}_1)\ddot{\zeta}_1 \\
&= \dot{V}_1 + \tanh(\dot{\zeta}_1)M_{11}^{-1}(\tau_1 - V_1 - G_1 - d_1 \\
&\quad - M_{12}\ddot{\zeta}_2 - M_{13}\ddot{\zeta}_3)
\end{aligned} \tag{6.53}
$$

From (6.22), we have

$$
\tau_1 = -M_{11}M_{31}^{-1}(C\ddot{\zeta}_2 + \mathcal{J}\ddot{\zeta}_3 + \mathcal{F} + \mathcal{K}) \tag{6.54}
$$

Integrating (6.54) into (6.53), we have

$$
\begin{aligned}
\dot{V}_2 &= \dot{V}_1 + \tanh(\dot{\zeta}_1)(-M_{31}^{-1}(C\ddot{\zeta}_2 + \mathcal{J}\ddot{\zeta}_3 + \mathcal{F} + \mathcal{K}) \\
&\quad - M_{11}^{-1}(V_1 + G_1 + d_1 + M_{12}\ddot{\zeta}_2 + M_{13}\ddot{\zeta}_3)) \\
&= \dot{V}_1 - \tanh(\dot{\zeta}_1)M_{31}^{-1}(\mathcal{F} + \mathcal{K}) \\
&\quad - \tanh(\dot{\zeta}_1)M_{11}^{-1}(V_1 + G_1 + d_1) \\
&\quad - \begin{bmatrix} \tanh(\dot{\zeta}_1)M_{31}^{-1}C + \tanh(\dot{\zeta}_1)M_{11}^{-1}M_{12} \\ \tanh(\dot{\zeta}_1)M_{31}^{-1}\mathcal{J} + \tanh(\dot{\zeta}_1)M_{11}^{-1}M_{13} \end{bmatrix}^T \begin{bmatrix} \ddot{\zeta}_2 \\ \ddot{\zeta}_3 \end{bmatrix}
\end{aligned}
$$

since $\| \tanh(\dot{\zeta}_1)\| \leq 1$, M_{31}^{-1}, M_{12} and M_{13} are all bounded. Let

$$
\left\| \begin{bmatrix} \tanh(\dot{\zeta}_1)M_{31}^{-1}C + \tanh(\dot{\zeta}_1)M_{11}^{-1}M_{12} \\ \tanh(\dot{\zeta}_1)M_{31}^{-1}\mathcal{J} + \tanh(\dot{\zeta}_1)M_{11}^{-1}M_{13} \end{bmatrix}^T \right\| \leq \varrho_1 \tag{6.55}
$$

and $\|M_{11}^{-1}\| \leq \varrho_2$, $\|M_{31}^{-1}\| \leq \varrho_3$, where ϱ_1, ϱ_2 and ϱ_3 are bounded constants. Considering Assumption 6.5 and Remark 6.4, we have

$$
\begin{aligned}
\dot{V}_2 &\leq -\frac{1}{2}\lambda_{min}(K_P)\|r\|^2 + \alpha^T\Delta + \frac{1}{4}\alpha^T\Sigma\Gamma^{-1}\alpha \\
&\quad + \varrho_1(\|\mu_d\| + \varsigma_2) + \varrho_2(L_{1\gamma}(\|\gamma_d\| + \varsigma_1) + L_{1f}) \\
&\quad + \varrho_3(L_{2\gamma}(\|\gamma_d\| + \varsigma_1) + L_{2f})
\end{aligned} \tag{6.56}
$$

Let $\rho_2 = \varrho_1(\|\mu_d\| + \varsigma_2) + \varrho_2(L_{1\gamma}(\|\gamma_d\| + \varsigma_1) + L_{1f}) + \varrho_3(L_{2\gamma}(\|\gamma_d\| + \varsigma_1) + L_{2f})$ and it is apparently bounded positive and we have $\dot{V}_2 \leq 0$, when $\|r\| \geq \sqrt{\frac{2\rho_1 + 2\rho_2}{\lambda_{min}(K_P)}}$. We can choose the proper K_P such that r can be arbitrarily small. Therefore, we can obtain stable internal dynamics with respect to the output $\dot{\zeta}_1$. Therefore, the ζ_1-subsystem (6.17) is asymptotically stable.

Theorem 6.1 *Consider the system (6.17)–(6.19) with Assumptions 6.3 and 6.2, under the action of control laws (6.37) and adaptation laws (6.38). For compact set Ω_1, where $(\zeta(0), \dot{\zeta}(0), \hat{a}(0)) \in \Omega_1$, the tracking errors r converges to the compact sets Ω_1 defined by (6.46), and all the signals in the closed loop system are bounded.*

Proof: From the results (6.47), it is clear that the tracking errors r_j converge to the compact set Ω_1 defined by (6.46). In addition, the signal $\tilde{\alpha}$ is bounded. Therefore, we can know e_2, $\dot{e}_2$, e_3, $\dot{e}_3$ are also bounded. From the boundedness of ζ_{2d}, ζ_{3d} in Assumption 6.2, we know that ζ_2, ζ_3 are bounded. Since $\dot{\zeta}_{2d}, \dot{\zeta}_{3d}$ are also bounded, it follows that $\dot{\zeta}_2, \dot{\zeta}_3$ are bounded. With α as constant, we know that $\dot{\alpha}$ is also bounded. From Lemma 6.2, we know that the ζ_1-subsystem (6.17) is stable, and $\zeta_1, \dot{\zeta}_1$ are bounded. This completes the proof.

The force control input τ_β is designed as

$$\tau_\beta = \lambda_d - K_f e_\lambda \tag{6.57}$$

where $e_\lambda = \lambda - \lambda_d$ is nonholonomic constraint force error vector and K_f is a constant matrix of proportional control feedback gains.

Substituting the control (6.14) and (6.57) into the reduced order dynamics (6.15) yields

$$
\begin{aligned}
(I + K_f)e_\lambda &= Z([M(q)\dot{R}(q) + C(q,\dot{q})R(q)]\dot{\zeta} + G + d(t) \\
&\quad -\tau_\alpha) = -ZR^{+T}M\ddot{\zeta}
\end{aligned}
\tag{6.58}
$$

Since $\ddot{\zeta}_2 \to \ddot{\zeta}_{2d}$, $\ddot{\zeta}_3 \to \ddot{\zeta}_{3d}$, and $\ddot{\zeta}_1 \to 0$ from Lemma 6.2, $\ddot{\zeta}$ is bounded, and $-ZR^{+T}M$ is also bounded; therefore, the size of e_λ can be adjusted by choosing the proper gain matrix K_f.

Since $\hat{r}$, ζ, $\dot{\zeta}$, ξ_r, $\dot{\xi}_r$, and e_λ are all bounded, it is easy to conclude that $\mathcal{U}_1$ is bounded from (6.37) and (6.57).

6.5   Simulation Studies

To verify the effectiveness of the proposed control algorithm, let us consider a wheeled mobile under-actuated manipulator shown in Fig. 6.1.

The following variables have been chosen to describe the vehicle (see also Fig. 6.1): τ_l, τ_r: the torques of two wheels; τ_1: the torques of the under-actuated joint, that is, $\tau_1 = 0$; θ_l, θ_r: the rotation angle of the left wheel and the right wheel of the mobile platform; v: the forward velocity of the mobile platform; θ: the direction angle of the mobile platform; ω: the rotation velocity of the mobile platform, and $\omega = \dot{\theta}$; θ_1: the joint angle of the under-actuated link; m_1, I_{z1}, l_1: the mass, the inertia moment, and the length for the link 1;

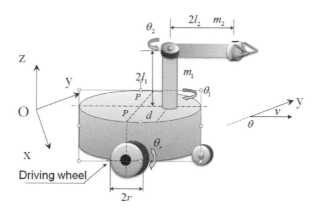

FIGURE 6.1

The mobile under-actuated manipulator in the simulation.

m_2, I_{z2}, l_2: the mass, the inertia moment , and the length for the link 2; r: the radius of the wheels; $2l$: the distance of the wheels; d: the distance between the manipulator and the driving center of the mobile base; m_p: the mass of the mobile platform; I_p: the inertia moment of the mobile platform; I_w: the inertia moment of each wheel; m_w: the mass of each wheel; g: gravity acceleration.

The mobile under-actuated manipulator is subject to the following constraint: $\dot{x}\cos\theta - \dot{y}\sin\theta = 0$. Using the Lagrangian approach, we can obtain the dynamic model with $q = [\theta_l, \theta_r, \theta_1]^T$, then we could obtain $M(q)\ddot{q} + C(q,\dot{q})\dot{q} + G(q) = B(q)\tau$, where $J^T\lambda$ is eliminated.

In the simulation, we assume the parameters $p_0 = 4.0kg \cdot m^2$, $p_1 = 0.5kg \cdot m^2$, $p_2 = 0.5kg \cdot m^2$, $p_3 = 1.0kg \cdot m^2$, $p_4 = 6.0kg \cdot m^2$, $q_0 = 2.0kg \cdot m^2$, $q_1 = 2.0kg \cdot m^2$, $q_2 = 1.0kg \cdot m^2$, $q_3 = 1.0kg \cdot m^2$, $q_4 = 0.5kg \cdot m^2$, $d = 0.5m$, $r = 0.5m$, $l = 0.5$, let $\dot{\zeta} = [\dot{\zeta}_1, \dot{\zeta}_2, \dot{\zeta}_3]^T = [\dot{\theta}, v, \dot{\theta}_1]^T$. The disturbances from environments on the system are introduced as $0.1\sin(t)$, $0.1\sin(t)$ and $0.1\sin(t)$ to the simulation model.

The control gains are selected as $K_P = \text{diag}[10.0]$, $\mathcal{B}_0 = \text{diag}[1.0]$, $\mathcal{M}_0 = \mathcal{V}_0 = \hat{\mathcal{V}}_0 = \mathcal{D}_0 = 0$, $b = 1.0$, $\Lambda_1 = \Lambda_2 = 1$, $\hat{a}(0) = [1.0, \ldots, 1.0]^T$, $\delta_i = \sigma_i = 1/(1+t)^2$, $\gamma_i = 0.05$. When $\dot{\zeta}$ is not measurable, a three-order high-gain observer is designed as follows: $\dot{\vartheta}_{j1} = \vartheta_{j2}/\epsilon$, $\dot{\vartheta}_{j2} = \vartheta_{j3}/\epsilon$, $\dot{\vartheta}_{j3} = (-\kappa_2\vartheta_{j3} -$

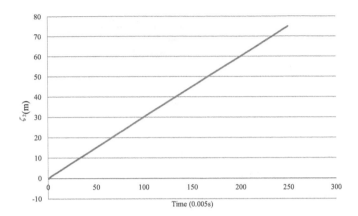

FIGURE 6.2
Tracking the desired position by output feedback control.

$\kappa_1 \vartheta_{j2} - \vartheta_{j1} + \zeta_j)/\epsilon$ with the parameters $\epsilon = 0.01$, $\kappa_2 = 3.0$, $\kappa_1 = 1.0$ and the initial condition $\vartheta_{j1}(0) = \zeta_j(0)$.

We provide two simulation studies: i) the desired trajectories are chosen as $\zeta_{2d} = 0.3t$ rad, $\zeta_2(0) = 0.1$m, $\theta(0) = -0.2$ rad, $\dot{\theta}(0) = 0.0$ rad/s, $\theta_{1d} = 0$ rad, $\theta_1(0) = -\pi/180$, $\dot{\theta}_1(0) = 0.0$ rad/s, the initial velocity for the mobile base is 0.0m/s; ii) the desired trajectories are chosen as $\zeta_{2d} = 0.3t$ rad and $\zeta_2(0) = 0.2$m, $\theta(0) = 0.2$ rad, $\dot{\theta}(0) = 0.0$ rad/s, $\theta_{1d} = 0.0$ rad, $\theta_1(0) = \pi/90$, $\dot{\theta}_1(0) = -0.2$ rad/s, the initial velocity for the mobile base is 0.0 m/s.

For the case i), the under-actuated joint (joint 1) is with zero initial velocity and we use the output feedback adaptive control to control the system. The position tracking using the output feedback adaptive control is shown in Figs. 6.2-6.3. The input torques are shown in Fig. 6.4. The final velocities for three variables are shown in Fig. 6.5. The produced trajectory is shown in Fig. 6.6. The nonholonomic constraint force of the mobile base is shown in Fig. 6.7. From these figures, we can see that v, and θ_1 are stable, and the motion on the mobile base is stable in a bounded region.

For the case ii), the under-actuated joint (joint 1) is with non-zero initial velocity. Similar to the above description, we utilize the output feedback adaptive control. The position tracking using the output feedback adaptive control

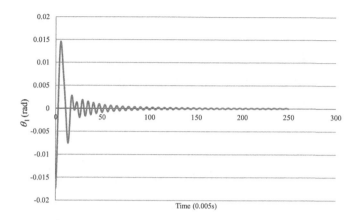

FIGURE 6.3

Tracking the desired angle of θ_1 by output feedback control.

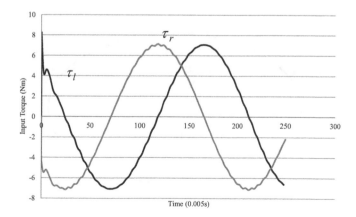

FIGURE 6.4

Input torques by output feedback control.

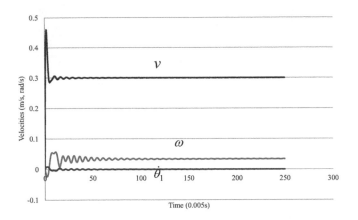

FIGURE 6.5

The velocities v, $\dot{\theta}$ and $\dot{\theta}_1$ produced by output feedback.

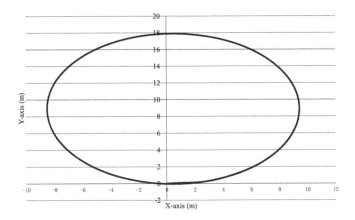

FIGURE 6.6

The produced bounded trajectory of the mobile manipulator.

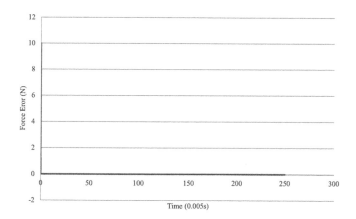

FIGURE 6.7

The nonholonomic constraint force.

is shown in Figs. 6.8-6.9. The input torques are shown in Fig. 6.10. The final velocities for three variables are shown in Fig. 6.11. The produced trajectory is shown in Fig. 6.12. The nonholonomic constraint force on the mobile base is shown in Fig. 6.13. From these figures, we can see that v and θ_1 are stable, and the motion of the mobile base is also stable in a bounded region.

From two examples, the simulation results show the proposed controls by the output feedback achieve good performance and are more realistic in practice, which validates the effectiveness of the control law in Theorem 6.1.

In both simulation examples, we do not know the boundedness of the system's parameters or the disturbances from the environment beforehand and do not reply on the derivative of states information by the additional sensors. However, it can be seen that the motion errors converge to zero and the zero dynamics of subsystem remains a bounded region simultaneously. Therefore, the proposed adaptive coupling control scheme attains good control performance even if the coupling dynamics is unknown beforehand. The tracking errors of control are small enough because of "adaptive" mechanisms from the simulation results. The simulations thus demonstrate the effectiveness of the proposed control in the presence of unknown nonlinear dynamic systems

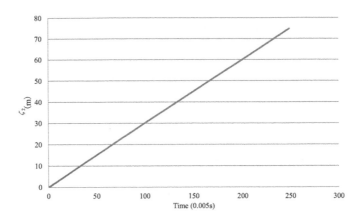

FIGURE 6.8

Tracking the desired position by output feedback control.

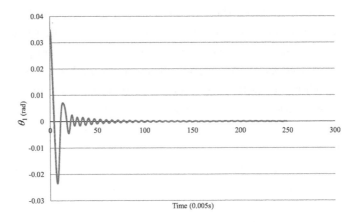

FIGURE 6.9

Tracking the desired angle of θ_1 by output feedback control.

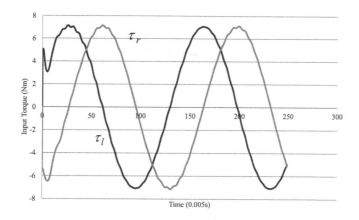

FIGURE 6.10

Input torques by output feedback control.

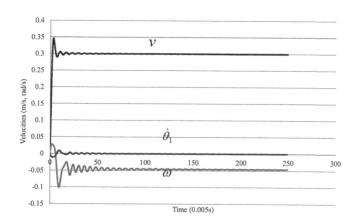

FIGURE 6.11

The velocities v, $\dot{\theta}$ and $\dot{\theta}_1$ produced by output feedback.

and environments. Different motion tracking performance can be achieved by adjusting control gains.

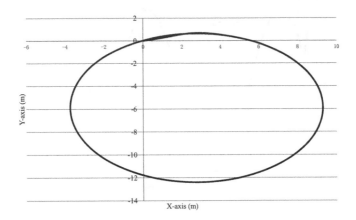

FIGURE 6.12

The produced bounded trajectory of the mobile manipulator.

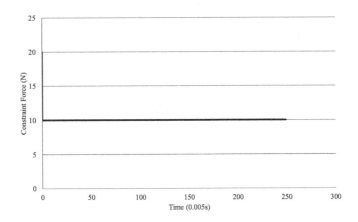

FIGURE 6.13

The nonholonomic constraint force.

6.6 Conclusion

In this chapter, adaptive motion/force control design by dynamic coupling and output feedback are first investigated for dynamic balance and desired trajectories tracking of mobile manipulator with kinematic nonholonomic constraints and dynamic nonholonomic constraints in the presence of unmodelled dynamics, or parametric/functional uncertainties. The controller is mathematically shown to guarantee semi-globally uniformly bounded stability, and the steady state compact sets to which the closed loop error signals for motion and force convergence are derived. The size of compact sets can be made small through appropriate choice of control design parameters. Simulation results demonstrate that the system is able to track reference signals satisfactorily, with all closed loop signals uniformly bounded.

7

Coordination Control

CONTENTS

Coordinated controls of multiple mobile manipulators have attracted the attention of many researchers [172], [174], [169]. Interest in multiple mobile manipulators stems from the better capability of the mobile manipulators in carrying out more complicated and dextrous tasks which may not be accomplished by a single mobile manipulator. It is an important technology for applying cooperative mobile manipulators to modern factories for transporting materials, and dangerous fields for dismantling bombs or moving nuclear-contaminated objects. Controls of multiple mobile manipulators present a significant increase in complexity over the single mobile manipulator case; moreover, they are much more difficult and challenging than the controls of multiple robotic manipulators [170, 171]. The difficulties of the control design lie in the fact that when multiple mobile manipulators coordinate each other and they are subject to kinematic constraints, they will form a closed kinematic chain mechanism that imposes a set of kinematic constraints on the position and velocity of coordinated mobile manipulators. As a result, the degrees of freedom of the whole system decrease, and internal forces are generated which need to be controlled. Therefore, for the above problems, we propose two coordination control approaches: centralized control and decentralized control in this chapter.

7.1 Centralized Coordination Control

7.1.1 Introduction

Multiple mobile manipulators in cooperation can complete tasks which are impossible for a single robot. For example, in [175], when a long and heavy object is carried, multiple cooperative robots reduce the weight and moment per robot. With the assumption of known dynamics, research on the control of coordination or cooperation has been carried out for multiple mobile manipulator systems. In [68], four methods previously developed for fixed base manipulation have been extended to mobile manipulation systems, including the operational space formulation, dextrous dynamic coordination of macro/mini structures, the augmented object model, the virtual linkage model, and then a decentralized control structure was proposed for cooperative tasks. In [172], for the tasks that require grasping and transporting large and flexible objects, decentralized coordination control was proposed for multiple mobile robots such that the robots can autonomously transport objects in a tightly controlled formation. In [173] and [174], modeling and centralized coordinating control were developed to handle deformable materials by multiple nonholonomic mobile manipulators operating in environments with obstacles. However, the internal forces existing in handling a rigid object are simplified. In [176], the same leader-follower type coordination motion control was proposed for multiple mobile robots engaging in cooperative tasks.

However, the above works seldom consider parametric uncertainties and unknown disturbances. Therefore, in this chapter, robust controls and robust adaptive controls [108, 65, 105, 66], will be investigated to deal with coordinated multiple mobile manipulators. We propose robust controls and robust adaptive controls for multiple mobile manipulators in cooperation carrying a common object in the presence of parametric uncertainties and unknown disturbances. A concise dynamics consisting of the dynamics of the mobile manipulators and the geometrical constraints between the end-effectors and

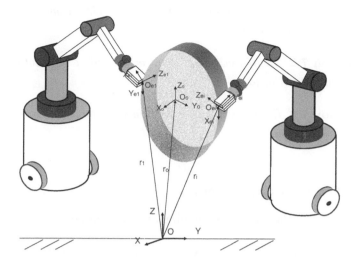

FIGURE 7.1

Two coordinated robots.

the object will be developed. Subsequently, robust controls and adaptive robust controls are designed for the coordinated multiple mobile manipulators to compensate for parametric uncertainties and suppress the bounded disturbances. The controls guarantee the outputs of the dynamic system to track some bounded hybrid signals which subsequently drive the kinematic system to the desired trajectory whereas the internal force tracking error remains bounded and can be made arbitrarily small. Feedback control design and stability analysis are performed via explicit Lyapunov techniques.

7.1.2   System Description and Assumptions

Consider m mobile manipulators holding a common rigid object in a task space. As shown in Fig. 7.1, different coordinate frames have been established for system modeling, in which $OXYZ$ is the inertial reference frame in which the position and orientation of the mobile manipulator end-effectors and the object are referred, $O_oX_oY_oZ_o$ is the object coordinate frame fixed at the center of mass of the object, and $O_{ie}X_{ie}Y_{ie}Z_{ie}$ is the end-effector frame of the ith manipulator located at the grasp point. To facilitate the dynamic formulation, the following assumptions are made.

Assumption 7.1 *All the end-effectors of the manipulators are rigidly at-*

tached to the common object so that there is no relative motion between the object and the end-effectors.

Assumption 7.2 *The object is rigid, that is, the object does not get deformed with the application of forces.*

Assumption 7.3 *Each manipulator is non-redundant and operating away from any singularity.*

7.1.3 Dynamics of System

The dynamics of the ith mobile manipulator in joint space are given by

$$M_i(q_i)\ddot{q}_i + C_i(q_i, \dot{q}_i)\dot{q}_i + G_i(q_i) + d_i = B_i(q_i)\tau_i + J_i^T f_i \tag{7.1}$$

where $q_i = [q_{i1}, \ldots, q_{in}]^T = [q_{iv}, q_{ia}]^T \in \mathbf{R}^n$ with $q_{iv} \in \mathbf{R}^{n_v}$ describing the generalized coordinates for the mobile platform and $q_{ia} \in \mathbf{R}^{n_a}$ denoting the generalized coordinates of the manipulator, and $n = n_v + n_a$. The symmetric positive definite inertia matrix $M_i(q_i) \in \mathbf{R}^{n \times n}$, the centripetal and Coriolis torques $C_i(\dot{q}_i, q_i) \in \mathbf{R}^{n \times n}$, the gravitational torque vector $G_i(q_i) \in \mathbf{R}^n$, the external disturbances $d_i(t) \in \mathbf{R}^n$ and the control inputs $\tau_i \in \mathbf{R}^k$ could be represented as

$$M_i(q_i) = \begin{bmatrix} M_{iv} & M_{iva} \\ M_{iav} & M_{ia} \end{bmatrix}, C_i(\dot{q}_i, q_i) = \begin{bmatrix} C_{iv} & C_{iva} \\ C_{iav} & C_{ia} \end{bmatrix}, G_i(q_i) = \begin{bmatrix} G_{iv} \\ G_{ia} \end{bmatrix}$$

$$d_i(t) = \begin{bmatrix} d_{iv}(t) \\ d_{ia}(t) \end{bmatrix}, \tau_i = \begin{bmatrix} \tau_{iv} \\ \tau_{ia} \end{bmatrix}, J_i = \begin{bmatrix} A_i & 0 \\ J_{iv} & J_{ia} \end{bmatrix}, f_i = \begin{bmatrix} f_{iv} \\ f_{ie} \end{bmatrix}$$

$B_i(q_i) = \text{diag}[B_{iv}, B_{ia}] \in \mathbf{R}^{n \times k}$ is a full rank input transformation matrix for the mobile platform and the robotic manipulator and assumed to be known because it is a function of fixed geometry of the system; $A_i = [A_{i1}^T(q_{iv}) \ldots, A_{il}^T(q_{iv})]^T : \mathbf{R}^{n_v} \to \mathbf{R}^{l \times n_v}$ is the kinematic constraint matrix which is assumed to have full rank l; $J_i^T \in \mathbf{R}^{n \times n}$ is a Jacobian matrix; and f_{iv} and f_{ie} are the constraint forces corresponding to the nonholonomic and holonomic constraints.

The mobile platform is subjected to nonholonomic constraints and the l non-integrable and independent velocity constraints can be expressed as

$$A_i(q_{iv})\dot{q}_{iv} = 0 \tag{7.2}$$

In the chapter, the mobile platform is assumed to be a completely nonholonomic constraint, and the holonomic constraint force is measured by the force sensor mounted on each mobile manipulator's end-effector. The l non-integrable and independent velocity constraints on the mobile platform can be viewed as restricting the dynamics on the manifold Ω_n

$$\Omega_n = \{(q_{iv}, \dot{q}_{iv}) | A_i(q_{iv})\dot{q}_{iv} = 0\} \tag{7.3}$$

Assume that the annihilator of the co-distribution spanned by the co-vector fields $A_{i1}(q_{iv})$, ..., $A_{il}(q_{iv})$ is an $(n_v - l)$-dimensional smooth nonsingular distribution Δ on $\mathbf{R}^{n_v}$. This distribution Δ is spanned by a set of $(n_v - l)$ smooth and linearly independent vector fields $H_{i1}(q_{iv})$, ..., $H_{i(n_v-l)}(q_{iv})$, i.e., $\Delta = \text{span}\{H_{i1}(q_{iv}), \ldots, H_{i(n_v-l)}(q_{iv})\}$. Thus, we have $H_i^T(q_{iv})A_i^T(q_{iv}) = 0$, $H_i(q_{iv}) = [H_{i1}(q_{iv}), \ldots, H_{i(n_v-l)}(q_{iv})] \in \mathbf{R}^{n_v \times (n_v-l)}$. Note that $H_i^T H_i$ is of full rank. Constraint (7.2) implies the existence of vector $\dot{\eta}_i \in \mathbf{R}^{n_v-l}$, such that

$$\dot{q}_{iv} = H_i(q_{iv})\dot{\eta}_i \tag{7.4}$$

Considering the nonholonomic constraints (7.2) and (7.4) and their derivative, the dynamics of a mobile manipulator (7.1) can be expressed as

$$M_i^1(\zeta_i)\ddot{\zeta}_i + C_i^1(\zeta_i, \dot{\zeta}_i)\dot{\zeta}_i + G_i^1 + d_i^1 = u_i + J_{ie}^T f_{ie} \tag{7.5}$$

where

$$M_i^1 = \begin{bmatrix} H_i^T M_{iv} H_i & H_i^T M_{iva} \\ M_{iav} H_i & M_{ia} \end{bmatrix}, \zeta_i = \begin{bmatrix} \eta_i \\ q_{ia} \end{bmatrix}, G_i^1 = \begin{bmatrix} H_i^T G_{iv} \\ G_{ia} \end{bmatrix},$$

$$C_i^1 = \begin{bmatrix} H_i^T M_{iv} \dot{H}_i + H_i^T C_{iv} H_i & H_i^T C_{iva} \\ M_{iav} \dot{H}_i + C_{iav} H_i & C_{ia} \end{bmatrix}, J_{ie} = \begin{bmatrix} 0 & 0 \\ J_{iv} H_i & J_{ia} \end{bmatrix}$$

$$u_i = B_i^1 \tau_i, B_i^1 = \begin{bmatrix} H_i^T B_{iv} & 0 \\ 0 & B_{ia} \end{bmatrix}, d_i^1 = \begin{bmatrix} H_i^T d_{iv} \\ d_{ia} \end{bmatrix}$$

The dynamics of m mobile manipulators from (7.5) can be expressed concisely as

$$M(\zeta)\ddot{\zeta} + C(\zeta, \dot{\zeta})\dot{\zeta} + G + D = U + J_e^T F_e \tag{7.6}$$

where $M(\zeta) = \text{block diag } [M_1^1(\zeta_1), \ldots, M_m^1(\zeta_m)] \in \mathbf{R}^{m(n-l) \times m(n-l)}$; $\zeta = [\zeta_1, \ldots, \zeta_m]^T \in \mathbf{R}^{m(n-l)}$; $U = [(B_1^1 \tau_1)^T, \ldots, (B_m^1 \tau_m)^T]^T \in \mathbf{R}^{m(n-l)}$; $G = [G_1^{1T}, \ldots, G_m^{1T}]^T \in \mathbf{R}^{m(n-l)}$; $F_e = [f_{1e}^T, \ldots, f_{me}^T]^T \in \mathbf{R}^{m(n-l)}$;

$C(\zeta, \dot\zeta) =$ block diag $[C_1^1(\zeta_1, \dot\zeta_1), \quad \ldots, \quad C_m^1(\zeta_m, \dot\zeta_m)] \in \mathbf{R}^{m(n-l) \times m(n-l)};$
$D = [d_1^{1T}, \quad \ldots, \quad d_m^{1T}]^T \in \mathbf{R}^{m(n-l)}; \quad J_e^T =$ block diag $[J_{1e}^T, \quad \ldots, \quad J_{me}^T]^T \in \mathbf{R}^{m(n-l) \times m(n-l)}.$

Let $x_o \in \mathbf{R}^{n_o}$ be the position/orientation vector of the object, the equation of motion of the object is written by the resultant force vector $F_o \in \mathbf{R}^{n_o}$ acting on the center of mass of the object, the symmetric positive definite inertial matrix $M_o(x_o) \in \mathbf{R}^{n_o \times n_o}$ of the object, the Corioli and centrifugal matrix $C_o(x_o, \dot x_o) \in \mathbf{R}^{n_o \times n_o}$, and the gravitational force vector $G_o(x_o) \in \mathbf{R}^{n_o}$ as

$$M_o(x_o)\ddot x_o + C_o(x_o, \dot x_o)\dot x_o + G_o(x_o) = F_o \tag{7.7}$$

Remark 7.1 *In this chapter, under Assumption 7.3, the mobile manipulators are non-redundant. As such, the degrees of freedom for each mobile manipulator must be equal to the dimension of the operational coordinate of the object, that is, $n_o = n - l$.*

Define $J_o(x_o) \in \mathbf{R}^{m(n-l) \times n_o}$ as $J_o(x_o) = [J_{1o}^T(x_o), \quad \ldots, \quad J_{mo}^T(x_o)]^T$ with the Jacobian matrix $J_{io}(x_o)$ from the object frame $O_o X_o Y_o Z_o$ to the ith mobile manipulator's end-effector frame $O_{ie} X_{ie} Y_{ie} Z_{ie}$. Then F_o can be written as

$$F_o = -J_o^T(x_o)F_e \tag{7.8}$$

Given the resultant force F_o, the end-effector force F_e satisfying (7.8) can be decomposed into two orthogonal components, one that contributes to the motion of the object, and one that produces the internal force. This is clearly represented by the following equation [177]

$$F_e = -(J_o^T(x_o))^+ F_o - F_I \tag{7.9}$$

where $(J_o^T(x_o))^+ \in \mathbf{R}^{m(n-l) \times n_o}$ is the pseudo-inverse matrix of $J_o^T(x_o)$, $F_I \in \mathbf{R}^{m(n-l)}$ is the internal force vector in the null space of $J_o^T(x_o)$, i.e., satisfying

$$J_o^T(x_o)F_I = 0 \tag{7.10}$$

and from [67], F_I can be parameterized by the vector of Lagrangian multiplier λ_I as

$$F_I = (I - (J_o^T(x_o))^+ J_o^T(x_o))\lambda_I \tag{7.11}$$

Let $\mathcal{J}^T = I - (J_o^T(x_o))^+ J_o^T(x_o)$, where $\mathcal{J}^T \in \mathbf{R}^{m(n-l) \times n_o}$ is Jacobian matrix for the internal force and satisfies

$$J_o^T(x_o)\mathcal{J}^T = 0 \tag{7.12}$$

Substituting (7.7) into (7.10), we have

$$F_e = -(J_o^T(x_o))^+ (M_o(x_o)\ddot{x}_o + C_o(x_o, \dot{x}_o)\dot{x}_o + G_o(x_o)) - \mathcal{J}^T \lambda_I \qquad (7.13)$$

Let $x_{ie} \in \mathbf{R}^{n-l}$ denote the position and orientation vector of the ith end-effector. Then x_{ie} is related to ζ_i, and the Jacobian matrix $J_{ie}(\zeta_i)$ as

$$\dot{x}_{ie} = J_{ie}(\zeta_i)\dot{\zeta}_i \qquad (7.14)$$

and the relationship between $\dot{x}_{ie}$ and $\dot{x}_o$ is given by

$$\dot{x}_{ie} = J_{io}(x_o)\dot{x}_o \qquad (7.15)$$

After combining (7.14) and (7.15), the following relationship between the joint velocity of the ith manipulator and the velocity of the object is obtained

$$J_{ie}(\zeta_i)\dot{\zeta}_i = J_{io}(x_o)\dot{x}_o \qquad (7.16)$$

As it is assumed that the manipulators work in a nonsingular region, the inverse of the Jacobian matrix $J_{ie}(\zeta_i)$ exists. Considering all the manipulators acting on the object at the same time yields

$$\dot{\zeta} = J_e^{-1}(\zeta)J_o(x_o)\dot{x}_o \qquad (7.17)$$

Differentiating (7.17) with respect to time t leads to

$$\ddot{\zeta} = J_e^{-1}(\zeta)J_o(x_o)\ddot{x}_o + \frac{d}{dt}(J_e^{-1}(\zeta)J_o(x_o))\dot{x}_o \qquad (7.18)$$

Using equations (7.17) and (7.18), the dynamics of multiple manipulators system (7.6), coupled with the object dynamics (7.7), are then given by

$$M_o(x_o)\ddot{x}_o + C_o(x_o, \dot{x}_o)\dot{x}_o + G_o(x_o) = F_o \qquad (7.19)$$

$$
\begin{aligned}
&M(\zeta)J_e^{-1}(\zeta)J_o(x_o)\ddot{x}_o \\
&+ \left(M(\zeta)\frac{d}{dt}(J_e^{-1}(\zeta)J_o(x_o)) + C(\zeta, \dot{\zeta})J_e^{-1}(\zeta)J_o(x_o) \right)\dot{x}_o \\
&+ G + D + J_e^T(\zeta)(J_o^T(x_o))^+ F_o + J_e^T(\zeta)\mathcal{J}\lambda_I = U
\end{aligned} \qquad (7.20)
$$

Combining (7.20) with (7.19) and multiplying both sides of (7.20) by $J_o^T(x_o)J_e^{-T}(\zeta)$, and using $J_o^T(\zeta)\mathcal{J}^T = 0$, the dynamics of multiple mobile manipulators system (7.6) with the object dynamics (7.7) are given by

$$\mathcal{M}\ddot{x}_o + \mathcal{C}\dot{x}_o + \mathcal{G} + \mathcal{D} = \mathcal{U} \qquad (7.21)$$

$$\lambda_I = Z(U - C^*\dot{x}_o - G^* - D) \qquad (7.22)$$

where

$$
\begin{aligned}
L &= J_e^{-1}(\zeta)J_o(x_o) \\
\mathcal{M} &= L^T M(\zeta)L + M_o(x_o) \\
\mathcal{C} &= L^T M(\zeta)\dot{L} + L^T C(\zeta,\dot{\zeta})L + C_o(x_o,\dot{x}_o) \\
\mathcal{G} &= L^T G(\zeta) + G_o(x_o) \\
\mathcal{D} &= L^T D \\
\mathcal{U} &= L^T U \\
M^* &= M(\zeta) + J_e^T(\zeta)(J_o^T(x_o))^+ M_o(x_o)(J_o(x_o))^+ J_e(\zeta) \\
Z &= (\mathcal{J}J_e(\zeta)(M^*)^{-1}J_e^T(\zeta)\mathcal{J}^T)^{-1}\mathcal{J}J_e(\zeta)(M^*)^{-1} \\
C^* &= M(\zeta)\dot{L} + C(\zeta,\dot{\zeta})L + J_e^T(\zeta)(J_o^T(x_o))^+ C_o(x_o,\dot{x}_o) \\
G^* &= G(\zeta) + J_e^T(\zeta)(J_o^T(x_o))^+ G_o(x_o)
\end{aligned}
$$

The dynamic equation (7.21) has the following structure properties, which can be exploited to facilitate the control system design.

Property 7.1 *The matrix $\mathcal{M}$ is symmetric, positive definite, and bounded, i.e., $\lambda_{min}(\mathcal{M})I \leq \mathcal{M} \leq \lambda_{max}(\mathcal{M})I$, where $\lambda_{min}(\mathcal{M})$ and $\lambda_{max}(\mathcal{M})$ denote the minimum and maximum eigenvalues of $\mathcal{M}$.*

Property 7.2 *The matrix $\dot{\mathcal{M}}-2\mathcal{C}$ is skew-symmetric, that is, $r^T(\dot{\mathcal{M}}-2\mathcal{C})r = 0, \forall r \in \mathbf{R}^{n_o}$.*

Property 7.3 *All Jacobian matrices are uniformly bounded and uniformly continuous if ζ and x_o are uniformly bounded and continuous, respectively.*

Property 7.4 *There exist some finite positive constants $c_j > 0$ $(1 \leq j \leq 4)$ and finite non-negative constants $c_j \geq 0$ $(j = 5)$ such that $\forall x_o \in \mathbf{R}^{n_o}$, $\forall \dot{x}_o \in \mathbf{R}^{n_o}$, $\|\mathcal{M}\| \leq c_1$, $\|\mathcal{C}\| \leq c_2 + c_3\|\dot{x}_o\|$, $\|\mathcal{G}\| \leq c_4$, and $sup_{t\geq 0}\|\mathcal{D}\| \leq c_5$.*

7.1.4   Robust Control Design

Given a desired motion trajectory $x_{od}(t)$, a desired internal force λ_I^d, since the system is inter-connected, we can obtain the desired motion trajectory $q(t)$ and $q_d(t)$. Therefore, the trajectory and internal force tracking control is to determine a control law such that for any $(x_o(0), \dot{x}_o(0)) \in \Omega$, $x_o, \dot{x}_o, \lambda_I$ converges to a manifold specified as Ω where

$$
\Omega_d = \{(x_o, \dot{x}_o, \lambda_I)|x_o = x_{od}, \dot{x} = \dot{x}_{od}, \lambda_I = \lambda_I^d\} \tag{7.23}
$$

Time-varying positive function δ is defined in Assumption 5.2, which converges to zero as $t \to \infty$ and satisfies

$$\lim_{t\to\infty} \int_0^t \delta(\omega)d\omega = \rho < \infty$$

with finite constant ρ.

Assumption 7.4 *The desired reference trajectories $x_{od}(t)$ are assumed to be bounded and uniformly continuous, and have bounded and uniformly continuous derivatives up to the third order. The desired internal force λ_I^d is also bounded and uniformly continuous.*

Let $e_o = x_o - x_{od}$, $\dot{x}_{or} = \dot{x}_{od} - K_o e_o$, $r = \dot{e}_o + K_o e_o$ with K_o diagonal positive definite and $e_I = \lambda_I - \lambda_I^d$.

Decoupled generalized position and constraint force are introduced separately. Considering the control input U as the form:

$$U = U_a + J_e^T(\zeta)\mathcal{J}^T U_b \tag{7.24}$$

then, (7.21) and (7.22) may be changed to

$$\mathcal{M}\ddot{x}_o + \mathcal{C}\dot{x}_o + \mathcal{G} + \mathcal{D} = L^T U_a \tag{7.25}$$

$$\lambda_I = Z(U_a - C^*\dot{x}_o - G^* - D) + U_b \tag{7.26}$$

Consider the following control laws:

$$L^T U_a = -K_p r - \frac{r\Phi^2}{\|r\|\Phi + \delta} \tag{7.27}$$

$$U_b = -\frac{\chi^2}{\chi + \delta}\ddot{x}_{od} + \lambda_I^d - K_f e_I \tag{7.28}$$

where

$$\Phi = C^T \Psi \tag{7.29}$$

$$\chi = c_1\|Z\|\|(L^T)^+\| \tag{7.30}$$

$$\Psi = [\|\ddot{x}_{or}\| \ \|\dot{x}_{or}\| \ \|\dot{x}_o\|\|\dot{x}_{or}\| \ 1 \ 1]^T \tag{7.31}$$

with $C = [c_1 \ c_2 \ c_3 \ c_4 \ c_5]^T$; K_p and K_f are positive definite, and $(L^T)^+ = J_e^T(\zeta)(J_o^T(x_o))^+$. Define $\nu = \lambda_{min}(K_p)/\lambda_{max}(\mathcal{M}) > 0$.

Theorem 7.1 *Considering the mechanical system described by (7.5) and using the control law (7.27) and (7.28), the following holds for any $(x_o(0), \dot{x}_o(0)) \in \Omega$:*

(i) r converges to a small set containing the origin with the convergence rate at least as fast as $e^{-\nu t}$;

(ii) e_o and $\dot{e}_o$ asymptotically converge to 0 as $t \to \infty$; and

(iii) e_I and τ are bounded for all $t \geq 0$.

Proof (i) Integrating (7.24) into (7.25), the closed-loop system dynamics can be rewritten as

$$\mathcal{M}\dot{r} = L^T U_a - (\mathcal{M}\ddot{x}_{or} + \mathcal{C}\dot{x}_{or} + \mathcal{G} + \mathcal{D}) - \mathcal{C}r \qquad (7.32)$$

Substituting (7.27) into (7.32), the closed-loop dynamic equation is obtained

$$\mathcal{M}\dot{r} = -K_p r - \frac{r\Phi^2}{\Phi||r|| + \delta} - \mu - C^2 r \qquad (7.33)$$

where $\mu = \mathcal{M}\ddot{x}_{or} + \mathcal{C}\dot{x}_{or} + \mathcal{G} + \mathcal{D}$.

Consider the Lyapunov function candidate as

$$V = \frac{1}{2} r^T \mathcal{M} r \qquad (7.34)$$

then

$$\dot{V} = r^T(\mathcal{M}\dot{r} + \frac{1}{2}\dot{\mathcal{M}}r) \qquad (7.35)$$

From Property 7.1, we have $\frac{1}{2}\lambda_{min}(\mathcal{M})r^T r \leq V \leq \frac{1}{2}\lambda_{max}(\mathcal{M})r^T r$. By using Property 7.2, the time derivative of V along the trajectory of (7.33) is

$$
\begin{aligned}
\dot{V} &= -r^T K_p r - r^T \mu - \frac{||r||^2 \Phi^2}{\Phi||r|| + \delta} \\
&\leq -r^T K_p r + ||r||\Phi - \frac{||r||^2 \Phi^2}{\Phi||r|| + \delta} \\
&\leq -r^T K_p r + \delta \\
&\leq -\lambda_{min}(K_p)||r||^2 + \delta
\end{aligned}
$$

Therefore, we arrive at $\dot{V} \leq -\nu V + \delta$. Thus, r converges to a set containing the origin with a rate at least at fast as $e^{-\nu t}$.

Integrating both sides of the above equation gives

$$V(t) - V(0) \leq -\int_0^t r^T K_p r ds + \rho < \infty \qquad (7.36)$$

Thus V is bounded, which implies that $r \in L_\infty^{n_o}$.

(ii) From (7.36), V is bounded, which implies that $x_o \in L_\infty^{n_o}$. We have $\int_0^t r^T K_p r ds \leq V(0) - V(t) + \rho$, which leads to $r \in L_2^{n_o}$. From $r = \dot{e}_o + K_o e_o$, it can be obtained that $e_o, \dot{e}_o \in L^{n_o}_\infty$. As we have established $e_o, \dot{e}_o \in L_\infty$, from Assumption 7.4, we conclude that $x_o(t), \dot{x}_o(t), \dot{x}_{or}(t), \ddot{x}_{or}(t) \in L^{n_o}_\infty$ and $\dot{q} \in L^{mn}_\infty$.

Therefore, all the signals on the right hand side of (7.32) are bounded, and we can conclude that $\dot{r}$ and therefore $\ddot{x}_o$ are bounded. Thus, $r \to 0$ as $t \to \infty$ can be obtained. Consequently, we have $e_o \to 0, \dot{e}_o \to 0$ as $t \to \infty$. It follows that $e_q, \dot{e}_q \to 0$ as $t \to \infty$.

(iii) Substituting the control (7.27) and (7.28) into the reduced order dynamic system model (7.26) yields

$$(I + K_f)e_I = Z(L^T)^+ \mathcal{M}\ddot{x}_o - \frac{\chi^2}{\chi + \delta}\ddot{x}_{od} \tag{7.37}$$

Since $\ddot{x}_o$ and Z are bounded, $x_o \to x_{od}$, $Z(L^T)^+\mathcal{M}\ddot{x}_o - \frac{\chi^2}{\chi+\delta}\ddot{x}_{od}$ is also bounded and the size of e_I can be adjusted by choosing the proper gain matrix K_f.

Since r, x_o, $\dot{x}_o$, x_{or}, $\dot{x}_{or}$, $\ddot{x}_{or}$, and e_I are all bounded, it is easy to conclude that τ is bounded from (7.27) and (7.28).

7.1.5 Robust Adaptive Control Design

In developing control laws (7.27) and (7.28), $c_j, 1 \le j \le 5$ are supposed to be known. However, in reality, these constants cannot be obtained easily. Therefore, we develop a control law which does not require the knowledge of $c_j, 1 \le j \le 5$.

Consider the robust adaptive control law as

$$L^T U_a = -K_p r - \sum_{j=1}^{5} \frac{r\hat{c}_j \Psi_j^2}{\|r\|\Psi_j + \delta_j} \tag{7.38}$$

$$U_b = -\frac{\hat{\chi}^2}{\hat{\chi} + \delta_1}\ddot{x}_{od} + \lambda_I^d - K_f e_I \tag{7.39}$$

where

$$\dot{\hat{c}}_j = -\omega_j \hat{c}_j + \sum_{j=1}^{5} \frac{\gamma_j \Psi_j^2 \|r\|^2}{\|r\|\Psi_j + \delta_j}, \quad j = 1, \ldots, 5 \tag{7.40}$$

$$\hat{\chi} = \hat{c}_1 \|\hat{Z}\| \|(L^T)^+\| \tag{7.41}$$

$\hat{Z} = (\mathcal{J}J_e(\zeta)(\hat{M}^*)^{-1}J_e^T(\zeta)\mathcal{J}^T)^{-1}\mathcal{J}J_e(\zeta)(\hat{M}^*)^{-1}$ with $\|L^T\hat{M}^*L\| = \hat{c}_1$, K_p

and K_f are positive definite; $\gamma_j > 0$; $\delta_j > 0$ and $\omega_j > 0$ satisfy Definition 5.1:

$$\int_0^\infty \delta_j(s)ds = \rho_{j\delta} < \infty \tag{7.42}$$

$$\int_0^\infty \omega_j(s)ds = \rho_{j\omega} < \infty \tag{7.43}$$

with the constants $\rho_{j\delta}$ and $\rho_{j\omega}$.

Theorem 7.2 *Considering the mechanical system described by (7.5), using the control law (7.38) and adaptation law (7.39), the following holds for any $(x_o(0), \dot{x}_o(0)) \in \Omega$:*

  (i) r converges to a small set containing the origin as $t \to \infty$;

  (ii) e_o and $\dot{e}_o$ converge a small set containing the origin as $t \to \infty$; and

  (iii) e_I and τ are bounded for all $t \geq 0$.

Proof (i). Substituting (7.38) and (7.40) into (7.32), the closed loop dynamic equation is obtained

$$\mathcal{M}\dot{r} = -K_p r - \sum_{j=1}^{5} \frac{r\hat{c}_j \Psi_j^2}{\|r\|\Psi_j + \delta_j} - \mu - C^2 r \tag{7.44}$$

Consider the Lyapunov candidate function

$$V = \frac{1}{2}r^T \mathcal{M} r + \frac{1}{2}\tilde{C}^T \Gamma^{-1} \tilde{C} \tag{7.45}$$

with $\tilde{C} = C - \hat{C}$. Its derivative is

$$\dot{V} = r^T(\mathcal{M}\dot{r} + \frac{1}{2}\dot{\mathcal{M}}r) + \tilde{C}^T \Gamma^{-1} \dot{\tilde{C}} \tag{7.46}$$

where $\Gamma = \text{diag}[\gamma_j] > 0$, $j = 1, \ldots, 5$.

From Property 7.1, we have $\frac{1}{2}\lambda_{min}\mathcal{M}r^T r \leq \frac{1}{2}r^T \mathcal{M}r \leq \frac{1}{2}\lambda_{max}\mathcal{M}r^T r$. By using Property 7.2, the time derivative of V along the trajectory of (7.44) is

$$
\begin{aligned}
\dot{V} &= -r^T K_p r - r^T \mu - r^T \sum_{j=1}^{5} \frac{r\hat{c}_j \Psi_j^2}{\|r\|\Psi_j + \delta_j} + \hat{C}^T \Omega \Gamma^{-1} \tilde{C} - \sum_{j=1}^{5} \frac{\|r\|^2 \tilde{c}_j \Psi_j^2}{\|r\|\Psi_j + \delta_j} \\
&\leq -r^T K_p r + \|r\|\Phi - \sum_{j=1}^{5} \frac{\|r\|^2 c_j \Psi_j^2}{\|r\|\Psi_j + \delta_j} + \hat{C}^T \Omega \Gamma^{-1} \tilde{C} \\
&\leq -r^T K_p r + C^T \Delta + \tilde{C}^T \Omega \Gamma^{-1} \hat{C} \\
&= -r^T K_p r + C^T \Delta + \tilde{C}^T \Omega \Gamma^{-1}(C - \tilde{C})
\end{aligned}
$$

$$= -r^T K_p r + C^T \Delta - \frac{1}{4} C^T \Omega \Gamma^{-1} C + \frac{1}{4} C^T \Omega \Gamma^{-1} C$$
$$+ \tilde{C}^T \Omega \Gamma^{-1} C - \tilde{C}^T \Omega \Gamma^{-1} \tilde{C}$$
$$= -r^T K_p r + C^T \Delta - (\frac{1}{2} C^T - \tilde{C}^T) \Omega \Gamma^{-1} (\frac{1}{2} C - \tilde{C}) + \frac{1}{4} C^T \Omega \Gamma^{-1} C$$
$$\leq -r^T K_p r + C^T \Delta + \frac{1}{4} C^T \Omega \Gamma^{-1} C$$

with $\Omega = \text{diag}[\omega_j]$ with $j = 1, \ldots, 5$, $\Delta = [\delta_1, \delta_2, \ldots, \delta_5]^T$.

Therefore, $\dot{V} \leq -\lambda_{min}(K_p) \|r\|^2 + C^T \Delta + \frac{1}{4} C^T \Omega \Gamma^{-1} C$. Since $C^T \Delta + \frac{1}{4} C^T \Omega \Gamma^{-1} C \to 0$ as $t \to \infty$. Noting (7.42) and (7.43) and $\dot{V} \leq 0$, from above, r converges to a small set containing the origin as $t \to \infty$.

(ii) Integrating both sides of the above equation gives

$$V(t) - V(0) \leq -\int_0^t r^T K_p r ds + \int_0^t (C^T \Delta + \frac{1}{4} C^T \Omega \Gamma^{-1} C) ds \quad (7.47)$$

Since C and Γ are constant, $\int_0^\infty \Delta ds = \rho_\delta = [\rho_{1\delta}, \ldots, \rho_{5\delta}]^T$, $\int_0^\infty \Omega ds = \rho_\omega = [\rho_{1\omega}, \ldots, \rho_{5\omega}]^T$, we can rewrite (7.47) as

$$V(t) - V(0) \leq -\int_0^t r^T K_p r ds + C^T \left(\int_0^t \Delta ds \right) + \frac{1}{4} C^T \left(\int_0^t \Omega ds \right) \Gamma^{-1} C ds$$
$$< -\int_0^t r^T K_p r ds + C^T \rho_\delta + C^T \rho_\omega \Gamma^{-1} C < \infty \quad (7.48)$$

Thus V is bounded, which implies that $r \in L_\infty^{n_o}$. From (7.48), we have

$$\int_0^t r^T K_p r ds < V(0) - V(t) + C_r^T \rho_\delta + C^T \rho_\omega \Gamma^{-1} C \quad (7.49)$$

which leads to $r \in L_2^{n_o}$. From $r = \dot{e}_o + K_o e_o$, it can be obtained that $e_o, \dot{e}_o \in L_\infty^{n_o}$. As we have established $e_o, \dot{e}_o \in L_\infty$, from Assumption 7.4, we conclude that $x_o(t), \dot{x}_o(t), \dot{x}_{or}(t), \ddot{x}_{or}(t) \in L_\infty^{n_o}$ and $\dot{q} \in L_\infty^{mn}$.

Therefore, all the signals on the right hand side of (7.44) are bounded, and we can conclude that $\dot{r}$ and therefore $\ddot{x}_o$ are bounded. Thus, $r \to 0$ as $t \to \infty$ can be obtained. Consequently, we have $e_o \to 0, \dot{e}_o \to 0$ as $t \to \infty$. It follows that $e_q, \dot{e}_q \to 0$ as $t \to \infty$.

(iii) Substituting the control (7.38) and (7.39) into the reduced order dynamics (7.26) yields

$$(I + K_f) e_I = Z(L^T)^+ \mathcal{M} \ddot{x}_o - \frac{\hat{\chi}^2}{\hat{\chi} + \delta_1} \ddot{x}_{od} \quad (7.50)$$

Since $\ddot{x}_o$ and Z are bounded, $x_o \to x_{od}$, $Z(L^T)^+ \mathcal{M} \ddot{x}_o - \frac{\hat{\chi}^2}{\hat{\chi} + \delta_1} \ddot{x}_{od}$ is also

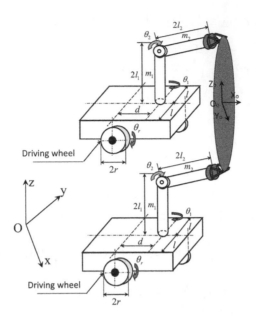

FIGURE 7.2

Two coordinated mobile manipulators.

bounded, the size of e_I can be adjusted by choosing the proper gain matrix K_f.

Since r, x_o, $\dot{x}_o$, x_{or}, $\dot{x}_{or}$, $\ddot{x}_{or}$, and e_I are all bounded, it is easy to conclude that τ is bounded from (7.38) and (7.39).

7.1.6   Simulation Studies

Let us consider two same 2-DOF mobile manipulators shown in Fig. 7.2. Each mobile manipulator is subjected to the following constraints:

$$\dot{x}_i \sin \theta_i - \dot{y}_i \cos \theta_i = 0$$

Using the Lagrangian approach, we can obtain the standard form (7.1) with $q_{iv} = [x_i \; y_i \; \theta_i]^T$, $q_{ia} = [\theta_{i1} \; \theta_{i2}]^T$, $q_i = [q_{iv} \; q_{ia}]^T$, $A_{iv} = [\sin \theta_i \; -\cos \theta \; 0.0]$, and $\zeta = [y_i \; \theta_i \; \theta_{i1} \; \theta_{i2}]^T$, $\dot{\zeta} = [\dot{y}_i \; \dot{\theta}_i \; \dot{\theta}_{i1} \; \dot{\theta}_{i2}]^T$ where y_i and $\dot{y}_i$ are the displacement and velocity of Y direction of the platform center of ith manipulator, respectively. The dynamics of the ith mobile manipulator are given by (9.22) in Section 9.1.

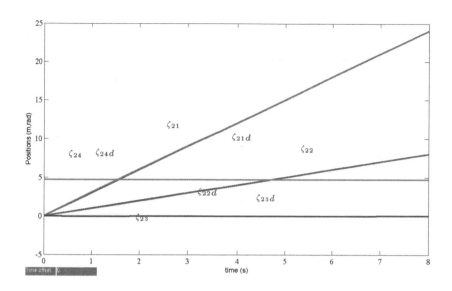

FIGURE 7.3

The joint positions of mobile manipulator I.

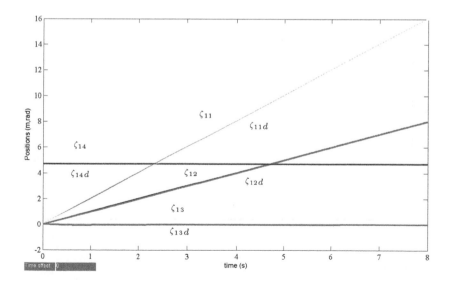

FIGURE 7.4

The joint velocities of mobile manipulator I.

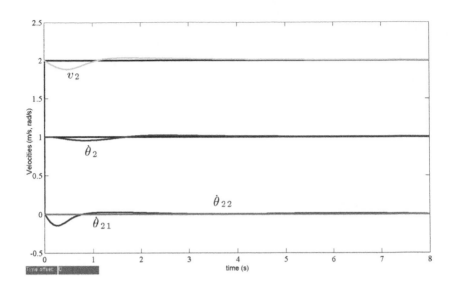

FIGURE 7.5

The joint torques of mobile manipulator I.

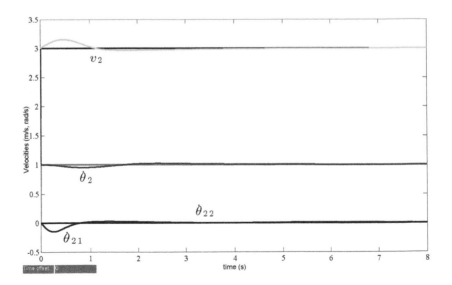

FIGURE 7.6

The joint positions of mobile manipulator II.

The position of end-effector can be given by

$$
\begin{aligned}
x_{ie} &= x_{if} - 2l_2 \sin\theta_{i2}\cos(\theta_i + \theta_{i1}) \\
y_{ie} &= y_{if} - 2l_2 \sin\theta_{i2}\sin(\theta_i + \theta_{i1}) \\
z_{ie} &= 2l_1 - 2l_2 \cos\theta_{i2} \\
\beta_{ie} &= \theta_i + \theta_{i1}
\end{aligned}
$$

where β_{ie} is the pitch angle for the ith end-effector.

For the mobile platform, we have

$$
\begin{bmatrix} \dot{x}_{if} \\ \dot{y}_{if} \\ \dot{\theta}_i \end{bmatrix} =
\begin{bmatrix} 1.0 & 0.0 & -d\sin\theta_i \\ 0.0 & 1.0 & d\cos\theta_i \\ 0.0 & 0.0 & 1.0 \end{bmatrix}
\begin{bmatrix} \dot{x}_i \\ \dot{y}_i \\ \dot{\theta}_i \end{bmatrix}
$$

and

$$
\begin{bmatrix} \dot{x}_i \\ \dot{y}_i \\ \dot{\theta}_i \end{bmatrix} =
\begin{bmatrix} \cos\theta_i & 0.0 \\ \sin\theta_i & 0.0 \\ 0.0 & 1.0 \end{bmatrix}
\begin{bmatrix} v_i \\ \dot{\theta}_i \end{bmatrix}
$$

Therefore, the mobile manipulator Jacobian matrix J_{ie} is given by

$$
\begin{bmatrix} \dot{x}_{ie} \\ \dot{y}_{ie} \\ \dot{z}_{ie} \\ \dot{\beta}_{ie} \end{bmatrix} =
\begin{bmatrix}
J_{i11} & J_{i12} & J_{i13} & J_{i14} \\
J_{i21} & J_{i22} & J_{i23} & J_{i24} \\
J_{i31} & J_{i32} & J_{i33} & J_{i34} \\
J_{i41} & J_{i42} & J_{i43} & J_{i44}
\end{bmatrix}
\begin{bmatrix} v_i \\ \dot{\theta}_i \\ \dot{\theta}_{i1} \\ \dot{\theta}_{i2} \end{bmatrix}
$$

where

$$
\begin{aligned}
J_{i11} &= \cos\theta_i,\ J_{i12} = -d\sin\theta_i + 2l_2 \sin\theta_{i2}\sin(\theta_i + \theta_{i1}) \\
J_{i13} &= 2l_2 \sin\theta_{i2}\sin(\theta_i + \theta_{i1}),\ J_{i14} = -2l_2 \cos\theta_{i2}\cos(\theta_i + \theta_{i1}) \\
J_{i21} &= \sin\theta_i,\ J_{i22} = d\cos\theta_i - 2l_2 \sin\theta_{i2}\cos(\theta_i + \theta_{i1}) \\
J_{i23} &= -2l_2 \sin\theta_{i2}\cos(\theta_i + \theta_{i1}),\ J_{i24} = -2l_2 \cos\theta_{i2}\sin(\theta_i + \theta_{i1}) \\
J_{i31} &= 0.0,\ J_{i32} = 0.0,\ J_{i33} = 0.0,\ J_{i34} = 2l_2 \sin\theta_{i2} \\
J_{i41} &= 0.0,\ J_{i42} = 1.0,\ J_{i43} = 1.0,\ J_{i44} = 0.0
\end{aligned}
$$

Let the position $x_o = [x_{1o}, x_{2o}, x_{3o}, x_{4o}]^T$ be positions to X axis, Y axis, Z axis and rotation angle to Z axis as shown in Fig. 7.2 and Z_o is parallel to the Z axis. The dynamic equation of the object is given by

$$
M_o(x_o)\ddot{x}_o + G_o(x_o) = J_{1o}^T(x_o)f_{1e} + J_{2o}^T(x_o)f_{2e}
$$

where

$$M_o(x_o) = \begin{bmatrix} m_o & 0.0 & 0.0 & 0.0 \\ 0.0 & m_o & 0.0 & 0.0 \\ 0.0 & 0.0 & m_o & 0.0 \\ 0.0 & 0.0 & 0.0 & I_o \end{bmatrix},$$

$$J_{1o}^T(x_o) = \begin{bmatrix} 1.0 & 0.0 & 0.0 & 0.0 \\ 0.0 & 1.0 & 0.0 & 0.0 \\ 0.0 & 0.0 & 1.0 & 0.0 \\ l_{c1}\sin x_{4o} & -l_{c1}\cos x_{4o} & 0.0 & 1.0 \end{bmatrix}, G_o(x_o) = \begin{bmatrix} 0.0 \\ 0.0 \\ -m_o g \\ 0.0 \end{bmatrix},$$

$$f_{ie} = \begin{bmatrix} f_{ix} \\ f_{iy} \\ f_{iz} \\ \tau_{i\beta} \end{bmatrix}, J_{2o}^T(x_o) = \begin{bmatrix} 1.0 & 0.0 & 0.0 & 0.0 \\ 0.0 & 1.0 & 0.0 & 0.0 \\ 0.0 & 0.0 & 1.0 & 0.0 \\ -l_{c2}\sin x_{4o} & l_{c2}\cos x_{4o} & 0.0 & 1.0 \end{bmatrix}.$$

Therefore, we could obtain $\mathcal{J}^T$ using $\mathcal{J}^T = I - (J_o^T(x_o))^+ J_o^T(x_o)$. The force vector F_e can be measured using force sensors mounted on the end-effectors of the mobile robots. Therefore, from (7.8), we can compute F_o. Subsequently, using (7.9), we can obtain F_I and using (7.11), we can obtain λ_I.

$$F_I = \begin{bmatrix} \cos x_{4o} & -\sin x_{4o} & 0.0 \\ \sin x_{4o} & \cos x_{4o} & 0.0 \\ 0.0 & 0.0 & 0.0 \\ 0.0 & l_{c1} & 1 \\ -\cos x_{4o} & \sin x_{4o} & 0.0 \\ -\sin x_{4o} & -\cos x_{4o} & 0.0 \\ 0.0 & 0.0 & 0.0 \\ 0.0 & l_{c2} & -1 \end{bmatrix} \begin{bmatrix} \lambda_{Ix} \\ \lambda_{Iy} \\ \lambda_{I\beta} \end{bmatrix},$$

where λ_{Ix}, λ_{Iy} and $\lambda_{I\beta}$ present components of compression force, shearing force and bending moment respectively.

The desired trajectory for the object and the desired internal force are chosen as

$$\begin{bmatrix} \dot{x}_{1od} \\ \dot{x}_{2od} \\ \dot{x}_{3od} \\ \dot{x}_{4od} \end{bmatrix} = \begin{bmatrix} \cos(\sin t + \frac{\pi}{2}) \\ \sin(\sin t + \frac{\pi}{2}) \\ 0.0 \\ \cos t + 0.01\cos t \end{bmatrix}, \lambda_{Id} = \begin{bmatrix} 5.0 \\ 0.0 \\ 0.0 \end{bmatrix}$$

We can obtain the desired trajectory of each mobile manipulator when θ_{i2} is fixed as $\frac{3\pi}{2}$

$$\dot{\zeta}_{id} = \begin{bmatrix} \dot{y}_{id} \\ \dot{\theta}_{id} \\ \dot{\theta}_{i1d} \\ \dot{\theta}_{i2d} \end{bmatrix} = \begin{bmatrix} \sin(\sin t + \frac{\pi}{2}) \\ \cos t \\ 0.01 \cos t \\ 0.0 \end{bmatrix}$$

The initial positions are chosen as $x_o(0) = [0.0, 0.0, 2l_1, 0.0]^T$ and $\zeta_i(0) = [0.0, \frac{\pi}{2}, 0.0, \frac{3\pi}{2}]^T$, the initial velocities are chosen as $\dot{x}_o(0) = \dot{\zeta}_i(0) = 0$.

Assume that the parameters are selected as $m_p = 6.0kg$, $m_1 = m_2 = 1.0kg$, $I_{zp} = 19kgm^2$, $I_{z1} = I_{z2} = 1.0kgm^2$, $d = 0.0$, $l = r = 1.0m$, $2l_1 = 1.0m$, $2l_2 = 0.6m$, the mass of the object $m_o = 1.0kg$, $I_o = 1.0kgm^2$, $l_{c1} = l_{c2} = 0.5m$. The disturbances on mobile manipulators are introduced into the simulation model as $d_1 = [0.1\sin t, -0.2\sin t, -0.05\sin t, 0, 0]^T$, $d_2 = [0.2\sin t, -0.2\sin t, 0, 0, 0]^T$. By Theorem 7.2, the control gains are selected as $K_p = \text{diag}[20.0]$, $K_o = \text{diag}[1.0]$, and $K_f = 8000$, $\hat{C}(0) = [1.0, 1.0, 1.0, 1.0, 1.0]^T$, $\delta_i = \omega_i = 1/(1+t)^2$, $i = 1, \ldots, 5$, and $\Gamma = \text{diag}[1.0]$. The simulation results for two mobile manipulators are shown in Figs. 7.3-7.9 [169]. The figures show the motion control for mobile manipulator I, and internal force is shown Fig. 7.9. Figs. 7.6-7.8 show the motion control for mobile manipulator II. The simulation results show that the trajectory and internal force tracking errors tend to the desired values, which validates the effectiveness of the control law in Theorem 7.2. Under the proposed control scheme, tracking of the desired trajectory and desired internal force is achieved and this is largely due to the "adaptive" mechanism. The simulation results demonstrate the effectiveness of the proposed adaptive control in the presence of parametric uncertainties and external disturbances. Although parametric uncertainties and external disturbances are both introduced into the simulation model, the force/motion control performance of system, under the proposed control, is not degraded. Different motion/force tracking performance can be achieved by adjusting parameter adaptation gains and control gains.

7.2 Decentralized Coordination

Most previous studies on the coordination of multiple mobile manipulators systems only deal with motion tracking control without the interacting with

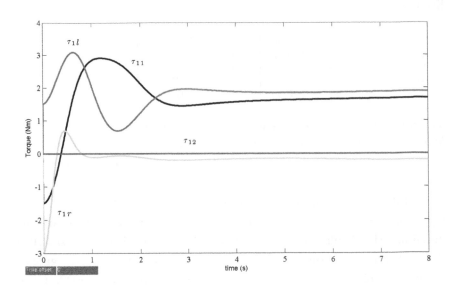

FIGURE 7.7

The joint velocities of mobile manipulator II.

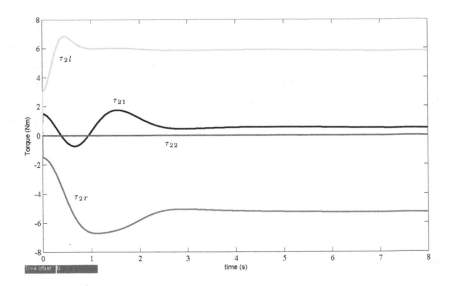

FIGURE 7.8

The joint torques of mobile manipulator II.

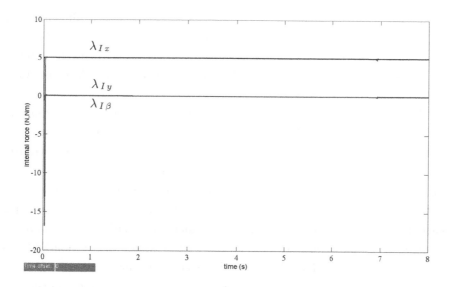

FIGURE 7.9
The internal force of mobile manipulator II.

the environment [172], [174], [175], [176], [169]. However, the large scale tasks, such as manufacturing and assembly in automatic factories and space explorations, often involve the situations where multiple robots are grasping an object in contact with environment, for example, scribing, painting, grinding, polishing, and contour following on a large scale. The purpose of controlling a coordinated system is to control the contact forces between the environment and object in the constrained directions and the motion of the object in unconstrained directions. The internal forces are produced within the grasped object, which do not contribute to system motion. The larger internal forces would damage the object so that we have to maintain internal forces at some desired values. Moreover, the model uncertainties and external disturbance of the robots and from the environment would disrupt the interaction. These problems will be investigated in this chapter.

In previous works, the coordination control of two mobile manipulators was investigated in [169], where the carried object did not interact with the non-rigid environments. The authors proposed the centralized control approach, by which a single controller is supposed to control all of the mobile manipulators in a centralized way. In [186], the centralized neuro-adaptive control was also proposed for robust force/motion tracking for multiple mobile manipulators in contact with a deformable working surface. However, it is difficult for a centralized controller to control many mobile manipulators in coordination. The centralized control creates a heavy computational burden, and limits a system's flexibility and reliability such that once the controller develops faults, the whole system is degraded.

The decentralized control system is believed to be more effective and suitable for multiple mobile robots cooperatively. In the literature, several decentralized control algorithms have been been reported using decentralized motion control [180], [181], [182] without introducing time-delay state information, in which each robot is controlled by its own controller. The interaction between the robots and environments can be found in many practical applications. Therefore, in this chapter, we consider the situations where multiple mobile manipulators are grasping an object in contact with a nonrigid surface, as shown in Fig. 7.10. The control objective of coordinated systems is to control the contact force between the environment and the object in the constrained direction, and the motion of the object in unconstrained directions while maintaining internal forces which do not contribute to system motion in some desired values.

A number of mobile manipulators in the real world are composed of a set of small interconnected subsystems. It is generally impossible to incorporate multiple feedback loops into the controller design, and it would be too costly even if they can be implemented. Therefore, decentralized control has been suggested to deal with the coordination of multiple mobile manipulators. First, we present the decentralize dynamics of an inter-connected system, which includes dynamics of each mobile manipulator as well as the object's dynamics; and the interaction between the object and the environment are presented. Decentralized adaptive robust control using local information is proposed for coordinated multiple mobile manipulators so as to suppress both local dynamic uncertainty and the nonlinear interconnections in each subsystem. The proposed schemes estimate unknown parameters of each subsystem, control object motion and contact force between object and the constraining surface, while maintaining the internal impedance forces in desired levels.

The presented control in this chapter could lead to a systematic robust decentralized design for each mobile manipulator to handle its own nonlinear parameterized dynamics. Given sufficiently smooth reference input, it can guarantee the convergence of the closed-loop system, and make the steady-state tracking error arbitrarily small.

7.2.1 System Description and Assumption

Consider m mobile manipulators holding a common rigid object in a task space with n degrees of freedom. As shown in Fig. 7.10, $OXYZ$ is the inertial reference frame. $O_oX_oY_oZ_o$ is the object coordinate frame fixed at the center of mass of the object. $O_{ei}X_{ei}Y_{ei}Z_{ei}$ is the end-effector frame of the ith manipulator located at the grasp point.

Assumption 7.5 *Each end-effector of the mobile manipulator shown in Fig. 7.11 is attached to the common object by a paired spring-damper system and a force sensor, and the distances and relative angles between the end-effectors can be obtained.*

Assumption 7.6 *Each manipulator is non-redundant and operating away from singularity.*

7.2.2 Dynamics of Interconnected System

The dynamics of the ith mobile manipulator in joint space are given by

$$M_i(q_i)\ddot{q}_i + C_i(q_i, \dot{q}_i)\dot{q}_i + G_i(q_i) + d_i(t) = B_i(q_i)(\tau_i - \tau_{if}(\dot{q})) + J_{ie}^T f_i \quad (7.51)$$

where $q_i = [q_{iv}^T, q_{ia}^T]^T \in \mathbb{R}^n$ with $q_{iv} \in \mathbb{R}^{n_v}$ describing the generalized coordinates for the mobile platform and $q_{ia} \in \mathbb{R}^{n_a}$ denoting the generalized coordinates of the manipulator, and $n = n_v + n_a$. The symmetric positive definite inertia matrix $M_i(q_i) \in \mathbb{R}^{n \times n}$, the centripetal and Coriolis torques $C_i(q_i, \dot{q}_i) \in \mathbb{R}^{n \times n}$, the gravitational torque vector $G_i(q_i) \in \mathbb{R}^n$, the external disturbances $d_i(t) \in \mathbb{R}^n$, the control inputs $\tau_i \in \mathbb{R}^{n_b}$, and the actuator friction force vector $\tau_{if}(\dot{q}) \in \mathbb{R}^{n_b}$ could be decoupled into the mobile base and the arm, and represented as

$$M_i(q) = \begin{bmatrix} M_{iv} & M_{iva} \\ M_{iav} & M_{ia} \end{bmatrix}, C_i(q_i, \dot{q}_i) = \begin{bmatrix} C_{iv} & C_{iva} \\ C_{iav} & C_{ia} \end{bmatrix}$$

$$G_i(q_i) = \begin{bmatrix} G_{iv} \\ G_{ia} \end{bmatrix}, d(t) = \begin{bmatrix} d_{iv}(t) \\ d_{ia}(t) \end{bmatrix}, \tau_i = \begin{bmatrix} \tau_{iv} \\ \tau_{ia} \end{bmatrix},$$

$$\tau_{if}(\dot{q}) = \begin{bmatrix} \tau_{ifv} \\ \tau_{ifa} \end{bmatrix} J_{ie} = \begin{bmatrix} A_i & 0 \\ J_{iv} & J_{ia} \end{bmatrix}, f_i = \begin{bmatrix} f_{iv} \\ f_{ia} \end{bmatrix}$$

$B_i(q_i) = \text{diag}[B_{iv}, B_{ia}] \in \mathbb{R}^{n \times n_b}$ is a full rank input transformation matrix for the mobile platform and the robotic manipulator and is assumed to be known because it is a function of fixed geometry of the system; $A_i = [A_{i1}^T(q_{iv}) \ \ldots, \ A_{il}^T(q_{iv})]^T : \mathbb{R}^{n_v} \rightarrow \mathbb{R}^{l \times n_v}$ is the kinematic constraint matrix which is assumed to have full rank l; $J_{ie}^T \in \mathbb{R}^{n \times n_J}$ is a Jacobian matrix; and f_{iv} and f_{ia} are the constraint forces corresponding to the nonholonomic and holonomic constraints.

Remark 7.2 *The mobile base is assumed to be completely nonholonomic, and the holonomic constraint force is measured by the force sensor mounted on each mobile manipulator's end-effector.*

Remark 7.3 *The friction force vector considered is Coulomb and viscous friction, given by*

$$\tau_{if} = s_i e^{-d_i \|q_i\|} + c_i \operatorname{sgn}(\dot{q}_i) + v_i \dot{q}_i \tag{7.52}$$

where $s_i > 0$, $d_i > 0$, $c_i > 0$ and v_i are constants related to the striction, Coulomb, and viscous friction, respectively. Obviously, there exist positive constants $f_c = \max_{i \in n}\{s_i + c_i\}$ and $f_v = \max_{i \in n}\{v_i\}$ such that $\|\tau_{if}(\dot{q}_i)\| \leq f_c + f_v \|\dot{q}_i\|$.

The mobile platform is subjected to nonholonomic constraints. The l non-integrable and independent velocity constraints can be expressed as

$$A_i(q_{iv})\dot{q}_{iv} = 0_{[l,1]}, \quad A_i(q_{iv}) \in \mathbb{R}^{l \times n_v} \tag{7.53}$$

Assume that the annihilator of the co-distribution spanned by the co-vector fields $A_{i1}(q_{iv})$, $\ldots$, $A_{il}(q_{iv})$ is an $(n_v - l)$-dimensional smooth nonsingular distribution Δ_i on $\mathbb{R}^{n_v}$. This distribution Δ_i is spanned by a set of $(n_v - l)$ smooth and linearly independent vector fields $H_1(q_{iv})$, $\ldots$, $H_{n_v-l}(q_{iv})$, i.e., $\Delta_i = \operatorname{span}\{H_{i,1}(q_{iv}), \ \ldots, \ H_{i,n_v-l}(q_{iv})\}$. Thus, we have

$$H_i^T(q_{iv})A_i^T(q_{iv}) = 0, \quad H_i(q_{iv}) = [H_{i,1}(q_{iv}), \ \ldots, \ H_{i,n_v-l}(q_{iv})] \in \mathbb{R}^{n_v \times (n_v-l)}$$

Note that $H^T H$ is of full rank. Constraint (7.53) implies the existence of vector $\dot{\eta}_i \in \mathbb{R}^{n_v-l}$, such that

$$\dot{q}_{iv} = H(q_{iv})\dot{\eta}_i \tag{7.54}$$

Considering the nonholonomic constraints (7.53) and (7.54) and their derivatives, the dynamics of a mobile manipulator (7.51) can be expressed as

$$M_i^1 \ddot{\zeta}_i + C_i^1 \dot{\zeta}_i + G_i^1 + d_i^1 = u_i - \tau_{if}^1 + J_{ie}^T f_{ie} \tag{7.55}$$

where $\zeta_i = [\eta_i^T, q_{ia}^T]^T$ and

$$M_i^1 = \begin{bmatrix} H_i^T M_{iv} H_i & H_i^T M_{iva} \\ M_{iav} H_i & M_{ia} \end{bmatrix}, B_i^1 = \begin{bmatrix} H_i^T B_{iv} & 0 \\ 0 & B_{ia} \end{bmatrix}$$

$$G_i^1 = \begin{bmatrix} H_i^T G_{iv} \\ G_{ia} \end{bmatrix}, \tau_{if}^1 = B_i^1 \tau_{if}(\dot{q}),$$

$$J_{ie}^T = \begin{bmatrix} 0 & 0 \\ J_{iv} H_i & J_{ia} \end{bmatrix}^T, d_i^1 = \begin{bmatrix} H_i^T d_{iv} \\ d_{ia} \end{bmatrix}, u_i = B_i^1 \tau_i,$$

$$C_i^1 = \begin{bmatrix} H_i^T M_{iv} \dot{H}_i + H_i^T C_{iv} H_i & H_i^T C_{iva} \\ M_{iav} \dot{H}_i + C_{iav} H_i & C_{ia} \end{bmatrix}$$

Remark 7.4 *From Assumption 7.5 and Fig. 7.11, the contact force between the ith mobile manipulator and object in (7.55) can be written as*

$$f_{ie} = K_{ie}(x_{ie} - x_{ie0}) + C_{ie}\dot{x}_{ie} \tag{7.56}$$

where K_{ie} is the ith spring coefficient and C_{ie} is the ith damper coefficient, $x_{ie} \in \mathbb{R}^{n_e}$ with $n_e = n - l$ is the end effector position, and x_{ie0} is the rest position of the spring.

Assumption 7.7 *In Remark 7.4, the parameters K_{ie} and C_{ie} are known beforehand.*

The dynamics of m mobile manipulators from (7.55) can be expressed in a compact form as

$$M\ddot{\zeta} + C\dot{\zeta} + G + D = u - \tau_f + J_e^T F_e \tag{7.57}$$

where $M = \text{diag}[M_i^1] \in \mathbb{R}^{m(n-l) \times m(n-l)}$; $\zeta = [\zeta_1^T, \ldots, \zeta_m^T]^T \in \mathbb{R}^{m(n-l)}$; $u = [u_1, \ldots, u_m]^T \in \mathbb{R}^{m(n-l)}$, $\tau_f = [\tau_{1f}, \ldots, \tau_{mf}]^T \in \mathbb{R}^{m(n-l)}$, with $B_1 = \text{diag}[B_i^1]$; $G = [G_1^{1T}, \ldots, G_m^{1T}]^T \in \mathbb{R}^{m(n-l)}$; $F_e = [f_{1e}^T, \ldots, f_{me}^T]^T \in \mathbb{R}^{m(n-l)}$; $C = \text{diag}[C_i^1] \in \mathbb{R}^{m(n-l) \times m(n-l)}$; $D = [d_1^1, \ldots, d_m^1]^T \in \mathbb{R}^{m(n-l)}$; and $J_e^T = \text{diag}[J_{ie}^T] \in \mathbb{R}^{m(n-l) \times m(n-l)}$.

The coordinate vector of object center of mass is denoted by $x_o \in \mathbb{R}^{n_o}$ and it is assumed that $\dot{x}_o$ and the object tip velocity vector $\dot{\chi}_t \in \mathbb{R}^{n_o}$ are related by

$$\dot{x}_o = R(\chi_t)\dot{\chi}_t \tag{7.58}$$

where $R(\chi_t) \in \mathbb{R}^{n_o \times n_o}$ is assumed invertible. The motion of the object is described by

$$M_o(x_o)\ddot{x}_o + C_o(x_o, \dot{x}_o)\dot{x}_o + G_o(x_o) = F_o + R^{-T} f_t \tag{7.59}$$

where $M_o(x_o) \in \mathbb{R}^{n_o \times n_o}$ is the symmetric positive definite inertial matrix of the object, $C_o(x_o, \dot{x}_o) \in \mathbb{R}^{n_o \times n_o}$ is the Coriolis and centrifugal matrix, and $G_o(x_o) \in \mathbb{R}^{n_o}$ is the gravitational force vector, $F_o \in \mathbb{R}^{n_o}$ is the resultant force vector acting on the center of mass of the object applied by the robots, and $f_t = J_t^T \lambda_t \in \mathbb{R}^{n_o}$ is the external force vector from the nonrigid surface, $J_t^T \in \mathbb{R}^{n_o \times n_t}$ is the constraint Jacobian matrix of the contact point and the Lagrange multiplier $\lambda_t = K_t(\chi_t - \chi_{t0}) + C_t \dot{\chi}_t \in \mathbb{R}^{n_t}$ with the spring initial position χ_{t0}, the known spring and damper coefficients K_t and C_t, respectively, which physically presents a impedance force from nonrigid contact. The constraint Jacobian matrix of the nonrigid surface J_t^T satisfies $J_t \dot{\chi}_t = 0$ so that there always exists a matrix $\mathcal{A} \in \mathbb{R}^{n_o \times (n_o - n_t)}$ satisfying $\mathcal{A}^T J_t^T = 0$ and moreover,

$$\dot{\chi}_t = \mathcal{A}\dot{x}_t \tag{7.60}$$

with the free motion sub-vector x_t of χ_t.

Considering (7.58), we can rewrite (7.60) as

$$\dot{x}_o = S\dot{x}_t \tag{7.61}$$

where $S = R(\chi_t)\mathcal{A} \in \mathbb{R}^{n_o \times (n_o - n_t)}$, integrating (7.61) into (7.59), and multiplying S^T, we have

$$\mathcal{M}_o(x_o)\ddot{x}_t + \mathcal{C}_o\dot{x}_t + \mathcal{G}_o(x_o) = \mathcal{F}_o \tag{7.62}$$

$$\lambda_t = Z_o(C_o\dot{x}_t + G_o - F_o) \tag{7.63}$$

where

$$\mathcal{M}_o(x_o) = S^T M_o(x_o)S$$
$$\mathcal{C}_o = S^T M_o(x_o)\dot{S} + S^T C_o(x_o, \dot{x}_o)S$$
$$Z_o = [J_t R^{-1} M_o^{-1} R^{-T} J_t^T]^{-1} J_t R^{-1} M_o^{-1}$$
$$\mathcal{G}_o(x_o) = S^T G_o(x_o)$$
$$\mathcal{F}_o = S^T F_o \tag{7.64}$$

Define $J_o(x_o) \in \mathbb{R}^{m(n-l) \times n_o}$ as $J_o(x_o) = [J_{1o}^T(x_o), \ \ldots, \ J_{mo}^T(x_o)]^T$, where $J_{io}(x_o) \in \mathbb{R}^{(n-l) \times n_o}$ is the Jacobian matrix from the object frame $O_o X_o Y_o Z_o$ to the ith mobile manipulator's end-effector frame $O_{ei} X_{ei} Y_{ei} Z_{ei}$. Then F_o is given by

$$F_o = -J_o^T(x_o)F_e \tag{7.65}$$

Considering the definition of $\mathcal{F}_o$, we can rewrite (7.65) as

$$\mathcal{F}_o = -S^T J_o^T(x_o) F_e \qquad (7.66)$$

Given the resultant force $\mathcal{F}_o$, the end-effector force F_e that satisfies (7.65) can be decomposed into two orthogonal components, one that contributes to the motion of the object, and one that produces the internal force. This is clearly represented by the following equation [177]

$$F_e = -(S^T J_o^T(x_o))^+ \mathcal{F}_o - \mathcal{F} \qquad (7.67)$$

where $(J_o^T(x_o))^+ \in \mathbb{R}^{m(n-l) \times n_o}$ is the pseudo-inverse matrix of $S^T J_o^T(x_o)$, $\mathcal{F} \in \mathbb{R}^{m(n-l)}$ is the internal force vector in the null space of $S^T J_o^T(x_o)$, i.e., satisfying

$$S^T J_o^T(x_o) \mathcal{F} = 0 \qquad (7.68)$$

and $\mathcal{F}$ can be parameterized by the vector of Lagrangian multiplier

$$\mathcal{F} = \mathbb{J}^T \lambda_I \qquad (7.69)$$

where $\mathbb{J}^T$ is Jacobian matrix for the internal force and satisfies $S^T J_o^T(x_o) \mathbb{J}^T = 0$. Substituting $\mathcal{F}_o$ of (7.62) into (7.67) yields

$$F_e = -(S^T J_o^T(x_o))^+ (\mathcal{M}_o(x_o)\ddot{x}_t + \mathcal{C}_o \dot{x}_t + \mathcal{G}_o(x_o)) - \mathcal{F} \qquad (7.70)$$

Let us put the end-effector position and orientation vector, $x_{ie} \in \mathbb{R}^{n_e}$, of every robot together to form a vector $x_e = [x_{1e}, \ldots, x_{me}]^T \in \mathbb{R}^{mn_e}$. Denote the Jacobian matrix as $J_{ie}(\zeta_i)$ which relates x_{ie} with ζ_i, e.g.,

$$\dot{x}_{ie} = J_{ie}(\zeta_i)\dot{\zeta}_i \qquad (7.71)$$

and the relationship between $\dot{x}_{ie}$ and $\dot{x}_o$ is given by

$$\dot{x}_{ie} = J_{io}(x_o)\dot{x}_o \qquad (7.72)$$

After combining (7.71) and (7.72), the following relationship between the joint velocity of the ith manipulator and the velocity of the object is obtained

$$J_{ie}(\zeta_i)\dot{\zeta}_i = J_{io}(x_o)\dot{x}_o \qquad (7.73)$$

We define $J_e(\zeta) = \text{diag}[J_{ie}]$ such that

$$\dot{x}_e = J_e(\zeta)\dot{\zeta} \qquad (7.74)$$

On the other hand, from (7.61), the relationship between $\dot{x}_e$ and $\dot{x}_t$ is given by

$$\dot{x}_e = J_o(x_o)S\dot{x}_t \tag{7.75}$$

After combining (7.74) and (7.75), the following relationship between the joint velocity of the ith manipulator and the velocity of the tip is obtained $J_e(\zeta)\dot{\zeta} = J_o(x_o)S\dot{x}_t$. As it is assumed that the manipulators work in a nonsingular region, the inverse of the Jacobian matrix $J_{ie}(\zeta_i)$ exists. Considering all the manipulators acting on the object at the same time yields

$$\dot{x}_t = (J_o(x_o)S)^+ J_e(\zeta)\dot{\zeta} \tag{7.76}$$

Differentiating (7.76) with respect to time leads to

$$\ddot{x}_t \;=\; (J_o(x_o)S)^+ J_e(\zeta)\ddot{\zeta} + \frac{d}{dt}[(J_o(x_o)S)^+ J_e(\zeta)]\dot{\zeta} \tag{7.77}$$

Let $\mathcal{J}_o(x_o) = J_o(x_o)S$, using equations (7.76) and (7.77). The dynamics of multiple manipulator systems (7.57), coupled with the object dynamics (7.62), are then given by

$$\mathcal{M}_o(x_o)(\mathcal{J}_o(x_o))^+ J_e(\zeta)\ddot{\zeta}$$
$$+ \left(\mathcal{M}_o \frac{d}{dt}[(\mathcal{J}_o(x_o))^+ J_e(\zeta)] + \mathcal{C}_o(\mathcal{J}_o(x_o))^+ J_e(\zeta) \right)\dot{\zeta}$$
$$+ \mathcal{G}_o = \mathcal{F}_o \tag{7.78}$$

$$M(\zeta)\ddot{\zeta} + C(\zeta,\dot{\zeta})\dot{\zeta} + G + D + J_e^T(\zeta)(\mathcal{J}_o^T(x_o))^+ \mathcal{F}_o + J_e^T(\zeta)\mathcal{F} = u - \tau_f \tag{7.79}$$

Let $L^T = J_e^T(\zeta)(\mathcal{J}_o^T(x_o))^+$. Integrating the above equations, we have

$$M_1\ddot{\zeta} + C_1\dot{\zeta} + G_1 + D_1 + J_e^T(\zeta)\mathcal{F} = u - \tau_f \tag{7.80}$$

where

$$M_1 = M(\zeta) + L^T \mathcal{M}_o L$$
$$G_1 = G + L^T \mathcal{G}_o$$
$$C_1 = C(\zeta,\dot{\zeta}) + L^T \left(\mathcal{M}_o \frac{d}{dt}[L] + \mathcal{C}_o(x_o,\dot{x}_o)L \right)$$
$$D_1 = D$$

Let $\Upsilon^T = \mathcal{J}_o^T(x_o)J_e^{-T}(\zeta)$. By introducing $\dot{\zeta} = \Upsilon\dot{\xi}$, we can rewrite (7.80) as

$$M_1\Upsilon\ddot{\xi} + (C_1\Upsilon + M_1\dot{\Upsilon})\dot{\xi} + G_1 + D_1 + J_e^T(\zeta)\mathcal{F} = u - \tau_f \tag{7.81}$$

Multiplying both sides of (7.81) by $\Upsilon^T = \mathcal{J}_o^T(x_o)J_e^{-T}(\zeta)$, and using (7.68), we have

$$M_2\ddot{\xi} + C_2\dot{\xi} + G_2 + D_2 = \Upsilon^T(u - \tau_f) \tag{7.82}$$

Considering (7.69), we have

$$\lambda_I = Z_1\left(C_1\dot{\zeta} + G_1 + D - (u - \tau_f)\right) \tag{7.83}$$

where $M_2 = \Upsilon^T M_1 \Upsilon$, $C_2 = \Upsilon^T(C_1\Upsilon + M\dot{\Upsilon})$, $G_2 = \Upsilon^T G_1$, $D_2 = \Upsilon^T D_1$, and $Z_1 = (\mathbb{J}J_e(\zeta)(M_1)^{-1}J_e^T(\zeta)\mathbb{J}^T)^{-1}J_e(\zeta)\mathbb{J}(M_1)^{-1}$.

The dynamics (7.82) have the following structure properties, which can be exploited to facilitate the control design.

Property 7.5 *The matrix M_1 is symmetric, positive definite, and is bounded from below and above, i.e., $\lambda_{min}(M_1)I \le M_1 \le \lambda_{max}(M_1)I$, where λ_{min} and λ_{max} denote the minimum and maximum eigenvalues of M_2.*

Property 7.6 *The matrix $\dot{M}_1 - 2C_1$ is skew-symmetric.*

Property 7.7 *All Jacobian matrices are uniformly bounded and uniformly continuous if ζ and x_o are uniformly bounded and continuous, respectively.*

Remark 7.5 *The matrix Z_1 is uniformly bounded and uniformly continuous from Property 7.5 and Property 7.7.*

Property 7.8 *There exists a finite positive vector $\mathbb{C} = [c_1, c_2, \ldots, c_7]^T$ with $c_i > 0$, such that $\forall \zeta, \dot{\zeta} \in \mathbb{R}^{m(n-l)}$, $\|M_1\| \le c_1$, $\|C_1\| \le c_2 + c_3\|\dot{\zeta}\|$, $\|G_1\| \le c_4$, $sup_{t \ge 0}\|D_1\| \le c_5$, and according to Remark 7.3, $\|\tau_f\| \le c_6 + c_7\|\dot{\zeta}\|$.*

7.2.3 Decentralized Adaptive Control

Assume a desired motion trajectory $x_t^d(t)$ and a bounded internal force $\mathcal{F}_d$. Since the system is inter-connected, we can obtain the desired corresponding motion trajectory $\zeta^d(t)$, $\dot{\zeta}^d(t)$ from x_t^d and $\dot{x}_t^d$; therefore, the trajectory control and internal force tracking control is to determine a control law such that for any $\zeta(0)$, $\dot{\zeta}(0)$ converges to a manifold specified as Ω where $\Omega_d = \{(\zeta, \dot{\zeta})|\zeta = \zeta^d, \dot{\zeta} = \dot{\zeta}^d\}$ and internal force $\mathcal{F}$ track $\mathcal{F}_d$ to the satisfaction of the design.

Assumption 7.8 *The desired reference trajectory $\xi^d(t)$ is bounded and uniformly continuous, and has bounded and uniformly continuous derivatives up to the third order.*

Let the control u take the form:

$$u = u_a - J_e^T(\zeta)\mathbf{J}^T u_b \qquad (7.84)$$

Then, (7.82) and (7.83) can be rewritten as

$$M_1\ddot{\zeta} + C_1\dot{\zeta} + G_1 + D_1 = u_a + \tau_f \qquad (7.85)$$

$$Z_1(u_a - C_1\dot{\zeta} - G_1 - D) - u_b = -\lambda_I \qquad (7.86)$$

To facilitate the robust control formulation, the following lemma and assumption are required.

Lemma 7.1 *For $x > 0$, $y > 0$ and $\delta \geq 1$, we have $\ln(\cosh(x)) + \delta \geq x$ and $\ln(\cosh(x)) - \ln(\cosh(y)) \geq x - y$.*

Proof: If $x \geq 0$, we have $\int_0^x \frac{2}{e^{2s}+1}ds < \int_0^x \frac{2}{e^{2s}}ds = 1 - e^{-2x} < 1$ Therefore, $\ln(\cosh(x)) + \delta \geq \ln(\cosh(x)) + \int_0^x \frac{2}{e^{2s}+1}ds$ with $\delta \geq 1$. Let $f(x) = \ln(\cosh(x)) + \int_0^x \frac{2}{e^{2s}+1}ds - x$, we have $\dot{f}(x) = \tanh(x) + \frac{2}{e^{2x}+1} - 1 = \frac{e^x-e^{-x}}{e^x+e^{-x}} + \frac{2}{e^{2x}+1} - 1 = 0$. From the mean value theorem, we have $f(x) - f(0) = \dot{f}(x)(x - 0)$. Since $f(0) = 0$, we have $f(x) = 0$, that is, $\ln(\cosh(x)) + \int_0^x \frac{2}{e^{2s}+1}ds = x$, then, we have $\ln(\cosh(x)) + \delta \geq x$. From the above, we have $\ln(\cosh(x)) - \ln(\cosh(y)) = \ln(\cosh(x)) + \delta - \ln(\cosh(y)) - \delta \geq \ln(\cosh(x)) + \delta - y \geq x - y$.

Let $e = \zeta - \zeta^d$, $\dot{\zeta}_r = \dot{\zeta}^d - \Lambda e$, $r = \dot{e} + \Lambda e$ with Λ being a diagonal positive definite constant matrix, we can rewrite (7.82) as

$$M_1\dot{r} + C_1 r = u_a - \Xi \qquad (7.87)$$

where $\Xi = M_1\ddot{\xi}_r + C_1\dot{\xi}_r + G_1 + D_1 - \tau_f$.

According to the definition of $\Xi \in \mathbb{R}^{m(n-l)}$, we denote Ξ_{ik}, $i = 1, 2, \ldots, m$, and $k = 1, 2, \ldots, n - l$ as the $((i - 1)(n - l) + k)$th elements of Ξ, which corresponds to the kth equation in the dynamics of the ith robot. Similarly, we denote r_{ik} as the $((i - 1)(n - l) + k)$th element of $r \in \mathbb{R}^{m(n-l)}$, and in addition, denote $r_i = [r_{i1}, r_{i2}, \ldots, r_{i(n-l)}]^T$. We define the jkth component of ith mobile manipulator in (7.87) as

$$\sum_{j=1}^{n-l} m_{ikj}\dot{r}_{ij} + \sum_{j=1}^{n-l} c_{ikj}(q, \dot{q})r_{ij} = u_{aik} - \Xi_{ik} \qquad (7.88)$$

Lemma 7.2 *Considering Property 7.8, the upper bound of Ξ satisfies*

$$\|\Xi_{ik}\| \leq \ln(\cosh(\|\Phi_{ik}\|)) + \delta \qquad (7.89)$$

where δ is a small constant, $\Phi_{ij} = \alpha_{ik}^T\varphi_i$ with $\varphi_i = [1, \sup\|r_i\|, \sup\|r_i\|^2]^T$ and $\alpha_{ik} = [\alpha_{ik1}, \alpha_{ik2}, \alpha_{ik3}]^T$ is a vector of positive constants defined below.

Proof: According to Property 7.8, the upper bound of Ξ_{ik} satisfies

$$
\begin{aligned}
\|\Xi_{ik}\| &\leq c_1\|\ddot{\zeta}_{ik}^d - \Lambda\dot{e}_{ik}\| + \left(c_2 + c_3\|\dot{\zeta}_{ik}^d + \dot{e}_{ik}\|\right)\|\dot{\zeta}_{ik}^d - \Lambda e_{ik}\| \\
&\quad + c_4 + c_5 + c_6 + c_7\|\dot{\zeta}_{ik}^d + \dot{e}_{ik}\| \\
&\leq c_1\|\ddot{\zeta}_{ik}^d\| + c_1\|\Lambda\|\|\dot{e}_{ik}\| + c_2\|\dot{\zeta}_{ik}^d\| + c_2\|\Lambda\|\|e_{ik}\| \\
&\quad + c_3\|\dot{\zeta}_{ik}^d\|^2 + c_3\|\dot{\zeta}_{ik}^d\|\|\Lambda\|\|e_{ik}\| + c_3\|\dot{e}_{ik}\|\|\dot{\zeta}_{ik}^d\| \\
&\quad + c_3\|\Lambda\|\|e_{ik}\|\|\dot{e}_{ik}\| + c_4 + c_5 + c_6 + c_7\|\dot{\zeta}_{ik}^d\| + c_7\|\dot{e}_{ik}\| \\
&\leq c_1\|\ddot{\zeta}_{ik}^d\| + c_2\|\dot{\zeta}_{ik}^d\| + c_3\|\dot{\zeta}_{ik}^d\|^2 + c_4 + c_5 + c_6 + c_7\|\dot{\zeta}_{ik}^d\| \\
&\quad + (c_2\|\Lambda\| + c_3\|\dot{\zeta}_{ik}^d\|\|\Lambda\|)\|e_{ik}\| + c_3\|\Lambda\|\|e_{ik}\|\|\dot{e}_{ik}\| \\
&\quad + (c_1\|\Lambda\| + c_3\|\dot{\zeta}_{ik}^d\| + c_7)\|\dot{e}_{ik}\| \\
&\leq \beta_1 + \beta_2\|e_{ik}\| + \beta_3\|\dot{e}_{ik}\| + \beta_4\|e_{ik}\|^2 + \beta_5\|\dot{e}_{ik}\|^2 \qquad (7.90)
\end{aligned}
$$

where $\beta_1 = c_1\|\ddot{\zeta}_{ik}^d\| + c_2\|\dot{\zeta}_{ik}^d\| + c_3\|\dot{\zeta}_{ik}^d\|^2 + c_4 + c_5 + c_6 + c_7\|\dot{\zeta}_{ik}^d\|$, $\beta_2 = c_2\|\Lambda\| + c_3\|\dot{\zeta}_{ik}^d\|\|\Lambda\|$, $\beta_3 = c_1\|\Lambda\| + c_3\|\dot{\zeta}_{ik}^d\| + c_7$, $\beta_4 = \frac{1}{2}c_3\|\Lambda\|$, $\beta_5 = \beta_4$.

Consider the linear system defined by $\dot{e}_{ik} = -\Lambda e_{ik} + r_{ik}$, $e_{ik}(0) = e_0$. Since the matrix $-\Lambda$ is Hurwitz, we see there exist constants a_1, a_2, a_3 and a_4 such that $\|e_{ik}(t)\| \leq a_1\|e_{ik0}\| + a_2 \sup\|r_{ik}\|$ and $\|\dot{e}_{ik}(t)\| \leq a_3\|e_{ik0}\| + a_4 \sup\|r_{ik}\|$. Substitute these two equations into (7.90) to finally obtain $\|\Xi_{ik}\| \leq \alpha_{ik1} + \alpha_{ik2} \sup\|r_{ik}\| + \alpha_{ik3} \sup\|r_{ik}\|^2$. We propose the following decentralized control for the multiple mobile manipulators

$$
u_{aik} = -k_{ik}r_{ik} - \ln(\cosh(\hat{\Phi}_{ik}))\operatorname{sgn}(r_{ik}) - \delta\operatorname{sgn}(r_{ik}) \qquad (7.91)
$$

where $\hat{\Phi}_{ik} = \hat{\alpha}_{ik}^T\varphi_{ik}$, $\dot{\hat{\alpha}}_{ik} = -\Sigma\hat{\alpha}_{ik} + \Gamma\varphi_{ik}\|r_{ik}\|$, $k_{ik} > 0$, $\delta > 1$, if $r_{ik} \geq 0$, $\operatorname{sgn}(r_{ik}) = 1$, else $\operatorname{sgn}(r_{ik}) = -1$, and $\Gamma = \operatorname{diag}[\gamma_{ik\iota}] > 0$, $\Sigma = \operatorname{diag}[\sigma_{ik1}, \sigma_{ik2}, \sigma_{ik3}]$ is a diagonal matrix whose each element $\sigma_{ik\iota}$ satisfies $\lim_{t\to\infty}\sigma_{ik\iota} = 0$, and $\lim_{t\to\infty}\int_0^t \sigma_{ik\iota}(\omega)d\omega = b_{ik\iota} < \infty$ with the finite constant $b_{ik\iota}$, $\iota = 1$, 2, 3. It is observed that the controller (7.91) only adopts the local feedback information.

Theorem 7.3 *Consider the mechanical system described by (7.82) and its decentralized dynamics model (7.88). Using the control law (7.91), the following hold for any $(\zeta(0), \dot{\zeta}(0))$:*

(i) r_{ik} converges to a set containing the origin as $t \to \infty$;

(ii) e_{ik} and $\dot{e}_{ik}$ converge to 0 as $t \to \infty$; and τ is bounded for all $t \geq 0$; and

(iii) $e_f = \lambda_I - \lambda_{Id}$ is bounded.

Proof: Consider a following Lyapunov function with $\tilde{\alpha}_{ik} = \alpha_{ik} - \hat{\alpha}_{ik}$ as

$$V = \frac{1}{2} r^T M_1 r + \sum_{i=1}^{m} \sum_{k=1}^{n-1} \sum_{\iota=1}^{3} \frac{1}{2\gamma_{ik\iota}} \tilde{\alpha}_{ik\iota} \tilde{\alpha}_{ik\iota} \qquad (7.92)$$

The derivative of V along (7.88) is given by

$$
\begin{aligned}
\dot{V}_1 &= \frac{1}{2} \left[r^T \dot{M}_1 r + \dot{r}^T M_1 r + r^T M_1 \dot{r} \right] + \sum_{i=1}^{m} \sum_{k=1}^{n-l} \sum_{\iota=1}^{3} \frac{1}{\gamma_{ij\iota}} \tilde{\alpha}_{ik\iota} \dot{\tilde{\alpha}}_{ik\iota} \\
&= \frac{1}{2} \sum_{i=1}^{m} \sum_{k=1}^{n-l} \sum_{j=1}^{n-l} r_{ik} \dot{m}_{ikj} r_{ij} + \sum_{i=1}^{m} \sum_{k=1}^{n-l} \sum_{j=1}^{n-l} r_{ik} m_{ikj} \dot{r}_{ij} \\
&\quad + \sum_{i=1}^{m} \sum_{k=1}^{n-l} \sum_{\iota=1}^{3} \frac{1}{\gamma_{ik\iota}} \tilde{\alpha}_{ik\iota} \dot{\tilde{\alpha}}_{ik\iota} \\
&= \sum_{i=1}^{m} \sum_{k=1}^{n-l} r_{ik} \left[\sum_{j=1}^{n-l} \frac{1}{2} \dot{m}_{ikj} r_{ij} - \sum_{j=1}^{n-l} c_{ikj} r_{ij} + u_{ik} - \Xi_{ik} \right] \\
&\quad + \sum_{i=1}^{m} \sum_{k=1}^{n-l} \sum_{\iota=1}^{3} \frac{1}{\gamma_{ik\iota}} \tilde{\alpha}_{ik\iota} \dot{\tilde{\alpha}}_{ik\iota} \qquad (7.93)
\end{aligned}
$$

Considering Property 7.6, and integrating (7.91) into (7.93), we have

$$
\begin{aligned}
\dot{V} &\leq \sum_{i=1}^{m} \sum_{k=1}^{n-l} r_{ik} \left[u_{ik} - \Xi_{ik} \right] + \sum_{i=1}^{m} \sum_{k=1}^{n-l} \sum_{\iota=1}^{3} \frac{1}{\gamma_{ik\iota}} \tilde{\alpha}_{ik\iota} \dot{\tilde{\alpha}}_{ik\iota} \\
&= \sum_{i=1}^{m} \sum_{k=1}^{n-l} r_{ik} [-k_{ik} r_{ik} - \ln(\cosh(\hat{\Phi}_{ik})) \operatorname{sgn}(r_{ik}) - \delta \operatorname{sgn}(r_{ik}) - \Xi_{ik}] \\
&\quad + \sum_{i=1}^{m} \sum_{k=1}^{n-l} \sum_{\iota=1}^{3} \tilde{\alpha}_{ik\iota}^T \gamma_{ik\iota}^{-1} \dot{\tilde{\alpha}}_{ik\iota} \\
&= -\sum_{i=1}^{m} \sum_{k=1}^{n-l} k_{ik} r_{ik}^2 - \sum_{i=1}^{m} \sum_{k=1}^{n-l} \delta \| r_{ik} \| + \sum_{i=1}^{m} \sum_{k=1}^{n-l} \sum_{\iota=1}^{3} \tilde{\alpha}_{ik\iota}^T \gamma_{ik\iota}^{-1} \dot{\tilde{\alpha}}_{ik\iota} \\
&\quad - \sum_{i=1}^{m} \sum_{k=1}^{n-l} r_{ik} \ln(\cosh(\hat{\Phi}_{ik})) \operatorname{sgn}(r_{ik}) - \sum_{i=1}^{m} \sum_{k=1}^{n-l} r_{ik} \Xi_{ik}
\end{aligned}
$$

$$\leq -\sum_{i=1}^{m}\sum_{k=1}^{n-l}k_{ik}r_{ik}^2 - \sum_{i=1}^{m}\sum_{k=1}^{n-l}\delta\|r_{ik}\|$$

$$-\sum_{i=1}^{m}\sum_{k=1}^{n-l}\|r_{ik}\|\ln(\cosh(\hat{\Phi}_{ik})) + \sum_{i=1}^{m}\sum_{k=1}^{n-l}\|r_{ik}\|\|\Xi_{ik}\|$$

$$+\sum_{i=1}^{m}\sum_{k=1}^{n-l}\sum_{\iota=1}^{3}\frac{\sigma_{ik\iota}}{\gamma_{ik\iota}}\tilde{\alpha}_{ik\iota}\hat{\alpha}_{ik\iota} - \sum_{i=1}^{m}\sum_{k=1}^{n-l}\sum_{\iota=1}^{3}\tilde{\alpha}_{ik\iota}\varphi_{ik\iota}\|r_{ik}\| \quad (7.94)$$

Considering Lemmas 7.1 and 7.2 and using $\tilde{\alpha}_{ik\iota}\hat{\alpha}_{ik\iota} = \hat{\alpha}_{ik\iota}(\alpha_{ik\iota} - \hat{\alpha}_{ik\iota}) = \frac{1}{4}\alpha_{ik\iota}^2 - (\frac{1}{2}\alpha_{ik\iota} - \hat{\alpha}_{ik\iota})^2 \leq \frac{1}{4}\alpha_{ik\iota}^2$, we have

$$\dot{V} \leq -\sum_{i=1}^{m}\sum_{k=1}^{n-l}k_{ik}r_{ik}^2 - \sum_{i=1}^{m}\sum_{k=1}^{n-l}\|r_{ik}\|\ln(\cosh(\hat{\Phi}_{ik}))$$

$$+\sum_{i=1}^{m}\sum_{k=1}^{n-l}\|r_{ik}\|\ln(\cosh(\Phi_{ik}) - \sum_{i=1}^{m}\sum_{k=1}^{n-l}\sum_{\iota=1}^{3}\tilde{\alpha}_{ik\iota}\varphi_{ik\iota}\|r_{ik}\|$$

$$+\sum_{i=1}^{m}\sum_{k=1}^{n-l}\sum_{\iota=1}^{3}\frac{\sigma_{ik\iota}}{\gamma_{ik\iota}}\tilde{\alpha}_{ik\iota}\hat{\alpha}_{ik\iota}$$

$$\leq -\sum_{i=1}^{m}\sum_{k=1}^{n-l}k_{ik}r_{ik}^2 + \sum_{i=1}^{m}\sum_{k=1}^{n-l}\sum_{\iota=1}^{3}\frac{\sigma_{ikl}}{\gamma_{ik\iota}}\tilde{\alpha}_{ik\iota}\hat{\alpha}_{ik\iota}$$

$$\leq -\sum_{i=1}^{m}\sum_{k=1}^{n-l}k_{ik}r_{ik}^2 + \frac{1}{4}\sum_{i=1}^{m}\sum_{k=1}^{n-l}\sum_{\iota=1}^{3}\frac{\sigma_{ik\iota}}{\gamma_{ikl}}\alpha_{ik\iota}^2 \quad (7.95)$$

Since $\frac{1}{4}\sum_{i=1}^{m}\sum_{k=1}^{n-l}\sum_{\iota=1}^{3}\frac{\sigma_{ik\iota}}{\gamma_{ik\iota}}\alpha_{ik\iota}^2$ is bounded and converges to zero as $t \to \infty$ by noting $\lim_{t\to\infty}\gamma_{ij\iota} = 0$, there exists $t > t_1$, $\frac{1}{4}\sum_{i=1}^{m}\sum_{k=1}^{n-l}\sum_{\iota=1}^{3}\frac{\sigma_{ik\iota}}{\gamma_{ik\iota}}\alpha_{ik\iota}^2 \leq \epsilon$ with a finite small constant ϵ, when $|r_{ik}| \geq \sqrt{\frac{\epsilon}{\lambda_{min}(k_{ik})}}$, $\dot{V} \leq 0$; then r_{ik} converges to a small set containing the origin as $t \to \infty$.

(ii) Integrating both sides of the above equation gives

$$V(t) - V(0) < -\int_{0}^{t}\sum_{i=1}^{m}\sum_{k=1}^{n-l}k_{ik}r_{ik}^2 ds + \frac{1}{4}\sum_{i=1}^{m}\sum_{k=1}^{n-l}\sum_{\iota=1}^{3}\frac{b_{ik\iota}}{\gamma_{ik\iota}}\alpha_{ik\iota}^2 \quad (7.96)$$

by noting $\lim_{t\to\infty}\sigma_{ik\iota} = 0$ and $\lim_{t\to\infty}\int_{0}^{t}\sigma_{ik\iota}(\omega)d\omega = b_{ik\iota} < \infty$. Thus V is bounded, which implies that $r \in L_\infty$. From $r = \dot{e} + \Lambda e$, it can be obtained that $e, \dot{e} \in L_\infty$. As we have established $e, \dot{e} \in L_\infty$ from Assumption 7.8, we conclude that $\zeta, \dot{\zeta}, \ddot{\zeta}, \varsigma, \dot{\varsigma}, \ddot{\varsigma} \in L_\infty$.

(iii) Substituting the control (7.91) into the reduced order dynamical sys-

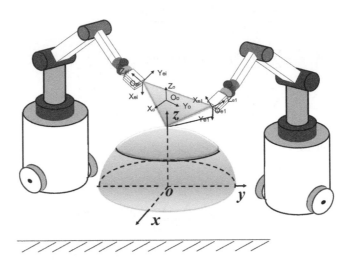

FIGURE 7.10
Coordinated operation of two robots.

tem yields

$$\begin{aligned}\lambda_I &= Z_1\left(C_1\dot\zeta + G_1 + D - u_a\right) + u_b \\ &= -Z_1 M_1 \Upsilon\ddot\zeta + u_b\end{aligned}\qquad(7.97)$$

We choose $u_b = \lambda_{Id} - K_b e_f$ with $e_f = \lambda_I - \lambda_{Id}$ such that we have

$$(I + K_b)e_f = -Z_1 M_1 \Upsilon\ddot\zeta \qquad(7.98)$$

Since $\ddot\zeta$ and Z_1 are bounded, $\zeta \to \zeta^d$, $-Z_1 M_1 \Upsilon\ddot\zeta$ is bounded and the size of λ_I can be adjusted by choosing the proper gain matrix K_b.

Since all the signals on the right hand side of (7.88) are bounded, it is easy to conclude that τ is bounded from (7.91).

7.2.4   Simulation Studies

Let us consider two same 2-DOF mobile manipulators shown in Fig. 7.12 of Section 7.1.6. The desired trajectory for the object, the desired internal force,

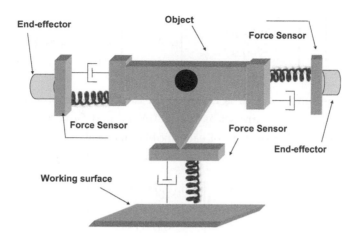

FIGURE 7.11
The relationship of end-effectors and object.

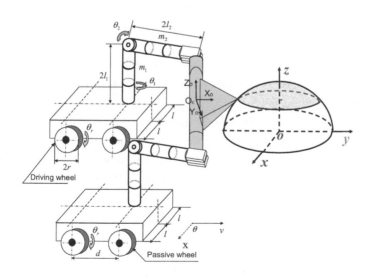

FIGURE 7.12
The 2-DOF mobile manipulators.

and the contact force are chosen as

$$
\begin{bmatrix} \dot{x}_{1od} \\ \dot{x}_{2od} \\ \dot{x}_{3od} \\ \dot{x}_{4od} \end{bmatrix} = \begin{bmatrix} \cos(0.5\sin t + \frac{\pi}{2}) \\ \sin(0.5\sin t + \frac{\pi}{2}) \\ 0.0 \\ \cos t + 0.01\cos t \end{bmatrix}, \lambda_{Id} = \begin{bmatrix} 5.0 \\ 0.0 \\ 0.0 \end{bmatrix}, \lambda_t = \begin{bmatrix} 0.0 \\ 0.0 \\ 10.0 \end{bmatrix}
$$

The spring-damper device is mounted on the tip of the object, and the parameters of the device are set as $K_t = 80N/m$, and $C_t = 0.01N/m^2$. We can obtain the desired trajectory of each mobile manipulator when θ_{i2} is fixed as $\frac{3\pi}{2}$,

$$
\begin{bmatrix} \dot{y}_{id} \\ \dot{\theta}_{id} \\ \dot{\theta}_{i1d} \\ \dot{\theta}_{i2d} \end{bmatrix} = \begin{bmatrix} \sin(0.5\sin t + \frac{\pi}{2}) \\ \cos t \\ 0.01\cos t \\ 0.0 \end{bmatrix}
$$

The initial positions and velocities are chosen as $x_o(0) = [0.0, 0.0, 2l_1, 0.0]^T$, $[y_{id}(0), \theta_{id}(0), \theta_{i1d}(0), \theta_{i2d}(0)]^T = [0.0, \frac{\pi}{2}, 0.0, \frac{3\pi}{2}]^T$, $\dot{x}_o(0) = \dot{y}_{id}(0) = \dot{\theta}_{id}(0) = \dot{\theta}_{i1d}(0) = \dot{\theta}_{i2d}(0) = 0.0$, respectively.

Assume that the parameters are selected as $m_p = 6.0kg$, $m_1 = m_2 = 1.0kg$, $I_{zp} = 19kgm^2$, $I_{z1} = I_{z2} = 1.0kgm^2$, $d = 0.0$, $l = r = 1.0m$, $2l_1 = 1.0m$, $2l_2 = 0.6m$, the mass of the object $m_o = 1.0kg$, $I_o = 1.0kgm^2$, $l_{c1} = l_{c2} = 0.5m$.

The friction vector considered is Coulomb and viscous friction, given by (7.52). However, the discontinuity of the friction characteristics at zero velocity is required for testing zero velocity; and when the velocity is zero or the system is stationary, the friction is indefinite and depends on the controlled torque. In the simulation, to improve the numerical efficiency, we adopt a revised friction model from [179]. For the ith joint motor, the structure for the revised friction model is described by the following mathematical model: $\tau_{if} = p_1 \operatorname{sgn}(\omega_i) + p_2\omega_i + (T_i - (p_1 \operatorname{sgn}(\omega_i) + p_2\omega_i))\exp(-(\frac{\omega_i}{\iota})^2)$ where τ_{if} is the revised friction and ι is a small positive scalar; T_i is given by

$$
T_i = \begin{cases} \Upsilon & \text{if } \tau_{si} > \Upsilon \\ \tau_{si} & \text{if } -\Upsilon \le \tau_{si} \le \Upsilon \\ \Upsilon & \text{if } \tau_{si} \le -\Upsilon \end{cases}
$$

where $\Upsilon = 0.12\tau_{max}$, ω_i is the motor angular velocity, τ_{si} is the motor torque, and τ_{max} is the maximum motor torque. In the simulation, we choose $\tau_{max} = 10Nm$, $p_2 = 0.0088Nm/rad/s$, and $p_1 = 1.2Nm/rad/s$. When the velocity is

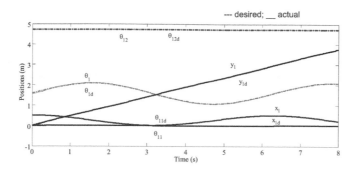

FIGURE 7.13

The joint positions of mobile manipulator I.

near zero, defined by $\iota = 0.001$, the friction is equal to the applied torque τ_{si}. When the velocity is greater than this, the third term in the above expression vanishes and the friction given by this revised model is equal to the friction given by (7.52).

The disturbances $d_1 = [0.1\sin t, -0.2\sin t, -0.05\sin t, 0, 0]^T$ and $d_2 = [0.2\sin t, -0.2\sin t, 0, 0, 0^T$ are introduced into the simulation model. By Theorem 7.3, the control gains are selected as $\Lambda = \text{diag}[0.5]$, $K_P = \text{diag}[150.0]$, and $K_f = 80000$, $\hat{C}(0) = [1.0, \ldots, 1.0]^T$, and $\Gamma = \text{diag}[1.0]$, $\Sigma = \text{diag}[\frac{1}{(1+t)^2}]$, $\delta = 1.5$.

The simulation results for two mobile manipulators are shown in Figs. 7.13-7.21. Figs. 7.13-7.15 show the motion control for mobile manipulator I, including the positions tracking in Fig. 7.13, the velocities in Fig. 7.14, and the input torques in Fig. 7.15. The internal force is shown in Fig. 7.17. Figs. 7.18 - Fig. 7.21 show the motion control for mobile manipulator II, similarly, including the position tracking in Fig. 7.18, the velocities in Fig. 7.19 and the input torques in Fig. 7.21. The contact force with the environments is shown in Fig. 7.22. Since the two mobile manipulators evenly share the loads, the motor torques of the joint 2 for every mobile manipulator are both shown in Fig. 7.16. The simulation results show that the trajectory and internal force tracking errors

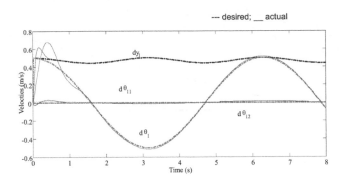

FIGURE 7.14

The joint velocities of mobile manipulator I.

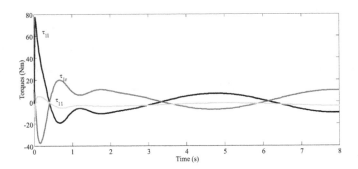

FIGURE 7.15

The input torque for mobile manipulator I.

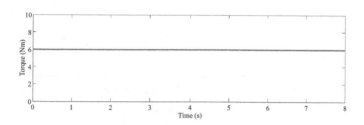

FIGURE 7.16

The input torque for joint 2 of mobile manipulator I.

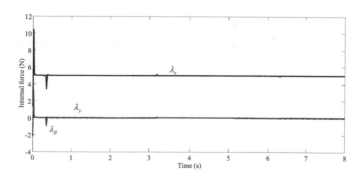

FIGURE 7.17

The internal force.

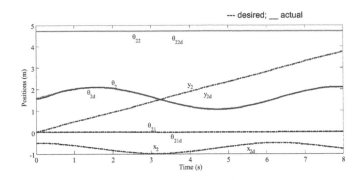

FIGURE 7.18

The joint positions of mobile manipulator II.

tend to the desired values, which validates the effectiveness of the control law
in Theorem 7.3. Under the proposed control scheme, tracking of the desired
trajectory and desired internal force is achieved and this is largely due to the
"adaptive" mechanism. The simulation results demonstrate the effectiveness of
the proposed adaptive control in the presence of parametric uncertainties and
external disturbances. Although parametric uncertainties and external disturbances are both introduced into the simulation model, the impedance force
and motion control performance of the system under the proposed control,
is not degraded. Different impedance force and motion tracking performance
can be achieved by adjusting parameter adaptation gains and control gains.

7.3   Conclusion

In this chapter, centralized robust adaptive centralized control strategies have
been first presented to control coordinated multiple mobile manipulators carrying a common object in the presence of uncertainties and disturbances. All
control strategies have been designed to drive the system motion to the de-

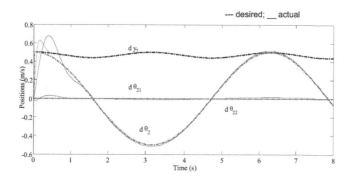

FIGURE 7.19

The joint velocities of mobile manipulator II.

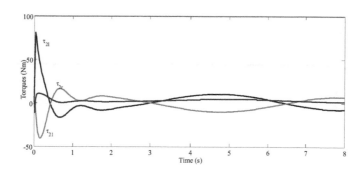

FIGURE 7.20

The input torques for mobile manipulator II.

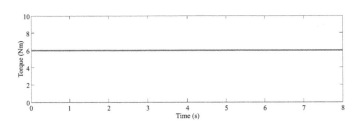

FIGURE 7.21

The input torque for joint 2 of mobile manipulator II.

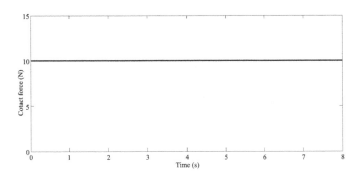

FIGURE 7.22

The contact force with the working surface.

sired manifold and at the same time guarantee the boundedness of the internal force. The proposed controls are nonregressor based and require no information on the system dynamics. Simulation results have shown the effectiveness of the proposed controls.

Then, a decentralized adaptive control version was presented systematically to control the coordinated multiple mobile manipulators interacting with a nonrigid surface in the presence of uncertainties and disturbances. All control strategies have been designed to drive the system's motion/force convergence to the desired manifold and at the same time guarantee the boundedness of the internal force. The proposed controls are nonregressor based and require no information on the system dynamics. Simulation studies have verified the effectiveness of the proposed controls.

8

Cooperation Control

CONTENTS

8.1 Introduction

The cooperation control of multiple mobile manipulators presents a significant increase in complexity over the single mobile manipulator case. Thus far, there are only some coordination schemes for multiple mobile manipulators in the literature: (i) hybrid position force control by a decentralized/centralized scheme, where the position of the object is controlled in a certain direction of the workspace, and the internal force of the object is controlled in a small range of the origin [68], [72], [174]; (ii) leader-follower control for a mobile manipulator, where one or a group of mobile manipulators or robotic manipulator play the role of the leader tracking a preplanned trajectory, and the rest of the mobile manipulators form the follower group moving in conjunction with the leader mobile manipulators [172], [75], [176].

However, in the hybrid position force control of constrained cooperating multiple mobile manipulator, [68], [72], [174], although the constraint object

Nomenclature

O_c	Contact point between the end-effector of mobile manipulator I and the object
O_h	Point where the end-effector of mobile manipulator II holds the object
O_o	Mass center of the object $O_c X_c Y_c Z_c$ frame fixed with the tool of mobile manipulator I with its origin at the contact point O_c
$O_h X_h Y_h Z_h$	Frame fixed with the end-effector or hand of mobile manipulator 2 with its origin at point O_h
$O_o X_o Y_o Z_o$	Frame fixed with the object with its origin at the mass centre O_o
$OXYZ$	World coordinates
r_c	Vector describing the posture of frame $O_c X_c Y_c Z_c$ with $r_c = [x_c^T, \theta_c^T]^T \in \mathbf{R}^6$
r_h	Vector describing the posture of frame $O_h X_h Y_h Z_h$ with $r_h = [x_h^T, \theta_h^T]^T \in \mathbf{R}^6$
r_o	Vector describing the posture of frame $O_o X_o Y_o Z_o$ with $r_o = [x_o^T, \theta_o^T]^T \in \mathbf{R}^6$
r_{co}	Vector describing the posture of frame $O_c X_c Y_c Z_c$ expressed in $O_o X_o Y_o Z_o$ with $r_{co} = [x_{co}^T, \theta_{co}^T]^T \in \mathbf{R}^6$
r_{ho}	Vector describing the posture of frame $O_h X_h Y_h Z_h$ expressed in $O_o X_o Y_o Z_o$ with $r_{ho} = [x_{ho}^T, \theta_{ho}^T]^T \in \mathbf{R}^6$
q_1	Vector of joint variables of mobile manipulator I
q_2	Vector of joint variables of mobile manipulator II
n_1	Degrees of freedom in mobile manipulator I
n_2	Degrees of freedom in mobile manipulator II
x_c	Position vector of O_c, the origin of frame $O_c X_c Y_c Z_c$
x_h	Position vector of O_h, the origin of frame $O_h X_h Y_h Z_h$
x_o	Position vector of O_o, the origin of frame $O_o X_o Y_o Z_o$
x_{co}	Position vector of O_c, the origin of frame $O_c X_c Y_c Z_c$ expressed in $O_o X_o Y_o Z_o$
x_{ho}	Position vector of O_h, the origin of frame $O_h X_h Y_h Z_h$ expressed in $O_o X_o Y_o Z_o$
θ_c	Orientation vector of frame $O_c X_c Y_c Z_c$
θ_h	Orientation vector of frame $O_h X_h Y_h Z_h$
θ_o	Orientation vector of frame $O_o X_o Y_o Z_o$
θ_{co}	Orientation vector of frame $O_c X_c Y_c Z_c$ expressed in $O_o X_o Y_o Z_o$
θ_{ho}	Orientation vector of frame $O_h X_h Y_h Z_h$ expressed in $O_o X_o Y_o Z_o$

---

is moving, it is usually assumed, for the ease of analysis, to be held tightly and thus has no relative motion with respect to the end-effectors of the mobile manipulators. These works have focused on dynamics based on pre-defined, fixed constraints among them. These works are not applicable to some applications which require both the motion of the object and its relative motion with respect to the end-effectors of the manipulators, such as sweeping tasks and cooperating assembly tasks by two or multiple mobile manipulators. The motion of the object with respect to the mobile manipulators can also be utilized to cope with the limited operational space and to increase task efficiency. Such tasks need simultaneous control of position and force in the given direction; therefore, impedance control like [172], [75], [176] may not be applicable.

In [15], possible kinds of cooperation for the industrial robotic systems were listed, including arc welding systems for complex contours, paint spraying of moving work pieces, belt picking and palletizing. In [14], a robotic system for arc welding was presented, where the cooperation movements are defined between the robot and the positioner for considerable efficiency at the robot station. In [13], the cooperation of a part positioning table and a manipulator for welding purpose was presented. The part positioning table manipulates the part into a position and orientation under the given task constraints and the manipulator produces the desired touch motion to complete the welding. Through this relative motion coordination, the welding velocity and the efficiency of the task can be significantly improved.

For space, undersea and modern manufacture robotic applications, there is a demand for performing tasks involving the assembly or disassembly of two objects without any special equipment, because the environment is not structured and not a priori known. Assembly and disassembly operations are decomposed into two types of tasks: independent and cooperative tasks. Independent tasks are characterized by the control of the absolute position and orientation of the robots to achieve separate but related goals. Cooperative tasks are characterized by the control of the relative position, orientation and contact force between the end-effectors. In this case, two robots can be used for assembling the objects in space, with each object being held by one robot [24]. It is necessary to develop a certain form of hybrid control scheme in order to control the relative motion/force between the objects and thus to carry out the task efficiently. The task of mating two sub-assemblies is a general example of a cooperative task that also requires control of the relative motion/force of the end-effectors.

In this chapter, we consider tasks for multiple mobile manipulators in which the following conditions may hold: a) the robots are kinematically constrained; and b) the robots are not physically connected, but work on a common object in completing a task, i.e., one robot manipulates a rigid payload while another spreads adhesive on the edges, with both robots in motion simultaneously. Conventional centralized and decentralized coordination schemes have not addressed cooperation tasks adequately, though the leader-follower scheme may be a solution. Therefore, there exists increasing demand for developing decentralized/centralized cooperation scheme by the mobility of multiple mobile manipulators that is applicable to all cooperation tasks mentioned above. Another motivation for developing a cooperation scheme is incorporating hybrid position and force control architecture with leader-follower coordination for easy and efficient implementation.

It should be noted that the success of the schemes [68], [69], [172], [173], [174] for coordinated controls of multiple mobile manipulators relies on knowledge of the complex dynamics of the robotic system. If there are parametric uncertainties in the dynamic model, such as the payload, the control designed without considering these uncertainties may lead to degraded performance and compromise the stability of the system. To deal with the uncertainties, adaptive and robust coordinated control schemes should be proposed.

Recently, some works have successfully incorporated adaptive controls to deal with dynamics uncertainty of a single mobile manipulator. In [115], adaptive neural network-based controls for the arm and the base had been proposed for the motion control of a mobile manipulator. Adaptive control was proposed for trajectory control of mobile manipulators subjected to nonholonomic constraints with unknown inertia parameters [103], which ensures the state of the system to asymptotically converge to the desired trajectory.

In this chapter, we shall investigate situations where one mobile robotic manipulator (referred as mobile manipulator I) performs the constrained motion on the surface of an object which is held tightly by another mobile robotic manipulator (referred as manipulator II) [104]. Mobile manipulator II is to be controlled in such a manner that the constraint object follows the planned motion trajectory, while mobile manipulator I is to be controlled such that its end-effector follows a planned trajectory on the surface with the desired contact force. We first present the dynamics of two mobile robotic manipulators manipulating an object with relative motion. This will be followed by centralized robust adaptive control to guarantee the convergence of the motion/force trajectories tracking of the constraint object under the parameter uncertainties and the external disturbances.

8.2   Description of Interconnected System

The system under study is schematically shown in Fig. 8.1. The object is held tightly by the end effector of mobile manipulator II and can be moved as required in space. The end effector of mobile manipulator I follows a trajectory on the surface of the object, and at the same time exerts a certain desired force on the object.

Assumption 8.1 *The surface of the object on which the end-effector of the mobile arm I moves is geometrically known.*

8.2.1   Kinematic Constraints of the System

The closed kinematic relationships of the system are given by the following equations [104]:

$$x_c = x_o + R_o(\theta_o)x_{co} \tag{8.1}$$

$$x_h = x_o + R_o(\theta_o)x_{ho} \tag{8.2}$$

$$R_c = R_o(\theta_o)R_{co}(\theta_{co}) \tag{8.3}$$

$$R_h = R_o(\theta_o) \tag{8.4}$$

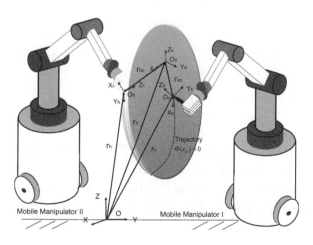

FIGURE 8.1

Cooperation operation of two robots

where $R_o(\theta_o) \in \mathbf{R}^{3\times3}$ and $R_{co}(\theta_{co}) \in \mathbf{R}^{3\times3}$ are the rotation matrices of θ_o and θ_{co} respectively; $R_c \in \{\mathbf{R}\}^{3\times3}$ and $R_h \in \{\mathbf{R}\}^{3\times3}$ are the rotation matrices of frames $O_cX_cY_cZ_c$ and $O_hX_hY_hZ_h$ with respect to the world coordinates, respectively. Differentiating the above equations with respect to time t and considering that the object is tightly held by manipulator II (accordingly, $\dot{x}_{ho} = 0$ and $\omega_{ho} = 0$), we have

$$\dot{x}_c = \dot{x}_o + R_o(\theta_o)\dot{x}_{co} - S(R_o(\theta_o)x_{co})\omega_o \qquad (8.5)$$

$$\dot{x}_h = \dot{x}_o - S(R_o(\theta_o)x_{ho})\omega_o \qquad (8.6)$$

$$\omega_c = \omega_o + R_o(\theta_o)\omega_{co} \qquad (8.7)$$

$$\omega_h = \omega_o \qquad (8.8)$$

with

$$S(u) := \begin{bmatrix} 0 & -u_3 & u_2 \\ u_3 & 0 & -u_1 \\ -u_2 & u_1 & 0 \end{bmatrix}$$

for a given vector $u = [u_1, u_2, u_3]^T$. Define $v_c = [\dot{x}_c^T, \omega_c^T]^T$, $v_h = [\dot{x}_h^T, \omega_h^T]^T$, $v_o = [\dot{x}_o^T, \omega_o^T]^T$, $v_{co} = [\dot{x}_{co}^T, \omega_{co}^T]^T$, and $v_{ho} = [\dot{x}_{ho}^T, \omega_{ho}^T]^T$. From (8.1) - (8.8), we have the following relationship:

$$v_c = Pv_o + R_Av_{co} \qquad (8.9)$$

$$v_h = Qv_o \qquad (8.10)$$

where

$$R_A = \begin{bmatrix} R_o(\theta_o) & 0 \\ 0 & R_o(\theta_o) \end{bmatrix}, P = \begin{bmatrix} I^{3\times3} & -S(R_o(\theta_o)x_{co}) \\ 0 & I^{3\times3} \end{bmatrix},$$

$$Q = \begin{bmatrix} I^{3\times3} & -S(R_o(\theta_o)x_{ho}) \\ 0 & I^{3\times3} \end{bmatrix}$$

Since $R_o(\theta_o)$ is a rotation matrix, $R_o(\theta_o)R_o^T(\theta_o) = I^{3\times3}$ and $R_AR_A^T = I^{6\times6}$. It is obvious that P and Q are of full rank.

From Assumption 8.1, suppose that the end-effector of the mobile manipulator I follows the trajectory $\Phi(r_{co}) = 0$ in the object coordinates. The contact force f_c is given by

$$f_c = R_A J_c^T \lambda_c \qquad (8.11)$$

$$J_c = \frac{\partial\Phi/\partial r_{co}}{\|\partial\Phi/\partial r_{co}\|} \qquad (8.12)$$

where λ_c is a Lagrange multiplier related to the magnitude of the contact force. The resulting force f_o due to f_c is thus derived as follows:

$$f_o = -P^T R_A J_c^T \lambda_c \tag{8.13}$$

8.2.2 Robot Dynamics

Consider two cooperating n-DOF mobile manipulators with nonholonomic mobile platforms as shown in Fig. 8.1. Combining (8.11) and (8.13), the dynamics of the constrained mobile manipulators can be described as

$$M_1(q_1)\ddot{q}_1 + C_1(q_1,\dot{q}_1)\dot{q}_1 + G_1(q_1) + d_1(t) = B_1\tau_1 + J_1^T\lambda_1 \tag{8.14}$$

$$M_2(q_2)\ddot{q}_2 + C_2(q_2,\dot{q}_2)\dot{q}_2 + G_2(q_2) + d_2(t) = B_2\tau_2 + J_2^T\lambda_2 \tag{8.15}$$

where

$$M_i(q_i) = \begin{bmatrix} M_{ib} & M_{iva} \\ M_{iab} & M_{ia} \end{bmatrix}, C_i(q_i,\dot{q}_i) = \begin{bmatrix} C_{ib} & C_{iba} \\ C_{iab} & C_{ia} \end{bmatrix}$$

$$G_i(q_i) = \begin{bmatrix} G_{ib} \\ G_{ia} \end{bmatrix}, d_i(t) = \begin{bmatrix} d_{ib}(t) \\ d_{ia}(t) \end{bmatrix} (i = 1,2)$$

$$J_1^T(q_1) = \begin{bmatrix} A_1^T & J_{1b}^T \\ 0 & J_{1a}^T \end{bmatrix} \begin{bmatrix} I & 0 \\ 0 & R_A J_c^T \end{bmatrix},$$

$$J_2^T(q_2) = \begin{bmatrix} A_2^T & J_{2b}^T \\ 0 & -J_{2a}^T P^T \end{bmatrix} \begin{bmatrix} I & 0 \\ 0 & R_A J_c^T \end{bmatrix}$$

$$\lambda_1 = \begin{bmatrix} \lambda_{1n} \\ \lambda_c \end{bmatrix}, \lambda_2 = \begin{bmatrix} \lambda_{2n} \\ \lambda_c \end{bmatrix}$$

$M_i(q_i) \in \{\mathbf{R}\}^{n_i \times n_i}$ is the symmetric bounded positive definite inertia matrix, $C_i(q_i,\dot{q}_i)\dot{q} \in \{\mathbf{R}\}^{n_i}$ denotes the centripetal and Coriolis forces; $G_i(q_i) \in \{\mathbf{R}\}^{n_i}$ are the gravitational forces; $\tau_i \in \{\mathbf{R}\}^{p_i}$ is the vector of control inputs; $B_i \in \{\mathbf{R}\}^{n_i \times p_i}$ is a full rank input transformation matrix and is assumed to be known because it is a function of the fixed geometry of the system; $d_i(t) \in \{\mathbf{R}\}^{n_i}$ is the disturbance vector; $q_i = [q_{ib}^T, q_{ia}^T]^T \in \{\mathbf{R}\}^{n_i}$, and $q_{ib} \in \{\mathbf{R}\}^{n_{iv}}$ describes the generalized coordinates for the mobile platform; $q_{ia} \in \{\mathbf{R}\}^{n_{ia}}$ are the coordinates of the manipulator, and $n_i = n_{iv} + n_{ia}$; $F_i = J_i^T\lambda_i \in \{\mathbf{R}\}^{n_i}$ denotes the vector of constraint forces; the $n_{iv} - m$ nonintegrable and independent velocity constraints can be expressed as $A_i\dot{q}_{ib} = 0$; $\lambda_i = [\lambda_{in}^T, \lambda_c^T]^T \in \{\mathbf{R}\}^{p_i}$ with λ_{in} being the Lagrangian multipliers with the nonholonomic constraints.

Assumption 8.2 *There is sufficient friction between the wheels of the mobile platforms and the surface such that the wheels do not slip.*

Proposition 8.1 *Consider the drift-free nonholonomic system*

$$\dot{q}_v = r_1(q_v)\dot{z}_1 + \ldots + r_m(q_v)\dot{z}_m$$

where $r_i(q_v)$ are smooth, linearly independent input vector fields. There exist state transformation $X = \mathcal{T}_1(q_v)$ and feedback $\dot{z} = \mathcal{T}_2(q_v)u_b$ on some open set $U \subset \{\mathbf{R}\}^n$ to transform the system into an $(m-1)$-chain, single-generator chained form, if and only if there exists a basis $f_1, \ldots, f_m$ for $\Delta_0 := \mathrm{span}\{r_1, \ldots, r_m\}$ which has the form

$$f_1 = (\partial/\partial q_{v1}) + \sum_{i=2}^{n_v} f_1^i(q_v)\partial/\partial q_{vi}$$

$$f_j = \sum_{i=1}^{n} f_j^i(q_v)\partial/\partial q_{vi}, \quad 2 \leq j \leq m$$

such that the distributions

$$G_j = \mathrm{span}\{\mathrm{ad}_{f_1}^i f_2, \ldots, \mathrm{ad}_{f_1}^i f_m : 0 \leq i \leq j\}, \quad 0 \leq j \leq n_v - 1$$

have constant dimensions on U and are all involutive, and G_{n_v-1} has dimensions $n_v - 1$ on U [84].

Under Assumption 8.2, we have $A_i \dot{q}_{ib} = 0$ with $A_i(q_{ib}) \in \{\mathbf{R}\}^{(n_{iv}-m) \times n_{iv}}$, and it is always possible to find a m rank matrix $H_i(q_{ib}) \in \{\mathbf{R}\}^{n_{iv} \times m}$ formed by a set of smooth and linearly independent vector fields spanning the null space of A_i, i.e.,

$$H_i^T(q_{ib})A_i^T(q_{ib}) = 0_{m \times (n_{iv}-m)} \tag{8.16}$$

Since $H_i = [h_{i1}(q_{ib}), \ldots, h_{im}(q_{ib})]$ is formed by a set of smooth and linearly independent vector fields spanning the null space of $A_i(q_{ib})$, we can define an auxiliary time function $v_{ib} = [v_{ib1}, \ldots, v_{ibm}]^T \in \{\mathbf{R}\}^m$ such that

$$\dot{q}_{ib} = H_i(q_{ib})v_{ib} = h_{i1}(q_{ib})v_{ib1} + \cdots + h_{im}(q_{ib})v_{ibm} \tag{8.17}$$

which is the so-called kinematics of a nonholonomic system. Let $v_{ia} = \dot{q}_{ia}$. One can obtain

$$\dot{q}_i = R_i(q_i)v_i \tag{8.18}$$

where $v_i = [v_{ib}^T, v_{ia}^T]^T$, and $R_i(q_i) = \mathrm{diag}[H_i(q_{ib}), I_{n_{ia} \times n_{ia}}]$.

Differentiating equation (8.18) yields

$$\ddot{q}_i = \dot{R}_i(q_i)v_i + R_i(q_i)\dot{v}_i \qquad (8.19)$$

Substituting (8.19) into (8.14) and (8.15), and multiplying both sides with $R_i^T(q_i)$ to eliminate λ_{in}, yields:

$$M_{i1}(q_i)\dot{v}_i + C_{i1}(q_i, \dot{q}_i)v_i + G_{i1}(q_i) + d_{i1}(t) = B_{i1}(q_i)\tau + J_{i1}^T\lambda_i \qquad (8.20)$$

where $M_{i1}(q_i) = R_i(q_i)^T M_i(q_i)R_i$, $C_{i1}(q_i, \dot{q}_i) = R_i^T(q_i)M_i(q_i)\dot{R}_i(q_i)$ $+R_i^T C_i(q_i, \dot{q}_i)R_i(q_i)$, $G_{i1}(q_i) = R_i^T(q_i)G_i(q_i)$, $d_{i1}(t) = R_i^T(q_i)d_i(t)$, $B_{i1} = R_i^T(q_i)B_i(q_i)$, $J_{i1}^T = R_i^T(q_i)J_i^T$, $\lambda_i = \lambda_c$.

Assumption 8.3 *There exists some diffeomorphic state transformation $T_2(q)$ for the class of nonholonomic systems considered in this paper such that the kinematic nonholonomic subsystem (8.18) can be globally transformed into the chained form.*

$$\begin{cases} \dot{\zeta}_{ib1} = u_{i1} \\ \dot{\zeta}_{ibj} = u_{i1}\zeta_{ib(j+1)}(2 \leq j \leq n_v - 1) \\ \dot{\zeta}_{ibn_v} = u_{i2} \\ \dot{\zeta}_{ia} = \dot{q}_{ia} = u_{ia} \end{cases} \qquad (8.21)$$

where

$$\zeta_i = [\zeta_{ib}^T, \zeta_{ia}^T]^T = T_1(q_i) = [T_{11}^T(q_{ib}), q_{ia}^T]^T \qquad (8.22)$$

$$v_i = [v_{ib}^T, v_{ia}^T]^T = T_2(q_i)u_i = [(T_{21}(q_{ib})u_{ib})^T, u_{ia}^T]^T \qquad (8.23)$$

with $T_2(q) = \text{diag}[T_{21}(q_{iv}), I]$ and $u = [u_{ib}^T, u_{ia}^T]^T$, $u_{ia} = \dot{q}_{ia}$.

Remark 8.1 *This assumption is reasonable and examples of a nonholonomic system which can be globally transformed into the chained form are the differentially driven wheeled mobile robot and the unicycle wheeled mobile robot [25]. A necessary and sufficient condition was given for the existence of the transformation $T_2(q)$ of the kinematic system (8.18) with a differentially driven wheeled mobile robot into this chained form (single chain) [10, 25]. For the other types of mobile platform (multichain case), the discussion on the existence condition of the transformation is given in Proposition 8.1.*

Considering the above transformations, the dynamic system (8.14) and (8.15) could be converted into the following canonical transformation, for $i = 1, 2$

$$M_{i2}(\zeta_i)\dot{u}_i + C_{i2}(\zeta_i, \dot{\zeta}_i)u_i + G_{i2}(\zeta_i) + d_{i2}(t) = B_{i2}\tau_i + J_{i2}^T\lambda_i \qquad (8.24)$$

where

$$M_{i2}(\zeta_i) = T_2^T(q_i)M_{i1}(q)T_2(q_i)|_{q_i=T_1^{-1}(\zeta_i)}$$
$$C_{i2}(\zeta_i,\dot{\zeta}_i) = T_2^T(q_i)[M_{i1}(q_i)\dot{T}_2(q_i) + C_{i1}(q_i,\dot{q}_i)T_2(q_i)]|_{q_i=T_1^{-1}(\zeta_i)}$$
$$G_{i2}(\zeta_i) = T_2^T(q_i)G_{i1}(q_i)|_{q_i=T_1^{-1}(\zeta_i)}$$
$$d_{i2}(t) = T_2^T(q_i)d_i(t)|_{q_i=T_1^{-1}(\zeta_i)}$$
$$B_{i2} = T_2^T(q_i)B_{i1}(q_i)|_{q_i=T_1^{-1}(\zeta_i)}$$
$$J_{i2}^T = T_2^T(q_i)J_{i1}^T|_{q=T_1^{-1}(\zeta_i)}$$

8.2.3   Reduced Dynamics

Assumption 8.4 *The Jacobian matrix J_{i2} is uniformly bounded and uniformly continuous, if q_i is uniformly bounded and uniformly continuous.*

Assumption 8.5 *Each manipulator is redundant and operating away from any singularity.*

Remark 8.2 *Under Assumptions 8.4 and 8.5, the Jacobian J_{i2} is of full rank. The vector $q_{ia} \in \{\mathbf{R}\}^{n_{ia}}$ can always be properly rearranged and partitioned into $q_{ia} = [q_{ia}^{1T}, q_{ia}^{2T}]^T$, where $q_{ia}^1 = [q_{ia1}^1, \ldots, q_{ia(n_{ia}-\kappa_i)}^1]^T$ describes the constrained motion of the manipulator and $q_{ia}^2 \in \{\mathbf{R}\}^{\kappa_i}$ denotes the remaining joint variables which make the arm redundant such that the possible breakage of contact could be compensated.*

Therefore, we have

$$J_{i2}(q_i) = [J_{i2b}, J_{i2a}^1, J_{i2a}^2] \tag{8.25}$$

Considering the object trajectory and relative motion trajectory as holonomic constraints, we can obtain

$$\dot{q}_{ia}^2 = -(J_{i2a}^2)^{-1}[J_{i2b}u_{ib} + J_{i2a}^1\dot{q}_{ia}^1] \tag{8.26}$$

and

$$u_i = \begin{bmatrix} u_{ib} \\ \dot{q}_{ia}^1 \\ -(J_{i2a}^2)^{-1}[J_{i2b}u_{ib} + J_{i2a}^1\dot{q}_{ia}^1] \end{bmatrix} = L_iu_i^1 \tag{8.27}$$

where

$$
L_i = \begin{bmatrix} I_{m \times m} & 0 \\ 0 & I_{(n_{ia}-\kappa_i) \times (n_{ia}-\kappa_i)} \\ -(J_{i2a}^2)^{-1} J_{i2b} & -(J_{i2a}^2)^{-1} J_{i2a}^1 \end{bmatrix}
$$

$$
u_i^1 = \begin{bmatrix} u_{ib} & \dot{q}_{ia}^1 \end{bmatrix}^T
$$

where $u_i^1 \in \{\mathbf{R}\}^{(n_{ia}+m-\kappa_i)}$, and $L_i \in \{\mathbf{R}\}^{(n_{ia}+m) \times (n_{ia}+m-\kappa_i)}$. From the definition of J_{i2} in (8.25) and L_i above, we have $L_i^T J_{i2}^T = 0$.

Combining (8.24) and (8.27), we can obtain the following compact dynamics:

$$
M \dot{u}^1 + C u^1 + G + d = B\tau + J^T \lambda \tag{8.28}
$$

where

$$
M = \begin{bmatrix} M_{12} L_1 & 0 \\ 0 & M_{22} L_2 \end{bmatrix}, L = \begin{bmatrix} L_1 & 0 \\ 0 & L_2 \end{bmatrix}
$$

$$
C = \begin{bmatrix} M_{12} \dot{L}_1 + C_{12} L_1 & 0 \\ 0 & M_{22} \dot{L}_2 + C_{22} L_2 \end{bmatrix}
$$

$$
G = \begin{bmatrix} G_{12} \\ G_{22} \end{bmatrix}, B = \begin{bmatrix} B_{12} & 0 \\ 0 & B_{22} \end{bmatrix}, \lambda = \lambda_c
$$

$$
d = \begin{bmatrix} d_{12}(t) \\ d_{22}(t) \end{bmatrix}, \tau = \begin{bmatrix} \tau_1 \\ \tau_2 \end{bmatrix}, J^T = \begin{bmatrix} J_{12}^T \\ J_{22}^T \end{bmatrix}
$$

Property 8.1 *Matrices $\mathcal{M} = L^T M$, $\mathcal{G} = L^T G$ are uniformly bounded and uniformly continuous if $\zeta = [\zeta_1, \zeta_2]^T$ is uniformly bounded and continuous, respectively. Matrix $\mathcal{C} = L^T C$ is uniformly bounded and uniformly continuous if $\dot{\zeta} = [\dot{\zeta}_1, \dot{\zeta}_2]^T$ is uniformly bounded and continuous.*

Property 8.2 *$\forall \zeta \in \{\mathbf{R}\}^{n_1+n_2}$, $0 < \lambda_{min} I \leq \mathcal{M}(\zeta) \leq \beta I$ where λ_{min} is the minimal eigenvalue of $\mathcal{M}$, and $\beta > 0$.*

8.3 Robust Adaptive Controls Design

8.3.1 Problem Statement and Control Diagram

Let $r_o^d(t)$ be the desired trajectory of the object, $r_{co}^d(t)$ be the desired trajectory on the object and $\lambda_c^d(t)$ be the desired constraint force. The first control objective is to drive the mobile manipulators such that $r_o(t)$ and $r_{co}(t)$ track their desired trajectories $r_o^d(t)$ and $r_{co}^d(t)$ respectively. Accordingly it is only necessary to make q track the desired trajectory $q^d = [q_1^{dT}, q_2^{dT}]^T$ since $q = [q_1^T, q_2^T]^T$ completely determines $r_o(t)$ and $r_{co}(t)$. Under Assumption 2.4, with the desired joint trajectory q^d, there exists a transformation $\dot{q}^d = R(q^d)v^d$, $\zeta^d = T_1(q^d)$ and $u_d = T_2^{-1}(q^d)v^d$ where $v^d = [v_1^{dT}, v_2^{dT}]^T$, $v = [v_1^T, v_2^T]^T$, $\zeta^d = [\zeta_1^{dT}, \zeta_2^{dT}]^T$, $\zeta = [\zeta_1^T, \zeta_2^T]^T$, $u_d = [u_{1d}^T, u_{2d}^T]^T$ and $u = [u_1^T, u_2^T]^T$. Therefore, the tracking problem can be treated as formulating a control strategy such that $\zeta \to \zeta^d$ and $u \to u_d$ as $t \to \infty$. The second control objective is to make $\lambda_c(t)$ track the desired trajectory $\lambda_c^d(t)$.

Definition 8.1 *Consider time-varying positive functions δ_k and α_ς which converge to zero as $t \to \infty$ and satisfy $\lim_{t\to\infty} \int_0^t \delta_k(\omega)d\omega = a_k < \infty$, $\lim_{t\to\infty} \int_0^t \alpha_\varsigma(\omega)d\omega = b_\varsigma < \infty$, with finite constants a_k and b_ς, where $k = 1, \ldots, 6$ and $\varsigma = 1, \ldots, 5$. There are many choices for δ_k and α_ς that satisfy the above condition, for example, $\delta_k = \alpha_\varsigma = 1/(1+t)^2$.*

8.3.2 Control Design

The complete model of the coordinated nonholonomic mobile manipulators consists of the two cascaded subsystems (8.21) and the combined dynamic model (8.28). As a consequence, the generalized velocity u cannot be used to control the system directly, as assumed in the design of controllers at the kinematic level. Instead, the desired velocities must be realized through the design of the control inputs τ (8.28). The above properties imply that the dynamics (8.28) retain the mechanical system structure of the original system (8.15), which is fundamental for designing the robust control law. In this section, we will develop a strategy so that the subsystem (8.21) tracks ζ^d through the design of a virtual control z, defined in (8.29) and (8.30) below, and at the same time, the output of mechanical subsystem (8.28) is controlled to track this desired signal. In turn, the tracking goal can be achieved.

For the given $\zeta^d = [\zeta_1^{dT}, \zeta_2^{dT}]^T$, the tracking errors are denoted as $e = \zeta - \zeta^d = [e_1^T, e_2^T]^T$, $e_i = [e_{ib}^T, e_{ia}^T]^T$, $e_{ib} = [e_{i1}, e_{i2}, \ldots, e_{in_v}]^T = \zeta_{ib} - \zeta_{ib}^d$, $e_{ia} = \zeta_{ia} - \zeta_{ia}^d$, and $e_\lambda = \lambda_c - \lambda_c^d$. Define the virtual control $z = [z_1^T, z_2^T]^T$ and $z_i = [z_{ib}^T, z_{ia}^T]^T$ as follows:

$$z_{ib} = \begin{bmatrix} u_{id1} + \eta_i \\ u_{id2} - s_{i(n_{iv}-1)}u_{id1} - k_{n_{iv}}s_{in_{iv}} \\ + \sum_{j=0}^{n_{iv}-3} \frac{\partial(e_{in_{iv}} - s_{in_{iv}})}{\partial u_{id1}^{(j)}} u_{id1}^{(j+1)} \\ + \sum_{j=2}^{n_{iv}-1} \frac{\partial(e_{in_{iv}} - s_{in_{iv}})}{\partial e_{ij}} e_{i(j+1)} \end{bmatrix} \tag{8.29}$$

$$z_{ia} = \dot{q}_{ia}^{1d} - K_{1a}(q_{ia}^1 - q_{ia}^{1d}) \tag{8.30}$$

$$s_i = \begin{bmatrix} e_{i1} \\ e_{i2} \\ \vdots \\ e_{in_{iv}} + s_{i(n_{iv}-2)} + k_{n_{iv}-1}s_{i(n_{iv}-1)}u_{id1}^{2l-1} \\ - \frac{1}{u_{id1}} \sum_{j=0}^{n_{iv}-4} \frac{\partial(e_{i(n_{iv}-1)} - s_{i(n_{iv}-1)})}{\partial u_{id1}^{(j)}} u_{id1}^{(j+1)} \\ - \sum_{j=2}^{n_{iv}-2} \frac{\partial(e_{i(n_{iv}-1)} - s_{i(n_{iv}-1)})}{\partial e_{ij}} e_{i(j+1)} \end{bmatrix} \tag{8.31}$$

$$\dot{\eta}_i = -k_0\eta_i - k_1 s_{i1} - \sum_{j=2}^{n_{iv}-1} s_{ij}\zeta_{i(j+1)}$$

$$+ \sum_{k=3}^{n_{iv}} s_{ik} \sum_{j=2}^{k-1} \frac{\partial(e_{ik} - s_{ik})}{\partial e_{ik}} \zeta_{i(k+1)} \tag{8.32}$$

and $l = n_{iv} - 2$, $u_{id1}^{(l)}$ is the lth derivative of u_{id1} with respect to t, and k_j is a positive constant and K_{ia} is diagonal positive.

Denote $\tilde{u} = [\tilde{u}_b, \tilde{u}_a]^T = [u_b - z_b, u_a - z_a]^T$ and define a filter tracking error

$$\sigma = \begin{bmatrix} u_b \\ \tilde{u}_a \end{bmatrix} + K_u \int_0^t \tilde{u}\,ds \tag{8.33}$$

with $K_u = \mathrm{diag}[0_{m \times m}, K_{u1}] > 0$, $K_{u1} \in \{\mathbf{R}\}^{(n_{ia}-\kappa_i) \times (n_{ia}-\kappa_i)}$. We could obtain $\dot{\sigma} = \begin{bmatrix} \dot{u}_b \\ \dot{\tilde{u}}_a \end{bmatrix} + K_u \tilde{u}$ and $u = \nu + \sigma$ with $\nu = \begin{bmatrix} 0 \\ z_a \end{bmatrix} - K_u \int_0^t \tilde{u}\,ds$.

We could rewrite (8.28) as

$$M\dot{\sigma} + C\sigma + M\dot{\nu} + C\nu + G + d = B\tau + J^T\lambda \tag{8.34}$$

If the system is certain, we could choose the control law given by

$$B\tau = M(\dot{\nu} - K_\sigma\sigma) + C(\nu + \sigma) + G + d - J^T\lambda_h \tag{8.35}$$

with diagonal matrix $K_\sigma > 0$ and the force control input λ_h as

$$\lambda_h = \lambda_d - K_\lambda\tilde{\lambda} - K_I \int_0^t \tilde{\lambda}\,dt \tag{8.36}$$

where $\tilde{\lambda} = \lambda_c - \lambda_c^d$, K_λ is a constant matrix of proportional control feedback gains, and K_I is a constant matrix of integral control feedback gains.

However, since $\mathcal{M}(\zeta)$, $\mathcal{C}(\zeta, \dot{\zeta})$, and $\mathcal{G}(\zeta)$ are uncertain, to facilitate the control formulation, the following assumption is required.

Assumption 8.6 *There exist some finite positive constants b, $c_\varsigma > 0$ ($1 \leq \varsigma \leq 4$), and finite non-negative constant $c_5 \geq 0$ such that $\forall \zeta \in \{\mathbf{R}\}^{2n}$, $\forall \dot{\zeta} \in \{\mathbf{R}\}^{2n}$, $\|\Delta M\| = \|\mathcal{M} - \mathcal{M}_0\| \leq c_1$, $\|\Delta C\| = \|\mathcal{C} - \mathcal{C}_0\| \leq c_2 + c_3\|\dot{\zeta}\|$, $\|\Delta G\| = \|\mathcal{G} - \mathcal{G}_0\| \leq c_4$, and $\sup_{t \geq 0}\|d_L(t)\| \leq c_5$, where M_0, C_0 and G_0 are nominal parameters of the system [12, 11].*

Let $\mathcal{B} = L^T B$, the proposed control for the system is given as

$$\mathcal{B}\tau = \mathrm{U}_1 + \mathrm{U}_2 \tag{8.37}$$

where U_1 is the nominal control

$$\mathrm{U}_1 = \mathcal{M}_0(\dot{\nu} - K_\sigma\sigma) + \mathcal{C}_0(\nu + \sigma) + \mathcal{G}_0 \tag{8.38}$$

and U_2 is designed to compensate for the parametric errors arising from estimating the unknown functions $\mathcal{M}$, $\mathcal{C}$ and $\mathcal{G}$ and the disturbance, respectively.

$$\mathrm{U}_2 = \mathrm{U}_{21} + \mathrm{U}_{22} + \mathrm{U}_{23} + \mathrm{U}_{24} + \mathrm{U}_{25} + \mathrm{U}_{26} \tag{8.39}$$

$$\mathrm{U}_{21} = -\frac{\beta}{\lambda_{min}} \frac{\hat{c}_1^2 \|K_\sigma\sigma - \dot{\nu}\|^2 \sigma}{\hat{c}_1\|K_\sigma\sigma - \dot{\nu}\|\|\sigma\| + \delta_1} \tag{8.40}$$

$$\mathrm{U}_{22} = -\frac{\beta}{\lambda_{min}} \frac{\hat{c}_2^2 \|\sigma + \nu\|^2 \sigma}{\hat{c}_2\|\sigma + \nu\|\|\sigma\| + \delta_2} \tag{8.41}$$

$$\mathrm{U}_{23} = -\frac{\beta}{\lambda_{min}} \frac{\hat{c}_3^2 \|\dot{\zeta}\|^2 \|\sigma + \nu\|^2 \sigma}{\hat{c}_3\|\dot{\zeta}\|\|\sigma + \nu\|\|\sigma\| + \delta_3} \tag{8.42}$$

$$\mathrm{U}_{24} = -\frac{\beta}{\lambda_{min}} \frac{\hat{c}_4^2 \sigma}{\hat{c}_4\|\sigma\| + \delta_4} \tag{8.43}$$

$$\mathrm{U}_{25} = -\frac{\beta}{\lambda_{min}} \frac{\hat{c}_5^2 \|L\|^2 \sigma}{\hat{c}_5\|L\|\|\sigma\| + \delta_5} \tag{8.44}$$

$$\mathrm{U}_{26} = -\beta\frac{\|\tilde{u}_b\|\|\Lambda\|^2 \sigma}{\|\Lambda\|\|\sigma\| + \delta_6} \tag{8.45}$$

where δ_k ($k = 1, \ldots, 6$) satisfies the conditions defined in Definition 8.1, and $\hat{c}_\varsigma$ denotes the estimate c_ς, which are adaptively tuned according to

$$\dot{\hat{c}}_1 = -\alpha_1\hat{c}_1 + \frac{\gamma_1}{\lambda_{min}}\|\sigma\|\|K_\sigma\sigma - \dot{\nu}\|, \hat{c}_1(0) > 0 \tag{8.46}$$

$$\dot{\hat{c}}_2 = -\alpha_2\hat{c}_2 + \frac{\gamma_2}{\lambda_{min}}\|\sigma\|\|\sigma + \nu\|, \hat{c}_2(0) > 0 \tag{8.47}$$

$$\dot{\hat{c}}_3 = -\alpha_3\hat{c}_3 + \frac{\gamma_3}{\lambda_{min}}\|\sigma\|\|\dot{\zeta}\|\|\sigma + \nu\|, \hat{c}_3(0) > 0 \qquad (8.48)$$

$$\dot{\hat{c}}_4 = -\alpha_4\hat{c}_4 + \frac{\gamma_4}{\lambda_{min}}\|\sigma\|, \ \hat{c}_4(0) > 0 \qquad (8.49)$$

$$\dot{\hat{c}}_5 = -\alpha_5\hat{c}_5 + \frac{\gamma_5}{\lambda_{min}}\|L\|\|\sigma\|, \ \hat{c}_5(0) > 0 \qquad (8.50)$$

with $\alpha_\varsigma > 0$ satisfying the condition in Definition 8.1 and $\gamma_\varsigma > 0$, ($\varsigma = 1, \ldots, 5$), and

$$\Lambda = \begin{bmatrix} \Lambda_1 & \Lambda_2 \end{bmatrix}^T \qquad (8.51)$$

$$\Lambda_i = \begin{bmatrix} k_1 s_{i1} + \sum_{j=2}^{n_v-1} s_{ij}\zeta_{i(j+1)} \\ -\sum_{j=3}^{n_v} s_j \sum_{k=2}^{j-1} \frac{\partial(e_{ik}-s_{ik})}{\partial e_{ik}}\zeta_{i(k+1)} \\ s_{in_v} \\ 0 \end{bmatrix} \qquad (8.52)$$

Remark 8.3 *The variables* $U_{21}, \ldots, U_{26}$ *are to compensate for the parametric errors arising from estimating the unknown functions* $\mathcal{M}$, $\mathcal{C}$ *and* $\mathcal{G}$ *and the disturbance. The choice of the variables in (8.40)-(8.45) is to avoid the use of sign functions which will lead to chattering. Based on the definition of* δ_k *in Definition 8.1, the denominators in (8.40)-(8.45) are non-negative and will only approach 0 when* $\delta_k \to 0$. *However, when* $\delta_k = 0$, *we can rewrite the equations in (8.40)-(8.45) as:*

$$U_{21} = -\frac{\beta}{\lambda_{min}}\hat{c}_1\|K_\sigma\sigma - \dot{\nu}\|\text{sgn}(\sigma)$$

$$U_{22} = -\frac{\beta}{\lambda_{min}}\hat{c}_2\|\sigma + \nu\|\text{sgn}(\sigma)$$

$$U_{23} = -\frac{\beta}{\lambda_{min}}\hat{c}_3\|\dot{\zeta}\|\|\sigma + \nu\|\text{sgn}(\sigma)$$

$$U_{24} = -\frac{\beta}{\lambda_{min}}\hat{c}_4\text{sgn}(\sigma)$$

$$U_{25} = -\frac{\beta}{\lambda_{min}}\hat{c}_5\|L\|\text{sgn}(\sigma)$$

$$U_{26} = -\beta\|\tilde{u}_b\|\|\Lambda\|\text{sgn}(\sigma)$$

From the above expressions, we can see that the variables $U_{21}, \ldots, U_{26}$ *are bounded when* $\hat{c}_\varsigma$, ζ, σ, v, $\dot{v}$, $\dot{\zeta}$, Λ *are bounded. As such, there is no division by zero in the control design.*

Remark 8.4 *Noting equations (8.40)-(8.45), and the corresponding adaptive laws (8.46)-(8.50), the signals required for the implementation of the adaptive* *robust control are σ, $\dot{\nu}$, ν, $\dot{\zeta}$ and Λ. Acceleration measurements are not required for the adaptive robust control.*

Remark 8.5 *For the computation of the control τ, we require the left inverse of the matrix $\mathcal{B}$ to exist such that $\mathcal{B}^+\mathcal{B} = \mathcal{B}^T(\mathcal{B}\mathcal{B}^T)^{-1}\mathcal{B} = I$. The matrix $\mathcal{B}$ can be written as*

$$\mathcal{B} = \begin{bmatrix} L_1^T T_2^T R_1^T B_1 & 0 \\ 0 & L_2^T T_2^T R_2^T B_2 \end{bmatrix}$$

The definition of L_i, in the equation below (30) is given by

$$L_i = \begin{bmatrix} I_{m \times m} & 0 \\ 0 & I_{(n_{ia}-\kappa_i) \times (n_{ia}-\kappa_i)} \\ -(J_{i2a}^2)^{-1} J_{i2b} & -(J_{i2a}^2)^{-1} J_{i2a}^1 \end{bmatrix} \in \{\mathbf{R}\}^{(n_{ia}+m) \times (n_{ia}+m-\kappa_i)}$$

L_i^T is full row ranked and the left inverse of L_i^T exists. The matrix R_i is defined as

$$R_i = \begin{bmatrix} H_i & 0 \\ 0 & I_{n_{ia} \times n_{ia}} \end{bmatrix} \in \{\mathbf{R}\}^{n_i \times (n_{ia}+m)}$$

Since $H_i \in \mathbb{R}^{n_{iv} \times m}$ is formed by a set of m smooth and linearly independent vector fields, R_i^T is full row ranked and the left inverse of R_i^T exists.
Since the matrices L_i^T and R_i^T are full row ranked, B_i is a full ranked input transformation matrix and T_2 is a diffeomorphism and there exists a left inverse of the matrix $\mathcal{B}$ such that $\mathcal{B}^+\mathcal{B} = \mathcal{B}^T(\mathcal{B}\mathcal{B}^T)^{-1}\mathcal{B} = I$.

Remark 8.6 *Application of sliding mode control generally leads to the introduction of the sgn function in the control laws, which would lead to the chattering phenomenon in the practical control [23]. To reduce the chattering phenomenon, we introduce positive time-varying functions δ_j, with properties described in Definition 8.1, in the control laws (8.40)-(8.45), such that the controls are continuous for $\delta_j \neq 0$.*

8.3.3   Control Stability

Theorem 8.1 *Consider the mechanical system described by (8.24), under Assumption 8.2. Using the control law (8.37), the following can achieved:*

(i) $e_\zeta = \zeta - \zeta_d$, $\dot{e}_\zeta = \dot{\zeta} - \dot{\zeta}_d$, $e_\lambda = \lambda_c - \lambda_c^d$ converge to a small set containing the origin as $t \to \infty$; and

(ii) all the signals in the closed-loop are bounded for all $t \geq 0$.

Proof 8.1 *Combining the dynamic equation (8.34) together with (8.31), (8.32) and (8.37), the close-loop system dynamics can be written as*

$$\dot{s}_{i1} = \eta_i + \tilde{u}_{i1} \tag{8.53}$$

$$\dot{s}_{i2} = (\eta_i + \tilde{u}_{i1})\zeta_{i3} + s_{i3}u_{id1} - k_2 s_{i2}u_{id1}^{2l} \tag{8.54}$$

$$\dot{s}_{i3} = (\eta_i + \tilde{u}_{i1})(\zeta_{i4} - \frac{\partial(e_{i3} - s_{i3})}{\partial e_{i2}}\zeta_{i3}) + s_{i4}u_{id1}$$
$$-s_{i2}u_{id1} - k_3 s_{i3}u_{id1}^{2l} \tag{8.55}$$

$$\vdots$$

$$\dot{s}_{i(n_{iv}-1)} = (\eta_i + \tilde{u}_{i1})\zeta_{in_{iv}} - (\eta_i + \tilde{u}_{i1})(\sum_{j=2}^{n_{iv}-2} \frac{\partial(e_{i(n_{iv}-1)} - s_{i(n_{iv}-1)})}{\partial e_{ji}}\zeta_{i(j+1)})$$
$$+s_{in_{iv}}u_{id1} - s_{i(n_{iv}-2)}u_{id1} - k_{(n_{iv}-1)}s_{i(n_{iv}-1)}u_{id1}^{2l} \tag{8.56}$$

$$\dot{s}_{in_{iv}} = (\eta_i + \tilde{u}_{i1}) \sum_{j=2}^{n_{iv}-2} \frac{\partial(e_{in_{iv}} - s_{in_{iv}})}{\partial e_{ij}}\zeta_{i(j+1)}$$
$$-k_{n_{iv}}s_{in_{iv}} - s_{i(n_{iv}-1)}u_{id1} + \tilde{u}_{i2} \tag{8.57}$$

$$\dot{\eta}_i = -k_0\eta_i - \Lambda_{i1} \tag{8.58}$$

$$M\dot{\sigma} = -M\dot{\nu} - C(\nu + \sigma) - G - d + B\tau + J^T\lambda \tag{8.59}$$

Let $\mathcal{D} = L^T d$. Multiplying L^T on both sides of (8.59), using (8.37), one can obtain

$$\mathcal{M}\dot{\sigma} = -\mathcal{M}_0 K_\sigma\sigma + (\mathcal{M}_0 - \mathcal{M})\dot{\nu} + (\mathcal{C}_0 - \mathcal{C})(\nu + \sigma) + (\mathcal{G}_0 - \mathcal{G}) - \mathcal{D} + \mathrm{U}_2$$
$$= -\mathcal{M}K_\sigma\sigma + (\mathcal{M} - \mathcal{M}_0)K_\sigma\sigma + (\mathcal{M}_0 - \mathcal{M})\dot{\nu}$$
$$+(\mathcal{C}_0 - \mathcal{C})(\nu + \sigma) + (\mathcal{G}_0 - \mathcal{G}) - \mathcal{D} + \mathrm{U}_2$$
$$= -\mathcal{M}K_\sigma\sigma + \Delta M(K_\sigma\sigma - \dot{\nu}) - \Delta C(\nu + \sigma) - \Delta G$$
$$-\mathcal{D} + \sum_{i=1}^{6}\mathrm{U}_{2i} \tag{8.60}$$

we have

$$\dot{\sigma} = -K_\sigma\sigma + \mathcal{M}^{-1}\Delta M(K_\sigma\sigma - \dot{\nu}) - \mathcal{M}^{-1}\Delta C(\nu + \sigma)$$
$$-\mathcal{M}^{-1}\Delta G - \mathcal{M}^{-1}\mathcal{D} + \mathcal{M}^{-1}\sum_{i=1}^{6}\mathrm{U}_{2i} \tag{8.61}$$

Let

$$\tilde{c}_\varsigma = \hat{c}_\varsigma - c_\varsigma \tag{8.62}$$

Consider the following positive definite functions:

$$V = V_1 + V_2 \tag{8.63}$$

$$V_1 = \frac{1}{2}\sum_{i=1}^{2}\sum_{j=2}^{n_{iv}} s_{ij}^2 + \frac{1}{2}\sum_{i=1}^{2} k_{i1}s_{i1}^2 + \frac{1}{2}\sum_{i=1}^{2}\eta_i^2$$

$$V_2 = \frac{1}{2}\sigma^T\sigma + \sum_{\varsigma=1}^{5}\frac{1}{2\gamma_\varsigma}\tilde{c}_\varsigma^2$$

Taking the time derivative of V_1 with (8.53)-(8.58) yields,

$$\begin{aligned}
\dot{V}_1 &= \sum_{i=1}^{2}\sum_{j=2}^{n_{iv}-1} s_{ij}\dot{s}_{ij} + \sum_{i=1}^{2} k_{i1}s_{i1}\dot{s}_{i1} + \sum_{i=1}^{2}\eta_i\dot{\eta}_i \\
&= -\sum_{i=1}^{2}\sum_{j=2}^{n_{iv}-1} k_{ij}s_{ij}^2 u_{id1}^{2l} - \sum_{i=1}^{2} k_{in_{iv}}s_{in_{iv}}^2 \\
&\quad - \sum_{i=1}^{2} k_0\eta_i^2 + \tilde{u}_b^T\Lambda
\end{aligned} \tag{8.64}$$

Taking the time derivative of V_2 and integrating (8.61) yields,

$$\begin{aligned}
\dot{V}_2 &= -\sigma^T K_\sigma\sigma \\
&\quad + \left[\sigma^T\mathcal{M}^{-1}\Delta M(K_\sigma\sigma - \dot{\nu}) + \sigma^T\mathcal{M}^{-1}\mathrm{U}_{21} + \frac{1}{\gamma_1}\tilde{c}_1\dot{\hat{c}}_1\right] \\
&\quad + \left[-\sigma^T\mathcal{M}^{-1}\Delta C(\sigma + \nu) + \sigma^T\mathcal{M}^{-1}\mathrm{U}_{22} + \frac{1}{\gamma_2}\tilde{c}_2\dot{\hat{c}}_2\right] \\
&\quad + \left[\sigma^T\mathcal{M}^{-1}\mathrm{U}_{23} + \frac{1}{\gamma_3}\tilde{c}_3\dot{\hat{c}}_3\right] \\
&\quad + \left[-\sigma^T\mathcal{M}^{-1}\Delta G + \sigma^T\mathcal{M}^{-1}\mathrm{U}_{24} + \frac{1}{\gamma_4}\tilde{c}_4\dot{\hat{c}}_4\right] \\
&\quad + \left[-\sigma^T\mathcal{M}^{-1}\mathcal{D} + \sigma^T\mathcal{M}^{-1}\mathrm{U}_{25} + \frac{1}{\gamma_5}\tilde{c}_5\dot{\hat{c}}_5\right] \\
&\quad + \sigma^T\mathcal{M}^{-1}\mathrm{U}_{26}
\end{aligned} \tag{8.65}$$

Considering Property 8.2, Assumption 8.6 and (8.40), the second right hand term of (8.65) is bounded by

$$\begin{aligned}
&\sigma^T\mathcal{M}^{-1}\Delta M(K_\sigma\sigma - \dot{\nu}) + \sigma^T\mathcal{M}^{-1}u_{21} + \frac{1}{\gamma_1}\tilde{c}_1\dot{\hat{c}}_1 \\
&\leq \frac{1}{\lambda_{min}}c_1\|K_\sigma\sigma - \dot{\nu}\|\|\sigma\| \\
&\quad - \frac{1}{\lambda_{min}}\hat{c}_1^2\frac{\|K_\sigma\sigma - \dot{\nu}\|^2\|\sigma\|^2}{\hat{c}_1\|K_\sigma\sigma - \dot{\nu}\|\|\sigma\| + \delta_1} + \frac{1}{\gamma_1}\tilde{c}_1\dot{\hat{c}}_1
\end{aligned}$$

$$= \frac{1}{\lambda_{min}}\hat{c}_1\|K_\sigma\sigma - \dot{\nu}\|\|\sigma\|$$

$$-\frac{1}{\lambda_{min}}\hat{c}_1^2\frac{\|K_\sigma\sigma - \dot{\nu}\|^2}{\hat{c}_1\|K_\sigma\sigma - \dot{\nu}\| + \delta_1}$$

$$+\tilde{c}_1\left[\frac{1}{\gamma_1}\dot{\hat{c}}_1 - \frac{1}{\lambda_{min}}\|K_\sigma\sigma - \dot{\nu}\|\|\sigma\|\right]$$

$$\leq \frac{1}{\lambda_{min}}\delta_1 - \frac{\alpha_1}{\gamma_1}\tilde{c}_1\hat{c}_1$$

$$\leq \frac{1}{\lambda_{min}}\delta_1 - \frac{\alpha_1}{\gamma_1}(\hat{c}_1 - \frac{1}{2}c_1)^2 + \frac{\alpha_1}{4\gamma_1}c_1^2 \qquad (8.66)$$

The last inequality arises from $-\tilde{c}_1\hat{c}_1 = -(\hat{c}_1 - \frac{1}{2}c_1)^2 + \frac{1}{4}c_1^2$.

Similarly, considering Property 8.2, Assumption 8.6 and (8.41), and (8.42), the third right hand term of (8.65) is bounded by

$$-\sigma^T M^{-1}\Delta C(\sigma + \nu) + \sigma^T M^{-1}u_{22} + \sigma^T M^{-1}u_{23}$$

$$+\frac{1}{\gamma_2}\tilde{c}_2\dot{\hat{c}}_2 + \frac{1}{\gamma_3}\tilde{c}_3\dot{\hat{c}}_3$$

$$\leq \frac{1}{\lambda_{min}}(c_2 + c_3\|\dot{\zeta}\|)\|\sigma + \nu\|\|\sigma\|$$

$$-\frac{1}{\lambda_{min}}\hat{c}_2^2\frac{\|\sigma + \nu\|^2\|\sigma\|^2}{\hat{c}_2\|\sigma + \nu\|\|\sigma\| + \delta_2}$$

$$+\frac{1}{\gamma_2}\tilde{c}_2\dot{\hat{c}}_2 - \frac{1}{\lambda_{min}}\hat{c}_3^2\frac{\|\dot{\zeta}\|^2\|\sigma + \nu\|^2\|\sigma\|^2}{\hat{c}_3\|\dot{\zeta}\|\|\sigma + \nu\|\|\sigma\| + \delta_2} + \frac{1}{\gamma_3}\tilde{c}_3\dot{\hat{c}}_3$$

$$= \frac{1}{\lambda_{min}}\hat{c}_2\|\sigma + \nu\|\|\sigma\| - \frac{1}{\lambda_{min}}\hat{c}_2^2\frac{\|\sigma + \nu\|^2\|\sigma\|^2}{\hat{c}_2\|\sigma + \nu\|\|\sigma\| + \delta_2}$$

$$+\tilde{c}_2\left[\frac{1}{\gamma_2}\dot{\hat{c}}_2 - \frac{1}{\lambda_{min}}\|\sigma + \nu\|\|\sigma\|\right]$$

$$+\frac{1}{\lambda_{min}}\hat{c}_3\|\dot{\zeta}\|\|\sigma + \nu\|\|\sigma\|$$

$$-\frac{1}{\lambda_{min}}\hat{c}_3^2\frac{\|\dot{\zeta}\|^2\|\sigma + \nu\|^2\|\sigma\|^2}{\hat{c}_3\|\dot{\zeta}\|\|\sigma + \nu\|\|\sigma\| + \delta_3}$$

$$+\tilde{c}_3\left[\frac{1}{\gamma_3}\dot{\hat{c}}_3 - \frac{1}{\lambda_{min}}\|\dot{\zeta}\|\|\sigma + \nu\|\|\sigma\|\right]$$

$$\leq \frac{1}{\lambda_{min}}\delta_2 - \frac{\alpha_2}{\gamma_2}\tilde{c}_2\hat{c}_2 + \frac{1}{\lambda_{min}}\delta_3 - \frac{\alpha_3}{\gamma_3}\tilde{c}_3\hat{c}_3$$

$$\leq \frac{1}{\lambda_{min}}\delta_2 - \frac{\alpha_2}{\gamma_2}(\hat{c}_2 - \frac{1}{2}c_2)^2 + \frac{\alpha_2}{4\gamma_2}c_2^2$$

$$+\frac{1}{\lambda_{min}}\delta_3 - \frac{\alpha_3}{\gamma_3}(\hat{c}_3 - \frac{1}{2}c_3)^2 + \frac{\alpha_3}{4\gamma_3}c_3^2 \qquad (8.67)$$

Similarly, considering Property 8.2 and Assumption 8.6 and (8.43), the *fourth right hand term of (8.65) is bounded by*

$$\sigma^T M^{-1} \Delta G + \sigma^T M^{-1} u_{24} + \frac{1}{\gamma_4} \tilde{c}_4 \dot{\hat{c}}_4$$

$$\leq \quad \frac{1}{\lambda_{min}} c_4 \|\sigma\| - \frac{1}{\lambda_{min}} \hat{c}_4^2 \frac{\|\sigma\|^2}{\hat{c}_4 \|\sigma\| + \delta_4} + \frac{1}{\gamma_4} \tilde{c}_4 \dot{\hat{c}}_4$$

$$= \quad \frac{1}{\lambda_{min}} \hat{c}_4 \|\sigma\| - \frac{1}{\lambda_{min}} \hat{c}_4^2 \frac{\|\sigma\|^2}{\hat{c}_4 \|\sigma\| + \delta_4}$$

$$+ \tilde{c}_4 \left[\frac{1}{\gamma_4} \dot{\hat{c}}_4 - \frac{1}{\lambda_{min}} \|\sigma\| \right]$$

$$\leq \quad \frac{1}{\lambda_{min}} \delta_4 - \frac{\alpha_4}{\gamma_4} \tilde{c}_4 \hat{c}_4$$

$$\leq \quad \frac{1}{\lambda_{min}} \delta_4 - \frac{\alpha_4}{\gamma_4} (\hat{c}_4 - \frac{1}{2} c_4)^2 + \frac{\alpha_4}{4\gamma_4} c_4^2 \qquad (8.68)$$

Similarly, considering Property 8.2, Assumption 8.6 and (8.44), the fifth right hand term of (8.65) is bounded by

$$\sigma^T \mathcal{M}^{-1} D + \sigma^T \mathcal{M}^{-1} u_{25} + \frac{1}{\gamma_5} \tilde{c}_5 \dot{\hat{c}}_5$$

$$\leq \quad \frac{1}{\lambda_{min}} c_5 \|L\| \|\sigma\| - \frac{1}{\lambda_{min}} \hat{c}_5^2 \frac{\|L\|^2 \|\sigma\|^2}{\hat{c}_5 \|L\| \|\sigma\| + \delta_5} + \frac{1}{\gamma_5} \tilde{c}_5 \dot{\hat{c}}_5$$

$$= \quad \frac{1}{\lambda_{min}} \hat{c}_5 \|L\| \|\sigma\| - \frac{1}{\lambda_{min}} \hat{c}_5^2 \frac{\|L\|^2 \|\sigma\|^2}{\hat{c}_5 \|L\| \|\sigma\| + \delta_5}$$

$$+ \tilde{c}_5 \left[\frac{1}{\gamma_5} \dot{\hat{c}}_5 - \frac{1}{\lambda_{min}} \|L\| \|\sigma\| \right]$$

$$\leq \quad \frac{1}{\lambda_{min}} \delta_5 - \frac{\alpha_5}{\gamma_5} \tilde{c}_5 \hat{c}_5$$

$$\leq \quad \frac{1}{\lambda_{min}} \delta_5 - \frac{\alpha_5}{\gamma_5} (\hat{c}_5 - \frac{1}{2} c_5)^2 + \frac{\alpha_5}{4\gamma_5} c_5^2 \qquad (8.69)$$

Combining (8.64) and (8.65), we obtain

$$\dot{V} \leq -\sum_{i=1}^{2} \sum_{j=2}^{n_{iv}-1} k_{ij} s_{ij}^2 u_{id1}^{2l} - \sum_{i=1}^{2} k_{in_{iv}} s_{in_{iv}}^2 - \sum_{i=1}^{2} k_0 \eta_i^2$$

$$+ \tilde{u}_b^T \Lambda - \sigma^T K_\sigma \sigma - \sum_{\varsigma=1}^{5} \frac{\alpha_\varsigma}{\gamma_\varsigma} (\hat{c}_\varsigma - \frac{1}{2} c_\varsigma)^2$$

$$+ \frac{1}{\lambda_{min}} \sum_{k=1}^{5} \delta_k + \sum_{\varsigma=1}^{5} \frac{\alpha_\varsigma}{4\gamma_\varsigma} c_\varsigma^2 + \sigma^T \mathcal{M}^{-1} U_{26} \qquad (8.70)$$

Considering Property 8.2 and (8.45), the fourth and ninth right-hand terms of (8.70) are bounded by

$$\tilde{u}_b^T \Lambda + \sigma^T \mathcal{M}^{-1} U_{26} \leq \|\tilde{u}_b\|\|\Lambda\| - \frac{\|\tilde{u}_b\|\|\Lambda\|^2\|\sigma\|^2}{\|\Lambda\|\|\sigma\|^2 + \delta_6}$$

$$\leq \delta_6 \tag{8.71}$$

Therefore, we can rewrite (8.70) as

$$\dot{V} \leq -\sum_{i=1}^{2} \sum_{j=2}^{n_{iv}-1} k_{ij} s_{ij}^2 u_{id1}^{2l} - \sum_{i=1}^{2} k_{in_v} s_{in_{iv}}^2 - \sum_{i=1}^{2} k_0 \eta_i^2$$

$$-\sigma^T K_\sigma \sigma - \sum_{\varsigma=1}^{5} \frac{\alpha_\varsigma}{\gamma_\varsigma}(\hat{c}_\varsigma - \frac{1}{2}c_\varsigma)^2 + \frac{1}{\lambda_{min}} \sum_{k=1}^{5} \delta_k$$

$$+\delta_6 + \sum_{\varsigma=1}^{5} \frac{\alpha_\varsigma}{4\gamma_\varsigma} c_\varsigma^2 \tag{8.72}$$

Noting Definition 8.1, we have $\mathcal{F} = \frac{1}{\lambda_{min}} \sum_{k=1}^{5} \delta_k + \sum_{\varsigma=1}^{5} \frac{\alpha_\varsigma}{4\gamma_\varsigma} c_\varsigma^2 + \delta_6 \to 0$ *as* $t \to \infty$.

We define $\mathcal{A} = \sum_{i=1}^{2} k_0 \eta_i^2 + \sum_{i=1}^{2} k_{in_{iv}} s_{in_{iv}}^2 + \sum_{i=1}^{2} \sum_{j=2}^{n_{iv}-1} k_{ij} s_{ij}^2 u_{id1}^{2l} + \lambda_{min}(K_\sigma)\|\sigma\|^2 + \sum_{\varsigma=1}^{5} \frac{\alpha_\varsigma}{\gamma_\varsigma}(\hat{c}_\varsigma - \frac{1}{2}c_\varsigma)^2$, *and from the definition, we have* $\mathcal{A} > 0$ $\forall \eta_i, s_{in_{iv}}, s_{ij}, u_{id1}, \sigma, c_\varsigma, i = 1, 2$ *and* $\varsigma = 1, \ldots, 5$.

Integrating both sides of (8.72) gives

$$V(t) - V(0) \leq -\int_0^t \mathcal{A}ds + \int_0^t \mathcal{F}ds$$

$$< -\int_0^t \mathcal{A}ds + \mathcal{C} \tag{8.73}$$

where $\mathcal{C} = \sum_{k=1}^{5} \frac{a_k}{\lambda_{min}} + \sum_{\varsigma=1}^{5} \frac{b_\varsigma}{4\gamma_\varsigma} c_\varsigma^2 + a_6$ *is a finite constant from Definition 8.1, we have* $V(t) < V(0) - \int_0^t \mathcal{A}ds + \mathcal{C}$. *Thus* V *is bounded and subsequently* $\eta_i, s_i, \sigma, \hat{c}_i, \nu$ *are bounded. From the definition of* s_i *in (8.31), it is concluded that* $[e_{i1}, e_{i2}, \ldots, e_{in_v}]^T$ *is bounded, which follows that* η *is bounded. From (8.73), we have* $s_{ij}u_{id1}, s_{in_{iv}}, \eta_i, \sigma \in L_2$, *which implies that* $\tilde{u}_b \in L_2^2$. *Since* $\sigma = u - z$ *is bounded and considering (8.22), (8.27), (8.30) and the definition of* $e_{ia}, \dot{e}_{ia} + K_{1a}e_{ia}$ *is bounded and can be rewritten as*

$$\dot{e}_{ia} \leq -K_{1a}e_{ia} + P$$

$$V_e = \frac{1}{2}e_{ia}^T e_{ia}$$

$$\dot{V}_e \leq -e_{ia}^T(K_{1a} - K_e)e_{ia} + \frac{1}{4}(n_{ia} - k_i)\lambda_{max}(K_e)\|p\|^2$$

where $P = [p, \ldots, p]^T \in \{\mathbf{R}\}^{n_{ia}-k_i}$ is a constant vector, $p > \|\sigma(t)\|$ $\forall$ t, $K_e \in \{\mathbf{R}\}^{n_{ia}-k_i \times n_{ia}-k_i}$ is a constant diagonal matrix chosen such that $\lambda_{\min}(K_{1a} - K_e) > 0$, $\lambda_{\max}(K_e)$ denotes the maximum eigenvalue of K_e, and $\lambda_{\min}(K_{1a} - K_e)$ denotes the minimum eigenvalue of $K_{1a} - K_e$. From the above equations, we can conclude that e_{ia} is bounded. Since q_{ia}^{1d}, the desired trajectory, is bounded, q_{ia}^1 and $\dot{q}_{ia}^1$ are bounded, which implies that ζ_{ia} and $\tilde{u}_{ia}$ are bounded as well. From (8.53)-(8.58), $d(s_{ij}u_{id1})/dt, \dot{s}_{iv}, \dot{\eta}_i, \dot{\tilde{u}}$ are bounded. Thus, from (8.33), $\dot{\nu}$ is bounded and $\dot{\sigma}$ is bounded as well. Therefore, from Remark 8.3, we can conclude that $\mathrm{U}_{21}, \ldots, \mathrm{U}_{26}$ are bounded.

Differentiating $u_{id1}^l \eta_i$ yields

$$\frac{d}{dt} u_{id1}^l \eta_i = -k_1 u_{id1}^l s_{i1} + l u_{id1}^{l-1} \dot{u}_{id1}^l \eta_i - k_0 u_{id1}^l \eta$$
$$- u_{id1}^l \left\{ \sum_{j=2}^{v-1} s_{ij} \zeta_{i(j+1)} - \sum_{j=3}^{v} s_{ij} \sum_{k=2}^{j-1} \frac{\partial(e_{ik} - s_{ik})}{\partial e_{ik}} \zeta_{i(k+1)} \right\}$$

where the first term is uniformly continuous and the other terms tend to zero. Since $\frac{d}{dt} u_{id1}^l \eta$ converges to zero [23], s_i and $\dot{s}_i$ converge to zero, $\zeta_i \to \zeta_{id}$ and $\dot{\zeta}_i \to \dot{\zeta}_{id}$ as $t \to \infty$.

Substituting the control (8.37) into the reduced order dynamics (8.28) yields

$$J^T[(K_\lambda + 1)e_\lambda + K_I \int_0^t e_\lambda dt]$$
$$= M(\dot{\sigma} + \dot{\nu}) + C(\nu + \sigma) + G + d$$
$$- L(L^T L)^{-1}(\mathrm{U}_1 + \mathrm{U}_2)$$

Since $\dot{\sigma}, \sigma, \dot{\nu}, \nu, c_i, \alpha_i, \dot{\zeta}, \gamma_i, \Lambda, \delta_i$ are all bounded, the right hand side of (8.74) is bounded, i.e., $J^T[(K_\lambda + 1)e_\lambda + K_I \int_0^t e_\lambda dt] = \Gamma(\dot{\sigma}, \sigma, \dot{\nu}, \nu, c_i, \alpha_i, \dot{\zeta}, \gamma_i, \Lambda, \delta_i)$, $\Gamma(*) \in L_\infty$.

Let $\int_0^t e_\lambda dt = E_\lambda$, then $\dot{E}_\lambda = e_\lambda$. By appropriately choosing $K_\lambda = \text{diag}[K_{\lambda,i}]$, $K_{\lambda,i} > -1$ and $K_I = \text{diag}[K_{I,i}]$, $K_{I,i} > 0$ to make $E_i(p) = \frac{1}{(K_{\lambda,i}+1)p+K_{I,i}}, p = d/dt$ a strictly proper exponential stable transfer function, it can be concluded that $\int_0^t e_\lambda dt \in L_\infty$, $e_\lambda \in L_\infty$, and the size of e_λ can be adjusted by choosing the proper gain matrices K_λ and K_I.

Since $\dot{\sigma}, \sigma, \dot{\nu}, \nu, c_i, \alpha_i, \dot{\zeta}, \gamma_i, \Lambda, \delta_i, e_\lambda$ and $\int_0^t e_\lambda dt$ are all bounded, we have τ bounded as well.

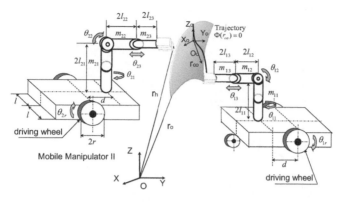

FIGURE 8.2

Cooperating 3-DOF mobile manipulators.

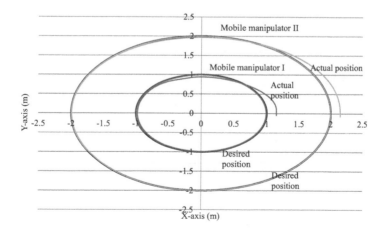

FIGURE 8.3

Tracking trajectories of both mobile platforms.

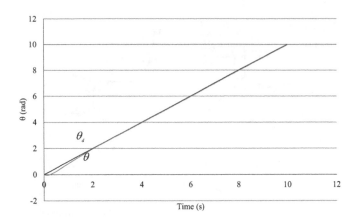

FIGURE 8.4

Tracking of θ for mobile manipulator I.

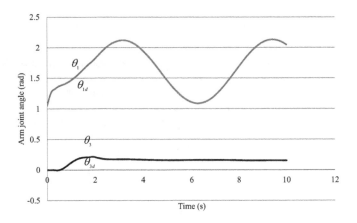

FIGURE 8.5

Tracking of arm joint angles of mobile manipulator I.

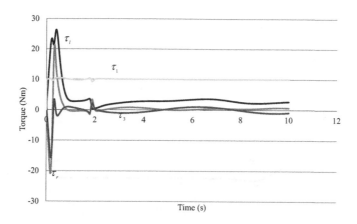

FIGURE 8.6

Input torques for mobile manipulator I.

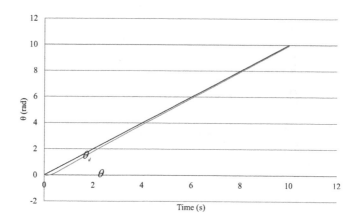

FIGURE 8.7

Tracking of θ for mobile manipulator II.

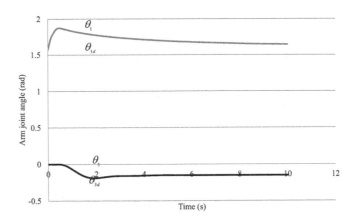

FIGURE 8.8

Tracking of arm joint angles of mobile manipulator II.

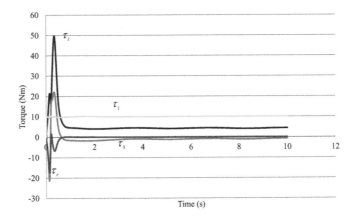

FIGURE 8.9

Torques of mobile manipulator II.

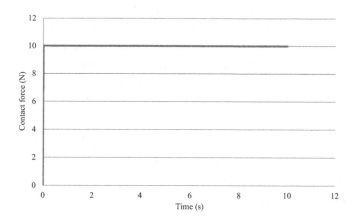

FIGURE 8.10

Contact force of relative motion.

8.4 Simulation Studies

To verify the effectiveness of the proposed control algorithm, we consider two similar 3-DOF mobile manipulators systems shown in Fig 8.2. Both mobile manipulators are subjected to the following constraint:

$$\dot{x}_i \cos \theta_i + \dot{y}_i \sin \theta_i = 0, \quad i = 1, 2$$

Using the Lagrangian approach, we can obtain the standard form for (8.14) and (8.15) with (9.51).

The parameters of the mobile manipulators used in this simulation are: $m_{1p} = m_{2p} = 5.0kg$, $m_{11} = m_{21} = 1.0kg$, $m_{12} = m_{22} = m_{13} = m_{23} = 0.5kg$, $I_{1w} = I_{2w} = 1.0kgm^2$, $I_{1p} = I_{2p} = 2.5kgm^2$, $I_{11} = I_{21} = 1.0kgm^2$, $I_{12} = I_{22} = 0.5kgm^2$, $I_{13} = I_{23} = 0.5kgm^2$, $d = l = r = 0.5m$, $2l_{11} = 2l_{21} = 1.0m$, $2l_{12} = 2l_{22} = 0.5m$, $2l_{13} = 0.05m$, $2l_{23} = 0.35m$. The mass of the object is $m_{obj} = 0.5kg$. The parameters are used for simulation purposes only, are assumed to be unknown and are not used in the control design. The desired trajectory of the object is $r_{od} = [x_{od}, y_{od}, z_{od}]^T$ where $x_{od} = 1.5\cos(t)$, $y_{od} = 1.5\sin(t)$, $z_{od} = 2l_1$. The corresponding desired trajectory of mobile manipulator II is

$q_{2d} = [x_{2d}, y_{2d}, \theta_{2d}, \theta_{21d}, \theta_{22d}]^T$ with $x_d = 2.0\cos(t)$, $y_d = 2.0\sin(t)$, $\theta_d = t$, $\theta_{22d} = \pi/2 rad$, and $\theta_{21d}, \theta_{23}$ to control force and compensate for task space errors. The end-effector holds tightly on the top point of the surface. The constraint relative motion by the mobile manipulator I is an arc with the center on the joint 2 of the mobile manipulator I, angle $= \pi/2 - \pi/6\cos(t)$, and constraint force set as $\lambda_c^d = 10.0N$. Therefore, from the constraint relative motion, we can obtain the desired trajectory of mobile manipulator I as $q_{1d} = [x_{1d}, y_{1d}, \theta_{1d}, \theta_{11d}, \theta_{12d}]^T$ with the corresponding trajectories $x_{1d} = 1.0\cos(t)$, $y_{1d} = 1.0\sin(t)$, $\theta_{1d} = t$, $\theta_{11d} = \pi/2 - \pi/6\cos(t)$, and $\theta_{12} = \pi/2$, θ_{13} used to compensate for position errors of the mobile platform.

For each mobile manipulator, by transformations similar to (8.22) and (8.23), $T_1(q)$ is defined as

$$
\begin{aligned}
\zeta_{i1} &= \theta_i \\
\zeta_{i2} &= x_i \cos\theta_i + y_i \sin\theta_i \\
\zeta_{i3} &= -x_i \sin\theta_i + y_i \cos\theta_i \\
\zeta_{i4} &= \theta_{i1} \\
\zeta_{i5} &= \theta_{i2} \\
\zeta_{i6} &= \theta_{i3} \\
u_{i1} &= v_{i2} \\
u_{i2} &= v_{i1} - (x_i \cos\theta_i + y_i \sin\theta_i)v_{i2} \\
u_{i3} &= \dot{\theta}_{i1} \\
u_{i4} &= \dot{\theta}_{i2} \\
u_{i5} &= \dot{\theta}_{i3}
\end{aligned}
$$

and we obtain the kinematic system in the chained form

$$
\begin{aligned}
\dot{\zeta}_{i1} &= u_{i1} \\
\dot{\zeta}_{i2} &= \zeta_{i3}u_{i1} \\
\dot{\zeta}_{i3} &= u_{i2} \\
\dot{\zeta}_{i4} &= u_{i3} \\
\dot{\zeta}_{i5} &= u_{i4} \\
\dot{\zeta}_{i6} &= u_{i5}
\end{aligned}
$$

The robust adaptive control (8.37) is used and the tracking errors for both mobile manipulators are, respectively,

$$
\begin{bmatrix} e_{11} \\ e_{12} \\ e_{13} \\ e_{14} \\ e_{15} \\ e_{16} \\ e_{\lambda_c} \end{bmatrix} = \begin{bmatrix} \zeta_{11} - \zeta_{1d1} \\ \zeta_{12} - \zeta_{1d2} \\ \zeta_{13} - \zeta_{1d3} \\ \zeta_{14} - \zeta_{1d4} \\ \zeta_{15} - \zeta_{1d5} \\ \zeta_{16} - \zeta_{1d6} \\ \lambda_c - \lambda_{cd} \end{bmatrix}
$$

$$
\begin{bmatrix} s_{11} \\ s_{12} \\ s_{13} \\ s_{14} \\ s_{15} \\ s_{16} \end{bmatrix} = \begin{bmatrix} e_{11} \\ e_{12} \\ e_{13} + k_{12}e_{12}u_{1d1} \\ 0.0 \\ 0.0 \\ 0.0 \end{bmatrix}
$$

$$
\begin{bmatrix} e_{21} \\ e_{22} \\ e_{23} \\ e_{24} \\ e_{25} \\ e_{26} \\ e_{\lambda_c} \end{bmatrix} = \begin{bmatrix} \zeta_{21} - \zeta_{2d1} \\ \zeta_{22} - \zeta_{2d2} \\ \zeta_{23} - \zeta_{2d3} \\ \zeta_{24} - \zeta_{2d4} \\ \zeta_{25} - \zeta_{2d5} \\ \zeta_{26} - \zeta_{2d6} \\ \lambda_c - \lambda_{cd} \end{bmatrix}
$$

$$
\begin{bmatrix} s_{21} \\ s_{22} \\ s_{23} \\ s_{24} \\ s_{25} \\ s_{26} \end{bmatrix} = \begin{bmatrix} e_{21} \\ e_{22} \\ e_{23} + k_{22}e_{22}u_{2d1} \\ 0.0 \\ 0.0 \\ 0.0 \end{bmatrix}
$$

The initial conditions selected for mobile manipulator I are $x_1(0) = 1.15m$, $y_1(0) = 0.0m$, $\theta_1(0) = 0.0rad$, $\theta_{11}(0) = 1.047rad$, $\theta_{12}(0) = \pi/2rad$, $\theta_{13}(0) = 0.0rad$, $\lambda(0) = 0.0N$ and $\dot{x}_1(0) = 0.5m/s$, $\dot{y}_1(0) = \dot{\theta}_1(0) = \dot{\theta}_{11}(0) = \dot{\theta}_{12}(0) = \dot{\theta}_{13(0)} = 0.0$, and the initial conditions selected for mobile manipulator II are $x_2(0) = 2.15m$, $y_2(0) = 0m$, $\theta_2(0) = 0.0rad$, $\theta_{21}(0) = 1.57rad$, $\theta_{22}(0) = \pi/2rad$, $\theta_{23}(0) = 0.0rad$, and $\dot{x}_2(0) = \dot{y}_2(0) = \dot{\theta}_2(0) = \dot{\theta}_{12}(0) = \dot{\theta}_{22}(0) = \dot{\theta}_{23}(0) = 0.0$.

In the simulation, the design parameters for each mobile manipulator are set as $k_0 = 5.0$, $k_1 = 180.0$, $k_2 = 5.0$, $k_3 = 5.0$, $\eta(0) = 0.0$, $K_{a1} = \text{diag}[2.0]$,

$K_\lambda = 0.3$, $K_I = 1.5$, $K_\sigma = \text{diag}[0.5]$, $K_u = \text{diag}[1.0]$. The design parameters in u_2 of (8.39) are $\gamma_i = 0.1$, $\alpha_i = \delta_i = 1/(1+t)^2$, and $\hat{c}_i(0) = 1.0$. The disturbances on each joint of each mobile manipulator are set to a time-varying form as $0.5\sin(t)$, $0.5\sin(t)$, $0.1\sin(t)$ and $0.1\sin(t)$, respectively. By using the control law (8.37) [167], we can obtain Fig. 8.3 to describe the trajectory of the mobile platforms of both mobile manipulators. Fig. 8.4, Fig. 8.7, Fig. 8.5 and Fig. 8.8 show the trajectory tracking ($\zeta - \zeta_d$) of the joints with the disturbances for both mobile manipulators, and the corresponding input torques for them are shown in Fig. 8.6 and Fig. 8.9. Fig. 8.10 shows the contact force tracking $\lambda_c - \lambda_c^d$. Since the joint 3 makes the manipulator redundant in the force space, from Fig. 8.10, we can see that the contact force is always more than zero, which means that the two mobile manipulators always keep in contact, and the force error quickly converges to 0 by adjusting K_λ and K_I. Therefore, the validity of the proposed controls is confirmed by these simulation results.

8.5   Conclusion

In this chapter, dynamics and control of two mobile robotic manipulators manipulating a constrained object have been investigated. In addition to the motion of the object with respect to the world coordinates, its relative motion with respect to the mobile manipulators is also taken into consideration. The dynamics of such system is established and its properties are discussed. Robust adaptive controls have been developed which can guarantee the convergence of positions and boundedness of the constraint force. The control signals are smooth and no projection is used in parameter update law. Simulation results showed that the proposed controls work quite well.

9

Appendix

CONTENTS

9.1 Example of 2-DOF Mobile Manipulator

For better understanding, an example mobile manipulator is used to show the derivation of the actual structures of robot dynamics. The robot models are used for simulation as well as experimental studies.

Consider a two-link manipulator with two revolute joints mounted on a two wheeled-driven mobile base as follows.

The following variables have been chosen to describe the vehicle (see also Fig. 9.1):

τ_l, τ_r: the torques of two wheels;

τ_1, τ_2: the torques of the joint 1 and the joint 2;

θ_l, θ_r: the rotation angle of the left wheel and the right wheel of the mobile platform;

v: the forward velocity of the mobile platform;

θ: the direction angle of the mobile platform;

ω: the rotation velocity of the mobile platform, and $\omega = \dot{\theta}$;

θ_1, θ_2: the joint angle of the link 1 and the link 2;

m_1, m_2: the mass of links of the manipulator;

I_{z1}, I_{z2}: the inertia moment of the link 1 and the link 2;

l_1, l_2: the link length of the manipulator;

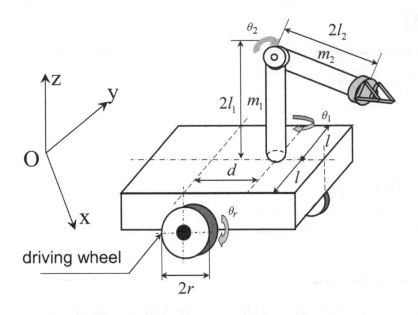

FIGURE 9.1

2-DOF mobile manipulator.

r: the radius of the wheels;

$2l$: the distance of the wheels;

d: the distance between the manipulator and the driving center of the mobile base;

m_p: the mass of the mobile platform;

I_p: the inertia moment of the mobile platform;

I_w: the inertia moment of each wheel;

m_w: the mass of each wheel;

g: gravity acceleration.

The positions of the mass center for the mobile base are given by x and y. which lead to the corresponding velocities as $\dot{x}$ and $\dot{y}$.

The positions of the mass center for two wheels are given by

$$x_r = x + l\sin\theta \tag{9.1}$$

$$y_r = y - l\cos\theta \tag{9.2}$$

$$x_l = x - l\sin\theta \tag{9.3}$$

$$y_l = y + l\cos\theta \tag{9.4}$$

which lead to the corresponding velocities as

$$\dot{x}_r = \dot{x} + l\dot{\theta}\cos\theta \tag{9.5}$$

$$\dot{y}_r = \dot{y} + l\dot{\theta}\sin\theta \tag{9.6}$$

$$\dot{x}_l = \dot{x} - l\dot{\theta}\cos\theta \tag{9.7}$$

$$\dot{y}_l = \dot{y} - l\dot{\theta}\sin\theta \tag{9.8}$$

The positions of the mass center for the 2-DOF manipulator are given by

$$x_1 = x + d\cos\theta \tag{9.9}$$

$$y_1 = y + d\sin\theta \tag{9.10}$$

$$x_2 = x + d\cos\theta - l_2\sin\theta_2\cos(\theta + \theta_1) \tag{9.11}$$

$$y_2 = y + d\sin\theta - l_2\sin\theta_2\sin(\theta + \theta_1) \tag{9.12}$$

$$z_2 = 2l_1 - l_2\cos\theta_2 \tag{9.13}$$

which lead to the corresponding velocities as

$$\dot{x}_1 = \dot{x} - d\dot{\theta}\sin\theta \tag{9.14}$$

$$\dot{y}_1 = \dot{y} + d\dot{\theta}\cos\theta \tag{9.15}$$

$$\dot{x}_2 = \dot{x} - d\dot{\theta}\sin\theta - l_2\dot{\theta}_2\cos\theta_2\cos(\theta + \theta_1)$$
$$+ l_2\sin\theta_2\sin(\theta_1 + \theta_2)(\dot{\theta} + \dot{\theta}_1) \tag{9.16}$$

$$\dot{y}_2 = \dot{y} + d\dot{\theta}\cos\theta - l_2\dot{\theta}_2\cos\theta_2\sin(\theta + \theta_1)$$
$$- l_2\sin\theta_2\cos(\theta + \theta_1)(\dot{\theta} + \dot{\theta}_1) \tag{9.17}$$

$$\dot{z}_2 = l_2\dot{\theta}_2\sin\theta_2 \tag{9.18}$$

The total kinetic energy is

$$K = k_p + k_1 + k_2 + k_r + k_l \tag{9.19}$$

where

$$k_p = \frac{1}{2}m_p(\dot{x}^2 + \dot{y}^2) + \frac{1}{2}I_p\dot{\theta}^2$$

$$k_1 = \frac{1}{2}m_1(\dot{x}_1^2 + \dot{y}_1^2) + \frac{1}{2}I_1(\dot{\theta} + \dot{\theta}_1)^2$$

$$k_2 = \frac{1}{2}m_2(\dot{x}_2^2 + \dot{y}_2^2 + (l_2\dot{\theta}^2 \sin\theta_2)^2) + \frac{1}{2}I_{z2}(\dot{\theta} + \dot{\theta}_1)^2 + \frac{1}{2}I_{y2}\dot{\theta}_2^2$$

$$k_r = \frac{1}{2}m_w(\dot{x}_r^2 + \dot{y}_r^2) + \frac{1}{2}I_{zw}\dot{\theta}^2 + \frac{1}{2}I_{yw}\dot{\theta}_r^2$$

$$k_l = \frac{1}{2}m_w(\dot{x}_l^2 + \dot{y}_l^2) + \frac{1}{2}I_{zw}\dot{\theta}^2 + \frac{1}{2}I_{yw}\dot{\theta}_l^2$$

Since the nonholonomic constraints exist, x, y, θ is not independent as follows

$$\begin{bmatrix} \dot{x} \\ \dot{y} \\ \dot{\theta} \end{bmatrix} = \begin{bmatrix} \cos\theta & 0 \\ \sin\theta & 0 \\ 0 & 1 \end{bmatrix} \begin{bmatrix} v \\ \omega \end{bmatrix} = \begin{bmatrix} \frac{r}{2}\cos\theta & \frac{r}{2}\cos\theta \\ \frac{r}{2}\sin\theta & \frac{r}{2}\sin\theta \\ \frac{r}{2l} & -\frac{r}{2l} \end{bmatrix} \begin{bmatrix} \dot{\theta}_r \\ \dot{\theta}_l \end{bmatrix} \quad (9.20)$$

we have

$$\begin{bmatrix} v \\ \omega \\ \dot{\theta}_1 \\ \dot{\theta}_2 \end{bmatrix} = \begin{bmatrix} \frac{r}{2} & \frac{r}{2l} & 0 & 0 \\ \frac{r}{2} & -\frac{r}{2l} & 0 & 0 \\ 0 & 0 & 1 & 0 \\ 0 & 0 & 0 & 1 \end{bmatrix} \begin{bmatrix} \dot{\theta}_r \\ \dot{\theta}_l \end{bmatrix} \quad (9.21)$$

Therefore, we choose $q = [\theta_r, \theta_l, \theta_1, \theta_2]^T$ as generalized coordinations. Considering (9.20), we could obtain the dynamics of the mobile manipulator as

$$M(q)\ddot{q} + C(q, \dot{q})\dot{q} + G(q) = B\tau \quad (9.22)$$

where

$$M(q) = \begin{bmatrix} m_{11} & m_{12} & m_{13} & m_{14} \\ m_{21} & m_{22} & m_{23} & m_{24} \\ m_{31} & m_{32} & m_{33} & m_{34} \\ m_{41} & m_{42} & m_{43} & m_{44} \end{bmatrix}, C(q, \dot{q}) = \begin{bmatrix} c_{11} & c_{12} & c_{13} & c_{14} \\ c_{21} & c_{22} & c_{23} & c_{24} \\ c_{31} & c_{32} & c_{33} & c_{34} \\ c_{41} & c_{42} & c_{43} & c_{44} \end{bmatrix}$$

$$m_{11} = p_0 - p_1 \cos\theta_1 \sin\theta_2 + p_2 \sin^2\theta_2 + p_3 \sin\theta_1 \sin\theta_2$$

$$m_{12} = q_0 + p_1 \cos\theta_1 \sin\theta_2 - p_2 \sin^2\theta_2$$

$$m_{13} = q_1 \sin\theta_1 \sin\theta_2 - q_2 \cos\theta_1 \sin\theta_2 + q_3^2 \sin\theta_2 + q_4$$

$$m_{14} = -q_1 \cos\theta_1 \cos\theta_2 - q_2 \sin\theta_1 \cos\theta_2$$

$$m_{21} = q_0 + p_1 \cos\theta_1 \sin\theta_2 - p_2 \sin^2\theta_2$$

$$m_{22} = p_0 - p_1 \cos\theta_1 \sin\theta_2 + p_2 \sin^2\theta_2 - p_3 \sin\theta_1 \sin\theta_2$$

$$m_{23} = q_1 \sin\theta_1 \sin\theta_2 + q_2 \cos\theta_1 \sin\theta_2 - q_3 \sin^2\theta_2 - q_4$$

$$m_{24} = -q_1 \cos\theta_1 \cos\theta_2 + q_2 \sin\theta_1 \cos\theta_2$$

$$m_{31} = q_1 \sin\theta_1 \sin\theta_2 - q_2 \cos\theta_1 \sin\theta_2 + q_3 \sin^2\theta_2 + q_4$$

$$m_{32} = q_1 \sin\theta_1 \sin\theta_2 + q_2 \cos\theta_1 \sin\theta_2 - q_3 \sin^2\theta_2 - q_4$$

$$m_{33} = I_{z1} + I_{z2} + m_2 l_2^2 \sin^2\theta_2$$

$$m_{34} = 0$$

$$m_{41} = -q_1 \cos\theta_1 \cos\theta_2 - q_2 \sin\theta_1 \cos\theta_2$$

$$m_{42} = -q_1 \cos\theta_1 \cos\theta_2 + q_2 \sin\theta_1 \cos\theta_2$$

$$m_{43} = 0$$

$$m_{44} = q_4$$

$$p_0 = \frac{1}{4}(m_p + m_1 + m_2 + 2m_w)r^2 + \frac{1}{4}(I_p + 2Iw + m_1 d^2 + m_2 d^2 + 2m_w l^2)r^2$$
$$+ (I_{z1} + I_{z2})r^2/4$$

$$p_1 = m_2 l_2 dr^2/2$$

$$p_2 = m_2 l_2^2 r^2/4$$

$$p_3 = m_2 l_2 r^2/2$$

$$q_0 = (m_p + m_1 + m_2 + 2m_w)r^2/4 - \frac{1}{4}(I_p + 2I_w + m_1 d^2 + m_2 d^2 + 2m_w l^2)r^2$$
$$- (I_{z1} + I_{z2})r^2/4$$

$$q_1 = m_2 l_2 r/2$$

$$q_2 = m_2 l_2 dr/2$$

$$q_3 = m_2 l_2 r^2/2$$

$$q_4 = (I_{z1} + I_{z2})r/2$$

$$c_{11} = (p_1 \sin\theta_1 \sin\theta_2 + p_3 \cos\theta_1 \sin\theta_2)\dot\theta_1/2$$
$$+ (2p_2 \sin\theta_2 \cos\theta_2 + p_3 \sin\theta_1 \cos\theta_2 - p_1 \cos\theta_1 \cos\theta_2)\dot\theta_2/2$$

$$c_{12} = -p_1 \sin\theta_1 \sin\theta_2 \dot\theta_1 + (p_1 \cos\theta_1 \cos\theta_2 - 2p_2 \sin\theta_2 \cos\theta_2)\dot\theta_2$$

$$c_{13} = (p_1 \sin\theta_1 \sin\theta_2 + p_3 \cos\theta_1 \sin\theta_2)\dot\theta_r/2 - p_3 \sin\theta_1 \sin\theta_2 \dot\theta_l$$
$$+ (q_1 \cos\theta_1 \sin\theta_2 + q_2 \sin\theta_1 \sin\theta_2)\dot\theta_1$$
$$+ (q_1 \sin\theta_1 \cos\theta_2 - q_2 \cos\theta_1 \cos\theta_2 + q_3 \sin\theta_2 \cos\theta_2)\dot\theta_2$$

$$c_{14} = (-p_1 \cos\theta_1 \cos\theta_2 + 2p_2 \sin\theta_2 \cos\theta_2 + p_3 \sin\theta_1 \cos\theta_2)\dot\theta_r/2$$
$$+ (p_1 \cos\theta_1 \cos\theta_2 - 2p_2 \sin\theta_2 \cos\theta_2)\dot\theta_l$$
$$+ (q_1 \sin\theta_1 \cos\theta_2 - q_2 \cos\theta_1 \cos\theta_2 + q_3 \sin\theta_2 \cos\theta_2)\dot\theta_1$$
$$+ (q_1 \cos\theta_1 \sin\theta_2 + q_2 \sin\theta_1 \sin\theta_2)\dot\theta_2$$

$$c_{21} = -p_1 \sin\theta_1 \sin\theta_2 \dot\theta_1/2 + (p_1 \cos\theta_1 \cos\theta_2 - 2p_2 \sin\theta_2 \cos\theta_2)\dot\theta_2/2$$

$$c_{22} = (p_1 \sin\theta_1 \sin\theta_2 - p_3 \cos\theta_1 \sin\theta_2 - p_3 \cos\theta_1 \sin\theta_2)\dot\theta_1/2$$

$$+(2p_2 \sin\theta \cos\theta_2 - p_3 \sin\theta_1 \cos\theta_2 - p_1 \cos\theta_1 \cos\theta_2)\dot\theta_2/2$$

$$c_{23} = -p_1 \sin\theta_1 \sin\theta_2 \dot\theta_r/2 + (p_1 \sin\theta_1 \sin\theta_2 - p_3 \cos\theta_1 \sin\theta_2)\dot\theta_l/2$$

$$+(q_1 \cos\theta_1 \sin\theta_2 - q_2 \sin\theta_1 sin\theta_2)\dot\theta_1$$

$$+(q_1 \sin\theta_1 \cos\theta_2 + q_2 \cos\theta_1 cos\theta_2 - q_3 \sin\theta_2 cos\theta_2)\dot\theta_2$$

$$c_{24} = (p_1 \cos\theta_1 \cos\theta_2 - 2p_2 \sin\theta_2 \cos\theta_2)\dot\theta_r/2$$

$$+(2p_2 \sin\theta_2 \cos\theta_2 - p_1 \cos\theta_1 \cos\theta_2 - p_3 \sin\theta_1 \cos\theta_2)\dot\theta_l/2$$

$$+(q_1 \sin\theta_1 \cos\theta_2 + q_2 \cos\theta_1 \cos\theta_2 - q_3 \sin\theta_2 \cos\theta_2)\dot\theta_1$$

$$c_{31} = -(p_1 \sin\theta_1 \sin\theta_2 + p_3 \cos\theta_1 \sin\theta_2)\dot\theta_r/2$$

$$+(p_1 \sin\theta_1 \sin\theta_2 - p_3 \cos\theta_1 \sin\theta_2)\dot\theta_l/2$$

$$+q_3 \sin\theta_2 \cos\theta_2 \dot\theta_2$$

$$c_{32} = (p_1 \sin\theta_1 \sin\theta_2 - p_3 \cos\theta_1 \sin\theta_2)\dot\theta_r/2 + (p_3 \cos\theta_1 \sin\theta_2 - p_1 \sin\theta_1 \sin\theta_2)\dot\theta_l/$$

$$-q_3 \sin\theta_2 \cos\theta_2 \dot\theta_2$$

$$c_{33} = 0$$

$$c_{34} = q_3 \sin\theta_2 \cos\theta_2 \dot\theta_r - q_3 \sin\theta_2 \cos\theta_2 \dot\theta_l + m_2 l_2^2 \sin\theta_2 \cos\theta_2 \dot\theta_1$$

$$c_{41} = p_1 \cos\theta_1 \cos\theta_2 - wp_2 \sin\theta_2 \cos\theta_2 - p_3 \sin\theta_1 \cos\theta_2)\dot\theta_r/2$$

$$-(p_1 \cos\theta_1 \cos\theta_2 - 2p_2 \sin\theta_2 \cos\theta_2)\dot\theta_l/2 + q_e \sin\theta_2 \cos\theta_2 \dot\theta_1$$

$$c_{42} = -(p_1 \cos\theta_1 \cos\theta_2 - 2p_2 \sin\theta_2 \cos\theta_2)\dot\theta_r/2$$

$$+(p_1 \cos\theta_1 \cos\theta_2 - I_{wp2} \sin\theta_2 \cos\theta_2 + p_3 \sin\theta_1 \cos\theta_2)\dot\theta_l/2$$

$$+q_3 \sin\theta_2 cos\theta_2 \dot\theta_1$$

$$c_{43} = q_3 \sin\theta_2 \cos\theta_2 \dot\theta_r + q_3 \sin\theta_2 \cos\theta_2 \dot\theta_l - m_2 l_2^2 \sin\theta_2 \cos\theta_2 \dot\theta_1$$

$$c_{44} = 0$$

$$G(q) = \begin{bmatrix} 0 & 0 & 0 & -m_2 g l_2 \sin\theta_2 \end{bmatrix}$$

$$B(q) = \begin{bmatrix} 1 & 0 & 0 & 0 \\ 0 & 1 & 0 & 0 \\ 0 & 0 & 1 & 0 \\ 0 & 0 & 0 & 1 \end{bmatrix}$$

From (9.21), it is easy to have

$$\begin{bmatrix} \dot\theta_r \\ \dot\theta_l \\ \dot\theta_1 \\ \dot\theta_2 \end{bmatrix} = \begin{bmatrix} \frac{1}{r} & \frac{l}{r} & 0 & 0 \\ \frac{1}{r} & -\frac{l}{r} & 0 & 0 \\ 0 & 0 & 1 & 0 \\ 0 & 0 & 0 & 1 \end{bmatrix} \begin{bmatrix} v \\ \omega \\ \dot\theta_1 \\ \dot\theta_2 \end{bmatrix} \tag{9.23}$$

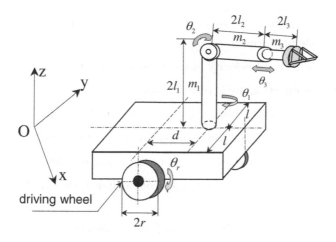

FIGURE 9.2

3-DOF robotic manipulator mounted on a mobile platform.

Since r is a fixed parameter for the system, R can be treated as a constant matrix.

Considering (9.23), let $\dot{\zeta} = [v, w, \dot{\theta}_1, \dot{\theta}_2]^T$ and we can rewrite the dynamics as

$$R^T M(q) R \ddot{\zeta} + R^T C(q, \dot{q}) R \dot{\zeta} + R^T G(q) = R^T B \tau \qquad (9.24)$$

with

$$R = \begin{bmatrix} \frac{1}{r} & \frac{1}{r} & 0 & 0 \\ \frac{1}{r} & -\frac{1}{r} & 0 & 0 \\ 0 & 0 & 1 & 0 \\ 0 & 0 & 0 & 1 \end{bmatrix} \qquad (9.25)$$

9.2   Example of 3-DOF Mobile Manipulator

Consider a 3-DOF robotic manipulator with two revolute joints and one prismatic joint mounted on a two-wheeled mobile platform shown in Fig. 9.2.

The following variables have been chosen to describe the system.

τ_l, τ_r: the torques of two wheels;

τ_1, τ_2, τ_3: the torques of the joint 1, the joint 2 and the joint 3;

θ_l, θ_r: the rotation angle of the left wheel and the right wheel of the mobile platform;

v: the forward velocity of the mobile platform;

θ: the direction angle of the mobile platform;

ω: the rotation velocity of the mobile platform, and $\omega = \dot{\theta}$;

$\theta_1, \theta_2, \theta_3$: the joint angle of the link 1, the link 2 and the link 3;

m_1, m_2, m_3: the mass of links of the manipulator;

I_{z1}, I_{z2}, I_{z3}: the inertia moment of the link 1, the link 2 and the link 3;

l_1, l_2, l_3: the fixed link length of the manipulator;

r: the radius of the wheels;

$2l$: the distance of the wheels;

d: the distance between the manipulator and the driving center of the mobile base;

m_p: the mass of the mobile platform;

I_p: the inertia moment of the mobile platform;

I_w: the inertia moment of each wheel;

m_w: the mass of each wheel;

g: gravity acceleration.

The positions of the mass center for the mobile base are given by x and y, which lead to the corresponding velocities as $\dot{x}$ and $\dot{y}$.

The positions of the mass center for two wheels are given by

$$x_r \;=\; x + l\sin\theta \tag{9.26}$$

$$y_r \;=\; y - l\cos\theta \tag{9.27}$$

$$x_l \;=\; x - l\sin\theta \tag{9.28}$$

$$y_l \;=\; y + l\cos\theta \tag{9.29}$$

which lead to the corresponding velocities as

$$\dot{x}_r \;=\; \dot{x} + l\dot{\theta}\cos\theta \tag{9.30}$$

$$\dot{y}_r \;=\; \dot{y} + l\dot{\theta}\sin\theta \tag{9.31}$$

$$\dot{x}_l \;=\; \dot{x} - l\dot{\theta}\cos\theta \tag{9.32}$$

$$\dot{y}_l \;=\; \dot{y} - l\dot{\theta}\sin\theta \tag{9.33}$$

The positions for the mass center for the 3-DOF manipulator are given by

$$x_1 = x + d\cos\theta \tag{9.34}$$

$$y_1 = y + d\sin\theta \tag{9.35}$$

$$x_2 = x + d\cos\theta - l_2\sin\theta_2\cos(\theta + \theta_1) \tag{9.36}$$

$$y_2 = y + d\sin\theta - l_2\sin\theta_2\sin(\theta + \theta_1) \tag{9.37}$$

$$z_2 = 2l_1 - l_2\cos\theta_2 \tag{9.38}$$

$$x_3 = x + d\cos\theta - (2l_2 + \theta_3)\sin\theta_2\cos(\theta + \theta_1) \tag{9.39}$$

$$y_3 = y + d\sin\theta - (2l_2 + \theta_3)\sin\theta_2\sin(\theta + \theta_1) \tag{9.40}$$

$$z_3 = 2l_1 - (2l_2 + \theta_3)\cos\theta_2 \tag{9.41}$$

which lead to the corresponding velocities as

$$\dot{x}_1 = \dot{x} - d\dot{\theta}\sin\theta \tag{9.42}$$

$$\dot{y}_1 = \dot{y} + d\dot{\theta}\cos\theta \tag{9.43}$$

$$\dot{x}_2 = \dot{x} - d\dot{\theta}\sin\theta - l_2\dot{\theta}_2\cos\theta_2\cos(\theta + \theta_1)$$
$$\quad + l_2\sin\theta_2\sin(\theta_1 + \theta_2)(\dot{\theta} + \dot{\theta}_1) \tag{9.44}$$

$$\dot{y}_2 = \dot{y} + d\dot{\theta}\cos\theta - l_2\dot{\theta}_2\cos\theta_2\sin(\theta + \theta_1)$$
$$\quad - l_2\sin\theta_2\cos(\theta + \theta_1)(\dot{\theta} + \dot{\theta}_1) \tag{9.45}$$

$$\dot{z}_2 = l_2\dot{\theta}_2\sin\theta_2 \tag{9.46}$$

$$\dot{x}_3 = \dot{x} - d\dot{\theta}\sin\theta - \dot{\theta}_3\sin\theta_2\cos(\theta + \theta_1) - (2l_2 + \theta_3)\dot{\theta}_2\cos\theta_2\cos(\theta + \theta_1)$$
$$\quad + (2l_2 + \theta_3)\sin\theta_2\sin(\theta + \theta_1)(\dot{\theta} + \dot{\theta}_1) \tag{9.47}$$

$$\dot{y}_3 = \dot{y} + d\dot{\theta}\cos\theta - \dot{\theta}_3\sin\theta_2\sin(\theta + \theta_1) - (2l_2 + \theta_3)\dot{\theta}_2\cos\theta_2\sin(\theta + \theta_1)$$
$$\quad - (2l_2 + \theta_3)\sin\theta_2\cos(\theta + \theta_1)(\dot{\theta} + \dot{\theta}_1) \tag{9.48}$$

$$\dot{z}_3 = -\dot{\theta}_3\cos\theta_2 + (2l_2 + \theta_3)\dot{\theta}_2\sin\theta_2 \tag{9.49}$$

The total kinetic energy is

$$K = k_p + k_1 + k_2 + k_3 + k_r + k_l \tag{9.50}$$

where

$$k_p = \frac{1}{2}m_p(\dot{x}^2 + \dot{y}^2) + \frac{1}{2}I_p\dot{\theta}^2$$

$$k_1 = \frac{1}{2}m_1(\dot{x}_1^2 + \dot{y}_1^2) + \frac{1}{2}I_1(\dot{\theta} + \dot{\theta}_1)^2$$

$$k_2 = \frac{1}{2}m_2(\dot{x}_2^2 + \dot{y}_2^2 + (l_2\dot{\theta}^2\sin\theta_2)^2) + \frac{1}{2}I_{z2}(\dot{\theta} + \dot{\theta}_1)^2 + \frac{1}{2}I_{y2}\dot{\theta}_2^2$$

$$k_3 = \frac{1}{2}m_3(\dot{x}_3^2 + \dot{y}_3^2 + \dot{z}_3^2) + \frac{1}{2}(I_{z3} + m_3\theta_3^2)(\dot{\theta} + \dot{\theta}_1)^2 + \frac{1}{2}(I_{y3} + m_3\theta_3^2)\dot{\theta}_2^2$$

$$k_r = \frac{1}{2}m_w(\dot{x}_r^2 + \dot{y}_r^2) + \frac{1}{2}I_{zw}\dot{\theta}^2 + \frac{1}{2}I_{yw}\dot{\theta}_r^2$$

$$k_l = \frac{1}{2}m_w(\dot{x}_l^2 + \dot{y}_l^2) + \frac{1}{2}I_{zw}\dot{\theta}^2 + \frac{1}{2}I_{yw}\dot{\theta}_l^2$$

which is subjected to the following constraints: $\dot{x}\cos\theta + \dot{y}\sin\theta = 0$. Using the Lagrangian approach as in the above section, we can choose $q = [\theta_r, \theta_l, \theta_1, \theta_2, \theta_3]^T$ as a generalized coordination as

$$M(q)\ddot{q} + C(q, \dot{q})\dot{q} + G(q) = B\tau \qquad (9.51)$$

where

$$M(q) = \begin{bmatrix} m_{11} & m_{12} & m_{13} & m_{14} & m_{15} \\ m_{21} & m_{22} & m_{23} & m_{24} & m_{25} \\ m_{31} & m_{32} & m_{33} & m_{34} & m_{35} \\ m_{41} & m_{42} & m_{43} & m_{44} & m_{45} \\ m_{51} & m_{52} & m_{53} & m_{54} & m_{55} \end{bmatrix},$$

$$C(q, \dot{q}) = \begin{bmatrix} c_{11} & c_{12} & c_{13} & c_{14} & c_{15} \\ c_{21} & c_{22} & c_{23} & c_{24} & c_{25} \\ c_{31} & c_{32} & c_{33} & c_{34} & c_{35} \\ c_{41} & c_{42} & c_{43} & c_{44} & c_{45} \\ c_{51} & c_{52} & c_{53} & c_{54} & c_{55} \end{bmatrix}$$

$$m_{11} = p_0 - p_1\cos\theta_1\sin\theta_2 + p_2\sin^2\theta_2 + p_3\sin\theta_1\sin\theta_2$$

$$m_{12} = q_0 + p_1\cos\theta_1\sin\theta_2 - p_2\sin^2\theta_2$$

$$m_{13} = q_1\sin\theta_1\sin\theta_2 - q_2\cos\theta_1\sin\theta_2 + q_3^2\sin\theta_2 + q_4$$

$$m_{14} = -q_1\cos\theta_1\cos\theta_2 - q_2\sin\theta_1\cos\theta_2$$

$$m_{15} = -q_5\cos\theta_1\cos\theta_2 - q_6\sin\theta_1\cos\theta_2$$

$$m_{21} = q_0 + p_1\cos\theta_1\sin\theta_2 - p_2\sin^2\theta_2$$

$$m_{22} = p_0 - p_1\cos\theta_1\sin\theta_2 + p_2\sin^2\theta_2 - p_3\sin\theta_1\sin\theta_2$$

$$m_{23} = q_1\sin\theta_1\sin\theta_2 + q_2\cos\theta_1\sin\theta_2 - q_3\sin^2\theta_2 - q_4$$

$$m_{24} = -q_1\cos\theta_1\cos\theta_2 + q_2\sin\theta_1\cos\theta_2$$

$$m_{25} = -q_5\cos\theta_1\cos\theta_2 + q_6\sin\theta_1\cos\theta_2$$

$$m_{31} = q_1 \sin\theta_1 \sin\theta_2 - q_2 \cos\theta_1 \sin\theta_2 + q_3 \sin^2\theta_2 + q_4$$

$$m_{32} = q_1 \sin\theta_1 \sin\theta_2 + q_2 \cos\theta_1 \sin\theta_2 - q_3 \sin^2\theta_2 - q_4$$

$$m_{33} = I_{z1} + I_{z2} + m_2 l_2^2 \sin^2\theta_2$$

$$m_{34} = 0$$

$$m_{35} = 0$$

$$m_{41} = -q_1 \cos\theta_1 \cos\theta_2 - q_2 \sin\theta_1 \cos\theta_2$$

$$m_{42} = -q_1 \cos\theta_1 \cos\theta_2 + q_2 \sin\theta_1 \cos\theta_2$$

$$m_{43} = 0$$

$$m_{44} = q_4$$

$$m_{45} = 0$$

$$m_{51} = m_{15}$$

$$m_{52} = m_{25}$$

$$m_{53} = m_{35}$$

$$m_{54} = m_{45}$$

$$m_{55} = q_7$$

$$p_0 = \frac{1}{4}(m_p + m_1 + m_2 + m_3 + 2m_w)r^2$$
$$+ \frac{1}{4}(I_p + 2Iw + m_1 d^2 + m_2 d^2 + m_3 d^2 + 2m_w l^2)r^2$$
$$+ (I_{z1} + I_{z2} + I_z 3 + m\theta_3^2)r^2/4$$

$$p_1 = m_2 l_2 dr^2/2$$

$$p_2 = m_2 l_2^2 r^2/4$$

$$p_3 = m_2 l_2 r^2/2$$

$$q_0 = (m_p + m_1 + m_2 + 2m_w)r^2/4 - \frac{1}{4}(I_p + 2I_w + m_1 d^2 + m_2 d^2 + 2m_w l^2)r^2$$
$$- (I_{z1} + I_{z2})r^2/4$$

$$q_1 = m_2 l_2 r/2$$

$$q_2 = m_2 l_2 dr/2$$

$$q_3 = m_2 l_2 r^2/2$$

$$q_4 = (I_{z1} + I_{z2})r/2$$

$$q_5 = m_3(2l_2 + \theta_3)r/2$$

$$q_6 = m_3(2l_2 + \theta_3)dr/2$$

$$q_7 = m_3$$

$$c_{11} = (p_1 \sin\theta_1 \sin\theta_2 + p_3 \cos\theta_1 \sin\theta_2)\dot\theta_1/2$$
$$+ (2p_2 \sin\theta_2 \cos\theta_2 + p_3 \sin\theta_1 \cos\theta_2 - p_1 \cos\theta_1 \cos\theta_2)\dot\theta_2/2$$

$$c_{12} = -p_1 \sin\theta_1 \sin\theta_2\dot\theta_1 + (p_1 \cos\theta_1 \cos\theta_2 - 2p_2 \sin\theta_2 \cos\theta_2)\dot\theta_2$$

$$c_{13} = (p_1 \sin\theta_1 \sin\theta_2 + p_3 \cos\theta_1 \sin\theta_2)\dot\theta_r/2 - p_3 \sin\theta_1 \sin\theta_2\dot\theta_l$$
$$+ (q_1 \cos\theta_1 \sin\theta_2 + q_2 \sin\theta_1 \sin\theta_2)\dot\theta_1$$
$$+ (q_1 \sin\theta_1 \cos\theta_2 - q_2 \cos\theta_1 \cos\theta_2 + q_3 \sin\theta_2 \cos\theta_2)\dot\theta_2$$

$$c_{14} = (-p_1 \cos\theta_1 \cos\theta_2 + 2p_2 \sin\theta_2 \cos\theta_2 + p_3 \sin\theta_1 \cos\theta_2)\dot\theta_r/2$$
$$+ (p_1 \cos\theta_1 \cos\theta_2 - 2p_2 \sin\theta_2 \cos\theta_2)\dot\theta_l$$
$$+ (q_1 \sin\theta_1 \cos\theta_2 - q_2 \cos\theta_1 \cos\theta_2 + q_3 \sin\theta_2 \cos\theta_2)\dot\theta_1$$
$$+ (q_1 \cos\theta_1 \sin\theta_2 + q_2 \sin\theta_1 \sin\theta_2)\dot\theta_2$$

$$c_{15} = (-p_1 \cos\theta_1 \cos\theta_2 + 2p_2 \sin\theta_2 \cos\theta_2 + p_3 \sin\theta_1 \cos\theta_2)\dot\theta_r/2$$
$$+ (p_1 \cos\theta_1 \cos\theta_2 - 2p_2 \sin\theta_2 \cos\theta_2)\dot\theta_l$$
$$+ (q_1 \sin\theta_1 \cos\theta_2 - q_2 \cos\theta_1 \cos\theta_2 + q_3 \sin\theta_2 \cos\theta_2)\dot\theta_1$$
$$+ (q_1 \cos\theta_1 \sin\theta_2 + q_2 \sin\theta_1 \sin\theta_2)\dot\theta_2$$

$$c_{21} = -p_1 \sin\theta_1 \sin\theta_2\dot\theta_1/2 + (p_1 \cos\theta_1 \cos\theta_2 - 2p_2 \sin\theta_2 \cos\theta_2)\dot\theta_2/2$$

$$c_{22} = (p_1 \sin\theta_1 \sin\theta_2 - p_3 \cos\theta_1 \sin\theta_2 - p_3 \cos\theta_1 \sin\theta_2)\dot\theta_1/2$$
$$+ (2p_2 \sin\theta \cos\theta_2 - p_3 \sin\theta_1 \cos\theta_2 - p_1 \cos\theta_1 \cos\theta_2)\dot\theta_2/2$$

$$c_{23} = -p_1 \sin\theta_1 \sin\theta_2\dot\theta_r/2 + (p_1 \sin\theta_1 \sin\theta_2 - p_3 \cos\theta_1 \sin\theta_2)\dot\theta_l/2$$
$$+ (q_1 \cos\theta_1 \sin\theta_2 - q_2 \sin\theta_1 sin\theta_2)\dot\theta_1$$
$$+ (q_1 \sin\theta_1 \cos\theta_2 + q_2 \cos\theta_1 cos\theta_2 - q_3 \sin\theta_2 cos\theta_2)\dot\theta_2$$

$$c_{24} = (p_1 \cos\theta_1 \cos\theta_2 - 2p_2 \sin\theta_2 \cos\theta_2)\dot\theta_r/2$$
$$+ (2p_2 \sin\theta_2 \cos\theta_2 - p_1 \cos\theta_1 \cos\theta_2 - p_3 \sin\theta_1 \cos\theta_2)\dot\theta_l/2$$
$$+ (q_1 \sin\theta_1 \cos\theta_2 + q_2 \cos\theta_1 \cos\theta_2 - q_3 \sin\theta_2 \cos\theta_2)\dot\theta_1$$

$$c_{25} = (p_1 \cos\theta_1 \cos\theta_2 - 2p_2 \sin\theta_2 \cos\theta_2)\dot\theta_r/2$$
$$+ (2p_2 \sin\theta_2 \cos\theta_2 - p_1 \cos\theta_1 \cos\theta_2 - p_3 \sin\theta_1 \cos\theta_2)\dot\theta_l/2$$
$$+ (q_1 \sin\theta_1 \cos\theta_2 + q_2 \cos\theta_1 \cos\theta_2 - q_3 \sin\theta_2 \cos\theta_2)\dot\theta_1$$

$$c_{31} = -(p_1 \sin\theta_1 \sin\theta_2 + p_3 \cos\theta_1 \sin\theta_2)\dot\theta_r/2$$
$$+ (p_1 \sin\theta_1 \sin\theta_2 - p_3 \cos\theta_1 \sin\theta_2)\dot\theta_l/2$$
$$+ q_3 \sin\theta_2 \cos\theta_2\dot\theta_2$$

$$c_{32} = (p_1 \sin\theta_1 \sin\theta_2 - p_3 \cos\theta_1 \sin\theta_2)\dot\theta_r/2 + (p_3 \cos\theta_1 \sin\theta_2 - p_1 \sin\theta_1 \sin\theta_2)\dot\theta_l$$
$$- q_3 \sin\theta_2 \cos\theta_2\dot\theta_2$$

$$c_{33} = 0$$

$$c_{34} = q_3 \sin\theta_2 \cos\theta_2 \dot\theta_r - q_3 \sin\theta_2 \cos\theta_2 \dot\theta_l + m_2 l_2^2 \sin\theta_2 \cos\theta_2 \dot\theta_1$$

$$c_{35} = q_3 \sin\theta_2 \cos\theta_2 \dot\theta_r - q_3 \sin\theta_2 \cos\theta_2 \dot\theta_l + m_2 l_2^2 \sin\theta_2 \cos\theta_2 \dot\theta_1$$

$$c_{41} = p_1 \cos\theta_1 \cos\theta_2 - w p_2 \sin\theta_2 \cos\theta_2 - p_3 \sin\theta_1 \cos\theta_2)\dot\theta_r/2$$
$$- (p_1 \cos\theta_1 \cos\theta_2 - 2p_2 \sin\theta_2 \cos\theta_2)\dot\theta_l/2 + q_e \sin\theta_2 \cos\theta_2 \dot\theta_1$$

$$c_{42} = -(p_1 \cos\theta_1 \cos\theta_2 - 2p_2 \sin\theta_2 \cos\theta_2)\dot\theta_r/2$$
$$+ (p_1 \cos\theta_1 \cos\theta_2 - I_{wp2} \sin\theta_2 \cos\theta_2 + p_3 \sin\theta_1 \cos\theta_2)\dot\theta_l/2$$
$$+ q_3 \sin\theta_2 cos\theta_2 \dot\theta_1$$

$$c_{43} = q_3 \sin\theta_2 \cos\theta_2 \dot\theta_r + q_3 \sin\theta_2 \cos\theta_2 \dot\theta_l - m_2 l_2^2 \sin\theta_2 \cos\theta_2 \dot\theta_1$$

$$c_{44} = 0$$

$$c_{45} = 0$$

$$G(q) = \begin{bmatrix} 0 & 0 & 0 & -m_2 g l_2 \sin\theta_2 & -m_3 g(2l_2 + \theta_3) \sin\theta_2 \end{bmatrix}$$

$$B(q) = \begin{bmatrix} 1 & 0 & 0 & 0 & 0 \\ 0 & 1 & 0 & 0 & 0 \\ 0 & 0 & 1 & 0 & 0 \\ 0 & 0 & 0 & 1 & 0 \\ 0 & 0 & 0 & 0 & 1 \end{bmatrix}$$

Bibliography

[1] http://robotics.stanford.edu/ ruspini/samm.html

[2] http://www.cs.washington.edu/homes/sisbot/HANP/

[3] http://www.dlr.de/rm/en/desktopdefault.aspx/tabid-5471/

[4] http://marsrovers.jpl.nasa.gov/home/index.html

[5] http://www.irobot.com

[6] http://mekabot.com/products/m1-mobile-manipulator/

[7] K. Liu and F. L. Lewis, "Decentralized continuous robust controller for mobile robots," In *Proceedings of IEEE International Conference Robotics and Automation*, pp. 1822-1827, 1990.

[8] J. Guldner and V. I. Utkin, "Stabilization of nonholonomic mobile robots using Lyapunov function for navigation and sliding mode control," in *Proceedings of the 33rd IEEE Conference on Decision & Control*, pp. 2967–2972, 1994.

[9] C. Samson, "Time-varying feedback stabilization of a nonholonomic wheeled mobile robot," *International Journal of Robotics Research*, vol. 12, pp. 55–66, 1993.

[10] R. Murray and S. Sastry, "Nonholonomic motion planning: steering using sinusoids," *IEEE Transactions on Automatic Control*, vol. 38, pp. 700–716, 1993.

[11] Z. Li, S. S. Ge, and A. Ming, "Adaptive Robust Motion/Force Control of Holonomic Constrained Nonholonomic Mobile Manipulators," *IEEE Trans. System, Man, and Cybernetics, Part B-Cybernetics*, vol. 37, no. 3, pp. 607-617, 2007.

[12] W. Dong, W. Xu, and W. Huo, "Trajectory tracking control of dynamic nonholonomic systems with unknown dynamics," *International Journal of Robust and Nonlinear Control*, vol. 9, pp. 905-922, 1999.

[13] S. Ahmad and S. Luo, "Coordinated motion control of multiple robotic devices for welding and redundancy coordination through constrained optimization in Cartesian space," *IEEE Trans. Robotics and Automation*, vol. 5, no. 4, 1989, pp. 409-417.

[14] G. S. Bolmsjo, "Programming robot systems for arc welding in small series production," *Robotics and Computer-Integrated Manufacturing*, vol. 5, no. 2-3, 1989, pp. 498-510.

[15] R. Weston, "Robot workplaces," *Microprocessors and Microsystems*, vol. 8, no. 5, 1984, pp. 245-248.

[16] Z. Sun, S. S. Ge, W. Huo, and T. H. Lee, "Stabilization of nonholonomic chained systems via nonregular feedback linearization," *System & Control Letters*, vol. 44, pp. 279–289, 2001.

[17] L. S. You and B. S. Chen, "Tracking control design for both holonomic and non-holonomic constrained mechanical systems: a unified viewpoint," *International Journal of Control*, vol. 58, no. 3, pp.587-612, 1993.

[18] J. Yuan, "Adaptive control of a constrained robot ensuring zero tracking and zero force errors," *IEEE Transactions on Automatic Control*, vol. 42, no. 12, pp. 1709-1714, 1997.

[19] Z. Wang, S. S. Ge, T. H, Lee, "Robust adaptive Neural Network Control of Uncertain Nonholonomic Systems with strong nonlinear drift," *IEEE Trans. System, Man, and Cybernetics, Part B: Cybernetics*, vol. 34, no. 5, pp.2048-2059 , 2004.

[20] C. Samson, "Control of chained systems: Application to path following and time-varying point-stabilization of mobile robots," *IEEE Transactions on Automatic Control*, vol. 40, pp. 64–77, 1995.

[21] C. Y. Su and Y. Stepanenko, "Robust motion/force control of mechanical systems with classical nonholonimic constraints," *IEEE Trans. Automatic Contr.*, vol. 39, no. 3, pp. 609–614, 1994.

[22] R. Murray, "Nilpotent bases for a class of nonintergrable distributions with applications to trajectory generation for nonholonomic systems," *California Inst. Tech., Pasadena, Tech. Memo, CIT-CDS 92-002, Oct.*, 1992.

[23] J. J. E. Slotine and W. Li, *Applied Nonlinear Control.* Englewood Cliffs, New Jersey: Prentice Hall, 1991.

[24] P. Dauchez, A. Fournier and R. Jourdan, "Hybrid control of a two-arm robot for complex tasks," *Robotics and Autonomous Systems*, vol. 5, pp. 323-332, 1989.

[25] W. Leroquais and B. d'Andréa-Novel, "Transformation of the kinematic models of restricted mobility wheeled mobile robots with a single platform into chain forms," *Proc. of the 34th Conference on Decision & Control*, New Orleans, LA, pp. 1443-1447, 1995.

[26] C. C. de Wit, B. Siciliano, and G. Bastin, *Theory of Robot Control*, New York: Springer, 1996.

[27] S. S. Ge, T. H. Lee and C. J. Harris, *Adaptive Neural Network Control of Robotic Manipulators*, World Scientific, London, December 1998.

[28] Z. Li and J. F. Canny, "Motion of two rigid bodies with rolling constraint," *IEEE Transactions on Robotics and Automation*, 6(1):62-72, February 1990.

[29] J.-P. Laumond, P.E. Jacobs, M. Taix, R.M. Murray , "A motion planner for nonholonomic mobile robots," *IEEE Transactions on Robotics and Automation*, vol. 10, no. 5, pp. 577 - 593, 1994.

[30] Z. Li, A. Ming, N. Xi, M. Shimojo, "Motion control of nonholonomic mobile underactuated manipulator," *IEEE International Conference on Robotics and Automation*, pp. 3512-3519, 2006.

[31] Z. Li, and J. Luo, "Adaptive Robust Dynamic Balance and Motion Controls of Mobile Wheeled Inverted Pendulums," *IEEE Transactions on Control Systems Technology*, vol. 17, no. 1, pp. 233-241.

[32] M. Zhang, and T. Tarn, "Hybrid Control of the Pendubot", *IEEE/ASME Trans. Mechatronics*, vol. 7, no. 1, pp. 79-86, 2002.

[33] G. Campion, G. Bastin, B. d'Andréa-Novel, " Structural properties and classification of kinematic and dynamic models of wheeled mobile robots," *IEEE Transactions on Robotics and Automation*, vol. 12, no.1, page(s): 47-62, 1996.

[34] R. L. Williams II , B. E. Carter , P. Gallina , and G. Rosati, "Dynamic Model with Slip for Wheeled Omni-Directional Robots," *IEEE Trans. Robotics and Automation*, vol. 18, no. 3, pp. 285-293, 2002.

[35] S. Wang, L. Lai, C. Wu, and Y. Shiue, "Kinematic Control of Omni-directional Robots for Time-optimal Movement between Two Configurations," *Journal of Intelligent and Robotic Systems*, vol. 49, no. 4, 2007, pp.397-410.

[36] D. Gracanin, K. P. Valavanis, N. C. Tsourveloudis, M. Matijasevic, "Virtual-environment-based navigation and control of underwater vehicles," *IEEE Robotics & Automation Magazine*, vol. 6, no. 2, 1999, pp. 53-63.

[37] F. Fahimi , "Sliding-Mode Formation Control for Underactuated Surface Vessels," *IEEE Trans. Robotics and Automation*, vol. 23, no. 3,pp. 617-622, 2007.

[38] S. S. Ge, Z. Sun, T. H. Lee, and M. W. Spong, "Feedback linearization and stabilization of second-order nonholonomic chained systems," *International Journal of Control*, vol. 74, no. 14, pp. 1383–1392, 2001.

[39] R. Tinos, M. H. Terra, and J. Y. Ishihara, "Motion and force control of cooperative robotic manipulators with passive joints," *IEEE Trans. Control Systems Technology*, vol. 14, no. 4, pp. 725-734, 2006.

[40] B. d'Andrea-Novel, G. Bastin, and G. Campion, "Modelling and control of non holonomic wheeled mobile robots," *Proceedings of 1991 International Conference on Robotics and Automation*, pages 1130-1135, Sacramento, CA, April 1991.

[41] N. J. Nilsson, *Principles of Artificial Intelligence*, Wellsboro, PA: Tioga, Jan. 1980.

[42] D. Kortenkamp, R. Bonasso, and R. Murphy, *Artificial Intelligence and Mobile Robots*, Cambridge, MA: MIT Press, 1998.

[43] L. Dorst and K. I. Trovato, "Optimal path planning by cost wave propagation in metric configuration space," *Proc. SPIE Mobile Robotics III*, 1989, vol. 1007, pp. 186-C197.

[44] J. C. Latombe, *Robot Motion Planning*, London, U.K.: Kluwer, 1991.

[45] O. Khatib, "Real-time obstacle avoidance for manipulators and mobile robots," *Int. J. Robot. Res.*, vol. 5, no. 1, pp. 90-98, 1986.

[46] D. Fox, W. Burgard, and S. Thrun, "The dynamic window approach to collision avoidance," *IEEE Robot. Autom. Mag.*, vol. 4, no. 1, pp. 23-33, Mar. 1997.

[47] R. Simmons, "The curvature-velocity method for local obstacle avoidance," *in Proc. IEEE Int. Conf. Robot. Autom.*, Apr. 1996, vol. 4, pp. 22-28.

[48] P. Moutarlier, B. Mirtich, and J. Canny, "Shortest paths for a robot to manifolds in configuration space," *Int. J. Robot. Res.*, vol. 15, no. 1, pp. 36-60, Feb. 1996.

[49] I. E. Paromtchik and C. Laugier, "Motion generation and control for parking an autonomous vehicle," *Proc. IEEE Int. Conf. Robotics Automation*, Minneapolis, MN, Apr. 1996, pp. 3117-7122.

[50] M. Khatib, H. Jaouni, R. Chatila, and J. P. Laumond, "Dynamic path modification for car-like nonholonomic mobile robots," *Proc. IEEE Int. Conf. Robotics Automation*, Albuquerque, NM, Apr. 1997, pp. 2920-2925.

[51] K. Jiang, L. D. Seneviratne, and S. W. E. Earles, "Time-optimal smooth-path motion planning for a mobile robot with kinematic constraints," *Robotica*, vol. 15, no. 5, pp. 547-553, 1997.

[52] L. Podsedkowski, "Path planner for nonholonomic mobile robot with fast replanning procedure," *in Proc. IEEE Int. Conf. Robotics Automation*, Leuven, Belgium, May 1998, pp. 3588-3593.

[53] S. Sekhavat, P. Svestka, J. P. Laumond, and M. H. Overmars , "Multi-level path planning for nonholonomic robots using semiholonomic subsystems," *Int. J. Robot. Res.*, vol. 17, no. 8, pp. 840-857, Aug. 1998.

[54] C. J. Ong and E. G. Gilbert, "Robot path planning with penetration growth distance," *Journal of Robotic Systems*, vol. 15, no. 2, pp. 57-74, 1998.

[55] L. M. Gambardella and C. Versino, "Robot motion planning integrating planning strategies and learning methods," *in Proc. 2nd Int. Conf. AI Planning Systems*, Chicago, IL, June 13-15, 1994.

[56] P. Svestka and M. H. Overmars, "Motion planning for carlike robots using a probabilistic approach," *Int. J. Robot. Res.*, vol. 16, no. 2, pp. 119-145, Apr. 1997.

[57] T. Fujii, Y. Arai, H. Asama, and I. Endo, "Multilayered reinforcement learning for complicated collision avoidance problems," *in Proc. IEEE International Conference Robotics Automation*, Leuven, Belgium, May 1998, pp. 2186-2191.

[58] V. Pavlov and A. Timofeyev, "Construction and stabilization of programmed movements of a mobile robot-manipulator," *Engineering Cybernetics*, vol. 14, no. 6, pp. 70-79, 1976.

[59] J. H. Chung and S. A. Velinsky, "Modeling and control of a mobile manipulator," *Robotica*, vol. 16, no. 6, pp. 607-613, 1998.

[60] K. Watanabe, K. Sato, K. Izumi and Y. Kunitake, "Analysis and control for an omnidirectional mobile manipulator," *Journal of Intelligent and Robotic Systems*, vol. 27, no. 1, pp. 3-20, 2000.

[61] J. Tan, N. Xi, and Y. Wang, "Integrated task planning and control for mobile manipulators," *International Journal of Robotics Research*, vol. 22, no. 5, pp. 337-354, 2003.

[62] O. Khatib, "Mobile manipulation: the robotic assistant," *Robotics and Autonomous Systems*, vol. 26, no. 2-3, pp. 175-183, 1999.

[63] B. Bayle, J. Y. Fourquet, and M. Renaud, "Manipulability of wheeled mobile manipulators: application to motion generation," *International Journal of Robotics Research*, vol. 22, no. 7-8, pp. 565-581, 2003.

[64] Z. Wang, C. Su, and S. S. Ge, "Adaptive control of mobile robots including actutor dynamics," *Autonomous Mobile Robots: Sensing, Control, Decision Making and Application*, S. S. Ge and F. L. Lewis, Eds., Boca Raton: Taylor & Francis Group, pp. 267-293, 2006.

[65] S. S. Ge, J. Wang, T. H. Lee and G. Y. Zhou, "Adaptive robust stabilization of dynamic nonholonomic chained systems", *Journal of Robotic Systems*, vol. 18, no. 3, pp. 119-133, 2001.

[66] Z. Wang, S. S. Ge, and T. H. Lee, "Robust motion/force control of uncertain holonomic/nonholonomic mechanical systems," *IEEE/ASME Trans. Mechatronics*, vol. 9, no. 1, pp. 118-123, 2004.

[67] N. Xi, T. J. Tarn, A. K. Bejczy, "Intelligent planning and control for multirobot coordination: an event-based approach," *IEEE Trans. Robotics and Automation*, vol. 12, no. 3, pp. 439-452, 1996.

[68] O. Khatib, K. Yokoi, K. Chang, D. Ruspini, R. Holmberg and A. Casal , "Coordination and decentralized cooperation of multiple mobile manipulators," *Journal of Robotic Systems*, vol. 13, no. 11 , pp. 755-764, 1996.

[69] Y. Yamamoto, and S. Fukuda, "Trajectory planning of multiple mobile manipulators with collision avoidance capability," *Proc. IEEE Int. Conf. on Robotics and Automation*, pp. 3565-3570, 2002.

[70] T. G. Sugar and V. Kumar, "Control of cooperating mobile manipulators," *IEEE Trans. Robotics and Automation*, vol. 18, pp. 94-103, 2002.

[71] H. G. Tanner, K. J. Kyriakopoulos, and N. J. Krikelis "Modeling of multiple mobile manipulators handling a common deformable object," *Journal of Robotic Systems*, vol. 15, pp. 599-623, 1998.

[72] Y. Yamamoto, Y. Hiyama, and A. Fujita, "Semi-autonomous reconfiguration of wheeled mobile robots in coordination," *Proc. IEEE Int. Conf. Robotics and Automation*, pp. 3456-3461, 2004.

[73] H. G. Tanner, S. Loizou, and K. J. Kyriakopoulos, "Nonholonomic navigation and control of cooperating mobile manipulators," *IEEE Trans. Robotics and Automation*, vol. 19, pp. 53-64, 2003.

[74] T. L. Huntsberger , A. Trebi-Ollennu, H. Aghazarian, P. S. Schenker, P. Pirjanian and H. D. Nayar "Distributed control of multi-robot systems engaged in tightly coupled tasks," *Autonomous Robots*, vol. 17, no. 1, pp. 929-5593, 2004.

[75] Y. Hirata, Y. Kume, T. Sawada, Z. Wang, and K. Kosuge, "Handling of an object by multiple mobile manipulators in coordination based

on caster-like dynamics," *Proc. IEEE International Conference Robotics and Automation* vol.26, pp. 807-812, 2004.

[76] E. Nakano, S. Ozaki, T. Ishida, and I. Kato, "Cooperational control of the anthropomorphous manipulator MELARM," *Proc 4th International Symposiums Industrial Robots*, pp. 251-260, 1974.

[77] S. Arimoto and F. Miyazaki and S. Kawamura, "Cooperative motion control of multiple robot arms or fingers," *Proc. IEEE International Conference Robotics and Automation*, pp. 1407-1412, 1987.

[78] Y. F. Zheng and J. Y. S. Luh, "Control of two coordinated robots in motion," *Proc. IEEE International Conference Robotics and Automation*, pp. 1761-1766, 1985.

[79] Y. H. Kim and F. L. Lewis, "Neural Network Output Feedback Control of Robot Manipulators," *IEEE Trans. Robotics and Automation*, vol. 15, no. 2, pp 301-309, 1999.

[80] S. S. Ge, C. C. Hang, and L. C. Woon, "Adaptive neural network control of robot manipulators in task space," *IEEE Trans. Industrial Electronics*, vol. 44, no. 6, pp. 746-752, 1997.

[81] F. Hong, S. S. Ge, F. L. Lewis, and T. H. Lee, "Adaptive neural-fuzzy control of nonholonomic mobile robots," *Autonomous Mobile Robots: Sensing, Control, Decision Making and Application*, S. S. Ge and F. L. Lewis Ed., Boca Raton: Taylor & Francis Group, pp. 229-265, 2006.

[82] N. H. McClamroch and D. Wang, "Feedback stabilization and tracking of constrained robots," *IEEE Trans. Automatic Control*, vol. 33, no. 5, pp. 419-426, 1988.

[83] A. Astol, "Discontinuous control of nonholonomic systems," *Systems & Control Letters*, vol. 27, pp. 37-45, 1996.

[84] G. C. Walsh and L. G. Bushnell, "Stabilization of multiple input chained form control systems," *System & Control Letters*, vol. 25, pp. 227-234, 1995.

[85] C. Samson, "Control of chained systems: application to path following and time-varying point-stabilization of mobile robots," *IEEE Trans. Automatic Control*, vol. 40, pp. 64-77, 1995.

[86] C. de Wit Canudas, H. Berghuis, and H. Nijmeijer, "Practical stabilization of nonlinear systems in chained form," *Proceedings of 33rd IEEE Conference on Decision & Control*, Lake Buena Vista, FL, USA, pp. 3475-3480, 1994.

[87] O. J. Sordalen, C. Canudas de Wit, "Exponential stabilization of nonholonomic chained systems," *IEEE Transactions on Automatic Control*, vol. 40, no. 1, pp. 35-49, 1995.

[88] I. Kolmanovsky and N. H. McClamroch, "Developments in nonholonomic control problems," *IEEE Control System Magazine*, vol. 15, no. 6, pp. 20-36, 1995.

[89] R. T. M'Closkey and R. M. Murray, "Convergence rate for nonholonomic systems in power form," *Proceedings of American Control Conference*, Chicago, USA, pp. 2489-2493, 1992.

[90] W. Huo and S. S. Ge, " Exponential stabilization of nonholonomic systems: an end approach," *International Journal of Control*, vol. 74, no. 15, pp. 1492-1500, 2001.

[91] S. S. Ge, Z. Sun, T. H. Lee, and M. W. Spong, " Feedback linearization and stabilization of second-order nonholonomic chained systems," *International Journal of Control*, vol. 74, no. 14, pp. 1383-1392, 2001.

[92] Z. P. Jiang and H. Nijmeijer, "A recursive technique for tracking control of nonholonomic systems in chained form," *IEEE Transactions on Automatic Control*, vol. 44, no. 2, 265-279, 1999.

[93] J. P. Hespanha, S. Liberzon, and A. S. Morse, "Towards the supervisory control of uncertain nonholonomic systems," *Proceedings of American Control Conference*, San Diego, CA, USA, pp. 3520-3524.

[94] K. Do and J. Pan, "Adaptive global stabilization of nonholonomic systems with strong nonlinear drifts," *Systems & Control Letters*, vol. 46, pp. 195-205, 2002.

[95] W. E. Dixon ,D. M. Dawson, E. Zergeroglu, and A. Behal, "Nonlinear control of wheeled mobile robots," *Lecture Notes in Control and Information Sciences*, Vol. 262. 2001.

[96] Z. P. Jiang, "Robust exponential regulation of nonholonomic systems with uncertainties," *Automatica*, vol. 36, no. 2, pp. 189-209, 2000.

[97] H. Seraji , "An on-line approach to coordinated mobility and manipulation," *Proceedings of International Conference on Robotics and Automation*, pages 28-35, Vol. 1, Atlanta, GA, May 1993.

[98] F. G. Pin and J. C. Culioli, "Multi-criteria position and configuration optimization for redundant platform/manipulator systems," *Proceedings of International Conference on Intelligent Robots and Systems*, pp. 103-107, July 1990.

[99] W. Miksch and D. Schroeder, "Performance-functional based controller design for a mobile manipulator," *Proceedings of International Conference on Robotics and Automation*, pp. 227-232, Nice, France, May 1992.

[100] K. Liu and F. L. Lewis, "Application of robust control techniques to a mobile robot system," *Journal of Robotic Systems*, vol. 9, no. 7, pp. 893-913, 1992.

[101] Y. Yamamoto and X. Yun, "Coordinating locomotion and manipulation of a mobile manipulator," *IEEE Trans. Automatic Control*, vol. 39, no. 6, pp. 1326-1332, 1994.

[102] Y. Yamamoto and X. Yun, "Effect of the dynamic interaction on coordinated control of mobile manipulators," *IEEE Trans. Robotics and Automation*, vol. 12, no. 5, pp. 816-824, 1996.

[103] W. Dong, "On trajectory and force tracking control of constrained mobile manipulators with parameter uncertainty," *Automatica*, vol. 38, no. 9, pp. 1475-1484, 2002.

[104] S. S. Ge and L. Huang and T. H. Lee, "Model-based and neural-network-based adaptive control of two robotic arms manipulating an object with relative motion," *Int. J. Sys. Sci.*, vol. 32, no. 1, pp. 9-23, 2001.

[105] S. S. Ge, Z. Wang, and T. H. Lee, "Adaptive stabilization of uncertain nonholonomic systems by state and output feedback," *Automatica*, vol. 39, no. 8, pp. 1451-1460, 2003.

[106] H. Arai and K. Tanie, "Nonholonomic control of a three-DOF planar underactuted manipulator," *IEEE Tran. Robotics and Automation*, vol. 14, no. 5, pp. 681-694, 1998.

[107] A. De Luca and G. Oriolo, "Trajectory planning and control for planar robots with passive last joint," *The International Journal of Robotics Research*, vol. 21, no. 5-6, pp. 575-590, 2002.

[108] S. S. Ge, J. Wang, T. H. Lee, and G. Y. Zhou, "Adaptive robust stabilization of dynamic nonholonomic chained systems," *Journal of Robotic Systems*, vol. 18, no. 3, pp. 119-133, 2001.

[109] M. Bergerman, C. Lee, and Y. Xu, "A dynamic coupling index for underactuated manipulators," *Journal of Robotic Systems*, vol. 12, no. 10, pp. 693-707, 1995.

[110] C. Su and Y. Stepanenko, "Robust motion/force control of mechanical systems with classical nonholonimic constraints," *IEEE Trans. Automatic Contr.*, vol. 39, no. 3, pp. 609-614, 1994.

[111] S. S. Ge, T. H. Lee, and C. J. Harris, *Adaptive Neural Network Control of Robot Manipulators*, River Edge, NJ: World Scientific, 1998.

[112] C. Kwan, F. L. Lewis, and D. M. Dawson, "Robust Neural-Network Control of Rigid-Link Electrically Driven Robots," *IEEE Trans. Neural Networks*, vol. 9, no. 4, pp. 581-588, 1998.

[113] R. Fierro and F. L. Lewis, "Control of a Nonholonomic Mobile Robot Using Neural Networks," *IEEE Trans. Neural Networks*, vol. 9, no. 4, pp. 589-600, 1998.

[114] L. Huang, S. S. Ge, and T. H. Lee, "Fuzzy unidirectional force control of constrained robotic manipulators," *Fuzzy Sets and Systems*, vol. 134, no. 1, pp. 135-146, 2003.

[115] S. Lin and A. A. Goldenberg, "Neural-network control of mobile manipulators," *IEEE Trans. Neural Network* , vol. 12, no. 5, pp. 1121-1133, 2001.

[116] S. S. Ge and C. C. Hang and T. H. Lee and T. Zhang, *Stable Adaptive Neural Network Control*, Boston, Kluwer Academic Publisher, 2002.

[117] H. K. Khalil, *Nonlinear Systems*, MacMillan, New York, 1992.

[118] J. C. Willems, *The Analysis of Feedback Systems*, MIT Press,Cambridge, MA, 1970.

[119] C. A. Desoer and M. Vidyasagar, *Feedback Systems: Input-Output Properties*, New York, Academic Press, 1975.

[120] K. S. Fu, R. C. Gonzalez, and C. S. G. Lee, *Robotics: Control, Sensing, Vision and Intelligence*, New York, McGraw-Hill, 1987.

[121] W. F. Carriker, P. K. Khosla, and B. H. Krogh, "Path planning for mobile manipulators for multiple task execution," *IEEE Transactions on Robotics and Automation*, vol. 7, no. 3, pp. 403-408, 1991.

[122] J. Lee and H. Cho, "Mobile manipulator motion planning for multiple tasks using global optimization approach," *Journal of Intelligent and Robotic Systems*, vol. 18, pp. 169-190, 1997.

[123] J. P. Desai and V. Kumar, "Nonholonomic motion planning for multiple mobile manipulators," *IEEE Conference on Robotics and Automation*, pp. 3409-3414, 1997.

[124] Y. Yamamoto and X. Yun, "Coordinated obstacle avoidance of a mobile manipulator," *IEEE Conference on Robotics and Automation*, pp. 2255-2260, 1995.

[125] H. Tanner and K. Kyriakopoulos, "Nonholonomic motion planning for mobile manipulators," *IEEE Conference on Robotics and Automation*, pp. 1233-1238, 2000.

[126] H. G. Tanner, S. G. Loizou, and K. J. Kyriakopoulos, "Nonholonomic navigation and control of cooperating mobile manipulators," *IEEE Transactions on Robotics and Automation*, vol. 19, no. 1, pp. 53-64, 2003.

[127] E. Papadopoulos, I. Poulakakis, and I. Papadimitriou, "On path planning and obstacle avoidance for nonholonomic platforms with manipulators: a polynomial approach," *International Journal of Robotics Research*, vol. 21, no. 4, pp. 367-383, 2002.

[128] G. Oriolo and C. Mongillo, "Motion planning for mobile manipulators along given end-effector paths," *IEEE Conference on Robotics and Automation*, pp. 2166-2172, 2005.

[129] J. B. Mbede, P. Ele, C. Mveh-Abia, Y. Toure, V. Graefe, and S. Ma, "Intelligent mobile manipulator navigation using adaptive neuro-fuzzy systems," *Information Science*, vol. 171, pp. 447-474, 2005.

[130] S. Furuno, M. Yamamoto, A. Mohri, "Trajectory Planning of Cooperative Multiple Mobile Manipulators,", *Proc. IEEE/RSJ Int. Conf. Intelligent Robots and Systems*, pp. 136-141, 2003.

[131] A. Yamashita, T. Arai, J. Ota and H. Asama, "Motion Planning of Multiple Mobile Robots for Cooperative Manipulation and Transportations," *IEEE Transactions on Robotics and Automation*, vol. 19, no. 2, pp. 223-237, 2003.

[132] H. G. Tanner, S. G. Loizou, and K. J. Kyriakopoulos, "Nonholonomic Navigation and Control of Cooperating Mobile Manipulators", *IEEE Transactions on Robotics and Automation*, vol. 19, no. 1, pp. 53-64, 2003.

[133] R. Fierro, L. Chaimowicz, and V. Kumar, "Multi-robot cooperation," *Autonomous Mobile Robots: Sensing, Control, Decision Making and Application*, S. S. Ge and F. L. Lewis Eds., Boca Raton: Taylor & Francis Group, pp. 417-459, 2006.

[134] J. Jean and L. Fu, "An Adaptive Control Scheme for Coordinated Multi-manipulator Systems," *IEEE Transactions on Robotics and Automation*, vol. 9, no. 2, pp. 226-231, 1993.

[135] R. W. Brockett, "Asymptotic stability and feedback stabilization, " *Differential Geometric Control Theory*, R.W. Brockett, R. S. Millman, and H. J. Sussmann, Eds., Boston, Birkhauser, pp. 181-191, 1983.

[136] M. W. Spong, "The swing up control problem for the Acrobot," *IEEE Contr. Syst.*, vol. 15, pp. 49-55, 1995.

[137] T. Takubo, H. Arai, and K. Tanie, "Control of mobile manipulator using a virtual impedance wall," *Proceedings of IEEE International Conference on Robotics and Automation*, pp. 3571-3576, 2002.

[138] Y. Liu and Y. Xu and M. Bergerman, "Cooperation control of multiple manipulators with passive joints," *IEEE Trans. Robotics and Automation*, vol. 15, no. 2, pp. 258-267, 1999.

[139] R. Tinos, M. H. Terra, and J. Y. Ishihara, "Motion and force control of cooperative robotic manipulators with passive joints," *IEEE Trans. Control Systems Technology*, vol. 14, no. 4, pp. 725-734, 2006.

[140] Y. C. Chang and B. S. Chen, "Robust tracking designs for both holonomic and nonholonomic constrained mechanical systems: adaptive fuzzy approach," *IEEE Trans. Fuzzy Syst.*, vol. 8, pp. 46-66, 2000.

[141] S. Behatsh, "Robust output tracking for nonlinear systems," *Int. J. Contr.*, vol. 51, no. 6, pp. 1381-1407, 1990.

[142] A. Isidori, L. Marconi, and A. Serrani. *Robust Autonomous Guidance: an Internal Model Approach.* New York, Springer, 2003.

[143] S. S. Ge, "Advanced control techniques of robot manipulator," *Proceedings of American Control Conference*, Pennsylvania, June, 1998 pp. 2185-2199.

[144] J. H. Yang, "Adaptive robust tracking control for compliant-joint mechanical arms with motor dynamics," *Proc. IEEE Conf. Decision and Control*, pp. 3394–3399, Dec 1999.

[145] R. Colbaugh and K. Glass, "Adaptive regulation of rigid-link electrically-driven manipulators," *Proc. IEEE Conf. Robotics and Automation*, pp. 293–299, 1995.

[146] C. Y. Su and Y. Stepanenko, "Hybrid adaptive/robust motion control of rigid-link electrically-driven robot manipulators," *IEEE Trans. Robotics and Automation*, vol. 11, no. 3, pp. 426-432, 1995.

[147] C. M. Anupoju, C. Y. Su, and M. Oya," Adaptive motion tracking control of uncertain nonholonomic mechanical systems including actuator dynamics," *IEE Proceedings Control Theory and Applications*, vol. 152, no. 5, pp. 575-580, 2005.

[148] M. A. Arteaga and R. Kelly, "Robot control without velocity measurements: new theory and experimental results," *IEEE Trans. Robotics and Automation*, vol. 20, no. 2, pp. 297-308, 2004.

[149] A. Astolfi, "On the stabilization of nonholonomic systems," *Proc. 33rd IEEE Conference on Decision & Control*, pp. 3481–3486, Lake Buena Vista, FL, USA, 1994.

[150] A. Astolfi, "Discontinuous control of nonholonomic systems," *Systems and Control Letters*, vol. 27, pp. 37–45, 1996.

[151] R. T. M'Closkey, and R. M. Murray, "Convergence rate for nonholonomic systems in power form," *Proceedings of the American Control Conference*, pp. 2489–2493, Chicago, USA, 1992.

[152] W. Huo, S. S. Ge, "Exponential stabilization of nonholonomic systems: an ENI approach," *International Journal of Control*, vol. 74, no. 15, pp. 1492–1500, 2001.

[153] R. Colbaugh, and R. Barany, and K. Glass, "Adaptive control of nonholonomic mechanical systems," *Proc. 35th IEEE Conf. Decision & Control*, pp. 1428–1434, Kobe, Japan, 1996.

[154] W. E. Dixon, and D. M. Dawson and E. Zergeroglu and A. Behal, *Nonlinear Control of Wheeled Mobile Robots*, Springer Verlag, London, 2001.

[155] Z. P. Jiang, "Robust exponential regulation of nonholonomic systems with uncertainties," *Automatica*, vol. 36, pp. 189–209, 2000.

[156] R. Fierro, and F. L. Lewis, "Control of a nonholonomic mobile robot: backstepping kinematics into dynamics," *Proc. 34th IEEE Conf. Decision & Control*, pp. 1722–1727, New Orleans, LA, USA, 1995.

[157] M. Krstic, and I. Kanellakopoulos, and P. V. Kokotovic, *Nonlinear and Adaptive Control Design*, Wiley, New York, 1995.

[158] I. Kosmatopoulos, and M. M. Polycarpou, and M. A. Christodoulou, and P. A. Ioannou, "High-order neural network structures for identification of dynamical systems," *IEEE Trans. Neural Networks*, vol. 6, pp. 422–431, 1995.

[159] R. M. Sanner, and J. E. Slotine, "Gaussian netowrks for direct adaptive control," *IEEE Trans. Neural Networks*, vol. 3, pp. 837–863, 1992.

[160] E. D. Sontag, "Smooth stabilization implies coprime factorization," *IEEE Trans. Automatic Control*, vol. 34, no. 4, pp. 435–443, 1989.

[161] V. I. Arnold, *Geometrical Methods in the Theory of Ordinary Differential Equations*, Springer, Berlin, 1987.

[162] K. D. Do and J. Pan, "Adptive global stabilization of nonholonomic systems with strong nonlinear drifts," *Systems & Control Letters*, vol. 46, pp. 195–205, 2002.

[163] S. S. Ge, and J. Wang and T. H. Lee, and G. Y. Zhou, "Adaptive robust stabilization of dynamic nonholonomic chained systems," *Journal of Robotic Systems*, vol. 18, no. 3, pp. 119–133, 2001.

[164] M. Fliess, and J. Levine, and P. Martin, and P. Rouchon, "Flatness and defect of nonlinear systems: introductory theory and examples," *International Journal of Control*, vol. 61, pp. 1327–1361, 1995.

[165] P. Morin, and J. -B. Pomet, and C. Samson, "Developments in time-varying feedback stabilization of nonlinear systems," *Preprints of Nonlinear Control Systems Design Symposium (NOLCOS'98)"*, pp. 587–594, Enschede, 1998.

[166] Z. Li, S. S. Ge, M. Adams, W. S. Wijesoma, "Robust Adaptive Control of Uncertain Force/Motion Constrained Nonholonomic Mobile Manipulators," *Automatica*, vol. 44, no. 3, pp. 776-784, 2008.

[167] Z. Li, S. S. Ge, M. D. Adams, W. S. Wijesoma, "Robust Adaptive Control of Cooperating Mobile Manipulators with Relative Motion," *IEEE Trans. System, Man, and Cybernetics, Part B-Cybernetics*, vol. 39, no. 1, pp. 103–116, 2009.

[168] Z. Li, S. S. Ge, M. Adams, and W. S. Wijesoma, "Adaptive Robust Output-feedback Motion/Force Control of Electrically Driven Nonholonomic Mobile Manipulators," *IEEE Transactions on Control Systems Technology*, Vol. 16, no. 6, pp. 1308–1315, 2008.

[169] Z. Li, S. S. Ge, Z. Wang, "Robust Adaptive Control of Coordinated Multiple Mobile Manipulators," *Mechatronics*, vol. 18, pp. 239–250, 2008.

[170] W. Gueaieb, F. Karray, and S. Al-Sharhan, "A robust hybrid intelligent position/force control scheme for cooperative manipulators," *IEEE/ASME Trans. Mechatronics*, vol. 12, no. 2, pp. 109-125, 2007.

[171] F. Caccavale, P. Chiacchio, A. Marino, and L. Villani, "Six-DOF impedance control of dual-arm cooperative manipulators," *IEEE/ASME Trans. Mechatronics*, vol. 13, no. 5, pp. 576-586, 2008.

[172] T. G. Sugar and V. Kumar, "Control of cooperating mobile manipulators," *IEEE Trans. Robotics and Automation*, vol. 18, pp. 94-103, 2002.

[173] H. G. Tanner, K. J. Kyriakopoulos, and N. J. Krikelis "Modeling of multiple mobile manipulators handling a common deformable object," *Journal of Robotic Systems*, vol. 15, pp. 599-623, 1998.

[174] H. G. Tanner, S. Loizou and K. J. Kyriakopoulos, "Nonholonomic navigation and control of cooperating mobile manipulators," *IEEE Trans. Robotics and Automation*, vol. 19, pp. 53-64, 2003.

[175] T. L. Huntsberger , A. Trebi-Ollennu, H. Aghazarian, P. S. Schenker, P. Pirjanian and H. D. Nayar "Distributed control of multi-robot systems engaged in tightly coupled tasks," *Autonomous Robots*, vol. 17, no. 1, pp. 929-993, 2004.

[176] R. Fierro, L. Chaimowicz, and V. Kumar,"Multi-robot cooperation," *Autonomous Mobile Robots: Sensing, Control, Decision Making and Application*, S. S. Ge and F. L. Lewis, Eds., Boca Raton: Taylor & Francis Group, pp. 417-459, 2006.

[177] J. Jean and L. Fu, "An adaptive control scheme for coordinated multi-manipulator systems," *IEEE Trans. Robotics and Automation*, vol. 9, no. 2, pp. 226-231, 1993.

[178] Z. Li, S. S. Ge, and A. Ming, "Adaptive robust motion/force control of holonomic-constrained nonholonomic mobile manipulator," *IEEE Trans. System, Man, and Cybernetics, Part B: Cybernetics*, vol. 37, no. 3, pp. 607-617, 2007.

[179] D. Karnopp, "Computer simulation of stick-strip friction in mechanical dynamic systems," *ASME J. Dynam. Syst., Meas., Contr.*, vol. 107, pp. 100-103, 1985.

[180] M. Hernandez and Y. Tang, "Adaptive output-feedback decentralized control of a class of second order nonlinear systems using recurrent fuzzy neural networks," *Neurocomputing*, vol. 73, pp. 461-467, 2009.

[181] S. Hsua and L. Fua, "A fully adaptive decentralized control of robot manipulators," *Automatica*, vol. 42, pp. 1761-1767, 2006.

[182] K. K. Tan , S. Huang, and T. H. Lee, " Decentralized adaptive controller design of large-scale uncertain robotic systems," *Automatica*, vol. 45, pp. 161-166, 2009.

[183] J. Ryu and S. K. Agrawal, "Planning and control of under-actuated mobile manipulators using differential flatness," *Auton Robot*, vol. 29, pp. 35–52, 2010.

[184] Z. Li, Y. Yang, and J. Li, "Adaptive Motion/Force Control of Mobile Under-actuated Manipulators with Dynamics Uncertainties by Dynamic Coupling and Output Feedback," *IEEE Trans. Control System Technology*, vol. 18, no. 5, pp. 1068-1079, 2010.

[185] M. W. Spong, S. Hutchinson, and M. Vidyasagar, *Robot Modeling and Control*, John Wiley & Sons, Inc., 2005.

[186] Z. Li, W. Chen, and J. Luo, "Adaptive Compliant Force-Motion Control of Coordinated Nonholonomic Mobile Manipulators Interacting With Unknown Non-rigid Environments," *Neurocomputing*, vol. 71, no. 7-9, pp. 1330-1344, 2008.

[187] M. Mariton, *Jump Linear Systems in Automatic Control*, New York, Marcel Dekker, 1990.

[188] E.K. Boukas, *Stochastic Hybrid Systems: Analysis and Design*, Birkhauser, Boston, 2005.

[189] A. A. G. Siqueira and M. H. Terra, "Nonlinear and Markovian controls of underactuated manipulators," *IEEE Transaction on Control Systems Technology*, vol. 12, no. 6, pp. 811-826, 2004.

[190] A. A. G. Siqueira and M. H. Terra, "A fault-tolerant manipulator robot based on H_2, H_∞, and mixed H_2/H_∞ Markovian controls," *IEEE/ASME Transactions on Mechatronics*, vol. 14, no. 2, pp. 257-263, 2009.

[191] Y. Y. Cao and L. James, "Robust H_∞ control of uncertain Markovian jump systems with time-delay," *IEEE Trans. Automat. Contr.*, vol. 45, pp. 77-83, 2000.

[192] J. Xiong, J. Lam, H. Gao, and D. W. C. Ho, "On robust stabilization of Markovian jump systems with uncertain switching probabilities," *Automatica*, vol. 41, no. 5, pp. 897-903, 2005.

[193] O. L. V. Costa, , J. B. R. Val, ,J. C. Geromel, "Continuous time state-feedback H2-control of Markovian jump linear system via convex analysis," *Automatica*, vol. 35, pp. 259-268, 1999.

[194] E. K. Boukas, P. Shi, K. Benjelloun, "On stabilization of uncertain linear systems with jump parameters," *International Journal of Control*, vol. 72, no. 9, pp. 842-850, 1999.

[195] S. Aberkane, D. Sauter, J. C. Ponsart and D, Theilliol, "H_∞ stochastic stabilization of active fault tolerant control systems: convex approach," *44th IEEE Conference on Decision and Control*, pp. 3783-3788, 2005.

[196] E. K. Boukas, "Robust constant gain feedback stabilization of stochastic systems," *IMA Journal of Mathematical Control and Information*, vol. 22, no. 3, pp. 334-349, 2005.

[197] H. J. Kushner, *Stochastic Stability and Control*, New York: Academic Press, 1967.

[198] A. V. Skorohod, *Asymptotic Methods in the Theory of Stochastic Differential Equations*, American Mathematical Society, Providence, RI, 1989.

[199] X. Feng, K. A. Loparo, Y. Ji and H. J. Chizeck, "Stochastic stability properties of jump linear systems," *IEEE Trans. Automat. Contr.*, vol. 37, pp. 38-53, 1992.

[200] L. Xie, "Output feedback H_∞ control of systems with parameter uncertainty," *Int. J. Cont.*, vol. 63, pp. 741-750, 1996.

[201] Y. S. Moon, P. Park and W. H. Kwon, "Delay-dependent robust stabilization of uncertain state-delayed systems," *Int. J. Cont.*, vol. 74, pp. 1447-1455, 2001.

[202] S. H. Esfahani, I. R. Petersen, "An LMI approach to output feedback-guaranteed cost control for uncertain time-delay systems," *International Journal of Robust and Nonlinear Control*, vol. 10, pp. 157-174, 2000.

[203] E. T. Jeung, J. H. Kim, H. B. Park, "H_∞-output feedback controller design for linear systems with time-varying delayed state," *IEEE Transaction of Automatic Control*, vol. 43, 971-974, 1998.

[204] M. Fukuda, M. Kojima, "Branch-and-cut algorithms for the bilinear matrix inequality eigenvalue problem," *Computational Optimization and Applications*, vol. 19, pp. 79-105, 2001.

[205] H. D. Tuan, P. Apkarian, "Low nonconvexity-rank bilinear matrix inequalities: algorithms and applications in robust controller and structure

designs," *IEEE Transactions on Automatic Control*, vol. 45, pp. 2111-2117, 2000.

[206] Z. Li, Y. Yang and S. Wang, "Adaptive dynamic coupling control of hybrid joints of human-symbiotic wheeled mobile manipulators with unmodelled dynamics," *International Journal of Social Robotics*, vol. 2, no. 2, pp. 109-120, 2010.

Index

Printed and bound by CPI Group (UK) Ltd, Croydon, CR0 4YY

18/10/2024

01776264-0007